Gifted Young
in Science

Potential Through
Performance

KRISTIN'S

Gifted Young
in Science

Potential Through
Performance

Paul F. Brandwein   A. Harry Passow
Editors

Deborah C. Fort   Gerald Skoog
Association Editor   Contributing Editor
                         Part IV

NATIONAL SCIENCE TEACHERS ASSOCIATION
WASHINGTON, D.C.

This book has been produced by Special Publications, the National Science Teachers Association, 1742 Connecticut Avenue, N. W., Washington, D. C., 20009. Phyllis R. Marcuccio, Director of Publications; Shirley L. Watt, Managing Editor of Special Publications.

Library of Congress Catalog Card Number: 88-63846

ISBN Number 0-87355-076-5

Printed in the United States of America
First edition

Contents

Part I

By Way of Beginning

Part II

Purpose and Principles

Part III

Programs: Certain Bases for Planning Curriculum and Instruction

Part IV

Personal Reflections: From Gifts to Talents

Part V

Part VI

Invitation to Our Readers

We bring you this book at a critical time. Indeed, the world is gathering up its resources to face new, great, unpredictable changes.

Thirty-four authors have joined to bring you this book; among them are natural and physical scientists, psychologists, historians, writers, scholars of curriculum and instruction, teachers of teachers, and teachers of the young. All have turned their scholarship, practical experience, and wisdom to probe the manner and mode by which elementary and high school students gifted in science could find opportunities to turn their intellectual resources and dispositions to contributions in science and technology.

The contributors probe many methodologies—among them, the place of unified concepts and processes; the interaction of the certainties of problem doing with the uncertainties of problem solving; the "hands-on" approaches coupled with the "brains-on" ones, as well as the much needed "hands-off" ones. Thus, the tools of the student are strengthened with the hard-won tools of the scientist. In short, the heart of this book opens a variety of opportunities to all who wish to help others press on with the arts of scientific investigation.

Clearly, all of us in the postindustrial era are now bound to a time in which science and technology will more than ever affect cultures, nations, societies, and personal lives. Yet the coming years, in spite of the uncertainties they bring, also offer great promise to the generations of young who use the resources of the past to mold the character of the future. You, the readers of this book, are profoundly concerned with the various environments of schooling facing teachers and students. Thus, you stand near the center of the vast new efforts necessary to offer the widest opportunity for all—teachers, administrators, parents, and those they nurture. Your overarching purpose: to fulfill the powers of the young in the pursuit of their personal and special capacities and excellences.

But we do not yet know who among the young during their school years will eventually turn to a lifework in science. We do know that there are wide opportunities in scientific and technological work available to the young with many interests, varied abilities, and personalities. We do know enough of the arts of teaching to create environments that nourish personal initiative in which the young may do works that demonstrate talent in science. And we are aware that, over the nation, teachers—whether in classrooms, in administrative offices, or in the library; whether as scientists in laboratories; whether as parents at home—are central, singly and together, to the creation of such environments.

The authors of *Gifted Young in Science: Potential Through Performance* explore the nature of work in science, in schooling, and in education, and thus explore in practice and in theory what is known and yet to be known. From the whole work, you, the readers—from the vantage point of your particular situations—may find models of the environments central to turning general and special gifts into particular talents in science.

I invite you to join the authors. For the book in hand advances your causes as teachers—in whatever capacity—as well as the causes of the young— particularly those young who daily engage in finding and creating their lives and their lifework.

Bill G. Aldridge
Executive Director
NSTA

Part I

By Way of Beginning

By Way of Beginning:
Seeding Destiny in an Open Society

We are living in a world at one of its strange turning points. Indeed, a number of students of intellectual history recognize the present as a major crisis in the history of the planet. In a curious way, we are catapulting into a new century a decade or so before its coming. Society mobilized to confront the Industrial Revolution a hundred years or so ago; we are now engaged in meeting the thrust of what has been called the Postindustrial Revolution. And as Daniel Bell* insists, on convincing data, the activity of this society will be centrally based in science and technology.[1]

Bell's analysis, as fresh today as it was when he wrote it, proceeds somewhat as follows: The preindustrial society interacts with nature; the industrial society with the machine. The preindustrial society is engaged mainly in agriculture; the industrial society induces machines to till the soil, to harvest the crops, and to produce its goods. In an industrial society, people are less in tune with the rhythms of nature, more with the rhythms of time. The work is not seasonal; it is technical, planned to harness energy—provided by fossil fuels, hydroelectric power, the atom, and the sun—for the varieties of activity in a technocratic society.

The bases of postindustrial society lie in many industries—those concerned with polymers, computers, electronics, genetics, solid-state physics, and a new "knowledge" industry, based in the unbelievable blink of computerized verbal and mathematical languages. These industries are based on the availability and participation of traditionally literate and numerate people who are also capable in the special symbolic languages and mathematics of physics, chemistry, geology, biology, space science, and the like. Thus, our labors lie in fostering the knowledges, attitudes, and skills bred in a society

---

*(1973) *The coming post-industrial society: A venture into social forecasting* (New York: Basic Books).

3

dependent for its present and future on science and technology. That is to say, the agricultural and manufacturing industries, without which our society could not survive, are now based on knowledges, on skills, on attitudes that conform to laboratory-based theory and practice in the sciences and their interdependent technologies.

The essential person in a postindustrial society may well be one educated to master the machines that can utilize the hoard of data produced by other educated individuals—the so-called professionals. Manufacturing technology in the modern industry of the developed world is geared to the supply of "knowledge workers" flocking, even now, to industries utilizing so-called "artificial intelligence"—that is, the artifices of the computer—and other data-gathering machines. Successful farmers now use computer-based data: They use electronically processed information on weather, soil, and crop management. Weather forecasters use data collected by satellite. The diagnoses of physicians are increasingly based on the analyses of medical technicians, who in turn use the services of machines to do microtechnical studies employing traces of radioactive substances.

Industry is modernizing at an astonishing rate: It often places at the center of manufacturing and in command of office management those experts in science, mathematics, and technology who can use the complex array of different data-processing machines to feed the overwhelming data bases the "memories" of modern computers can accommodate. Traditional manufacturing may not compete; entire plants will need to be redesigned. Postindustrial society is thus mainly a service economy. Still, *all* citizens will be required to understand enough of the technical matters and detail enhancing and coercing modern life and living to achieve some degree of psychological safety. Witness the anxieties induced by an old disease, cancer; a new one, AIDS; a nuclear disaster at Chernobyl; and one in the offing, the "greenhouse effect."

To reemphasize, by the year 2000 it is possible that 90 out of 100 workers will be engaged in service industries, leaving a labor force of 10 percent or so engaged in manufacturing and farming. Even now, only 2 percent of the population is engaged in agriculture; in 1900 there were some 70–80 percent so employed. Thus, the science-based revolution is already moving at a fast pace; recent reports indicate that more than half of new jobs are in the service industries. The thrust is intense; it affects every facet of life.

It is apparent we stand witness to all three worlds at once: Even as the television screen brings us visions of *preindustrial* society, as we ponder the pictures of men and women in the backbreaking work of contending with nature in underdeveloped agrarian societies, we also observe *industrial* societies contending with and attending to the machine. And then, as we watch, we are bemused to see our young taking to the new machines (processors, computers, not to forget the new arts of television) that have swallowed the

4

knowledge of times past and can organize data past, present, and future. The knowledge workers using the new technologies spew out new information at an inconceivable rate; thus, scientific and technological data is estimated as doubling every 8 to 10 years rather than every 15 years, as happened a decade ago.

This, then, is the meaning of *postindustrial* and of science-based; for Western society the new age no longer means contending either with nature or with brute machines. Now, often with effective surprise, we realize that informed mind contends with informed mind, aided and pressed by information gathered by not-so-artificial intelligence, the latter, to be sure, under the guidance of *human intelligence*. Hence, an incredible task before us: to transform schooling and meld human intelligence so that it proceeds confidently and with skill to manage increasingly complex machines. But, even as the young undertake the newer modes of instruction in *schooling,* the *education* of our workers proceeds at an accelerating pace. In other words, the modern workplace now requires young prepared in the newer modes of conserving and transmitting information derived from raw data and in the more complex modes of conceptualizing the data into systematic knowledge. A word, however, on the use (and misuse) of these significant terms: schooling and education.[2]

The briefest definition of the function of each would explain schooling as an attempt to transmit the concepts, values, and skills prized by a community acting under the constraints of its customs, rules, and law (local, state, and federal). Education, on the other hand, concerns itself with the broad environment—indeed, an ecology of achievement of the widest kind—that affects all modes of life in school and out: It brings about changes in individuals and the culture, whether these changes act upon predispositions, purposes, and practices, or character and intellect. In brief, in one way or another, education affects all modes of life and living.

Education is not necessarily related to graduations, stages, or exits from any form of endeavor, public or private. Thus, education occurs in situations and institutions outside of school—at home and with family; at church, synagogue, temple, mosque, lamasery; with nonchurchgoers as well; with peers and with groups; in games and in sports; in chance encounters everywhere; from TV, books, newspapers, and magazines; definitely from a job, from work; in personal contemplation; or in retreat; from experience of all sorts. In any event, while schooling ought *not* to be equated with all of education, both do interact. In fact, educational experiences at home precede and then parallel schooling and hence are both prelude to, and synergetic with, achievement in school. However, serious misunderstanding occurs when problems, say, of drug addiction, or spread of diseases, or deviant social behaviors, are thought to be curable solely through *schooling.* Indeed, schooling may contribute to the amelioration of such conditions but

5

obviously and necessarily in combination with education, beginning with and supported, sustained, and continued by the home, community, peers, diet, medical aid, psychological aid, and the like. But, however these terms are defined, however they are attacked or defended, schooling and education form the major part of a special environment—that of society—affecting the development of the young. Having said this by way of introduction, we now turn to the special problem in schooling and education with which we are concerned.

It is understating by far to propose that the young now in our schools will be central to the development of knowledge in the 21st century. From these young should come a considerable number of the scientists and technologists who, together with those now active in science, will be vital in advancing the culture. Some 60 years ago Alfred North Whitehead (1929),* distinguished mathematician and philosopher of science, voiced his own fears about societies that do not look to the education of their young:

> When one considers in its length and its breadth the importance of this question of the education of the nation's young, the broken lives, the defeated hopes, the national failures, which result from the frivolous inertia with which it is treated, it is difficult to restrain within oneself a savage rage. In the conditions of modern life the rule is absolute, the race which does not value trained intelligence is doomed. Not all your heroism, not all your social charm, not all your wit, not all your victories on land or at sea, can move back the finger of fate. Today we maintain ourselves. Tomorrow science will have moved forward yet one more step, and there will be no appeal from the judgment which will then be pronounced on the uneducated. (p. 26)

A Necessary Thrust

Jonas Salk, who developed the vaccine for immunization against polio, has stated some of his beliefs about schooling and education in *Anatomy of Reality* (1983).‡ Knowing that teachers acknowledge as their superordinate task their duty to advance the causes of the young, he presses upon us an additional purpose; we must not only meet the needs of all, but also,

> In the face of the magnitude of our problems, we are in deep need of recognizing extraordinary human beings . . . If this is so, then it is possible that a new human preoccupation may be in the process of emergence, i.e., a preoccupation with human creativity. (p. 84)

---

*Alfred North Whitehead (1929) *The aims of education* (New York: Mentor Books).

‡New York: Columbia University Press.

By Way of Beginning

And F. James Rutherford, Director of Science Education for the American Association for the Advancement of Science (AAAS), writes in the introduction to a study he is heading:

> The first phase of this study *[Project 2061]* will be conducted during 1986, the year in our lifetimes, as it turns out, in which the most famous of all comets will be nearest to the earth. The children born that year will, on average: enter school in 1991, graduate from high school in 2004, enter the job market between 2005 and 2015, have children who start school in the 2020s, run things for two or three decades, retire from work in the 2050s, and live to see Halley's Comet when it returns in 2061.
>
> What we do as a nation during the next 5 to 10 years to reform education will affect an entire life span. Project 2061 will try to identify the content and character of education in science, mathematics, and technology for those children soon to be born, soon to start their education, soon to be responsible for the future of civilization. (Preface, 1985)*

The literature of current reform movements supports different theses on the nature of the current crisis, but all agree that the education of the generation "soon to be responsible for the future of civilization" is of critical importance. Scientists and technologists have indeed affected the future of civilization; and, as so many predict and as we're beginning to observe, our lives will be even more deeply affected by science and technology in the years ahead. Thus, our pressing need to define the kind of schooling and education in science and technology our young require.

No matter how we try, in time to come, we cannot now set aside one responsibility of schooling and education—courted, neglected, and embraced, then set aside once again—that is, our responsibility to "recogniz[e] extraordinary human beings" in our schools. But, as we shall see, however "extraordinary" these people are, first as children and then as adults, they still *require* the sustenance of colleagues and coworkers and the support of vast systems, not only of funds for research but also of a considerable number of scientists and technicians to assist their work. As we shall see, "extraordinary," talented contributors in science depend greatly on a good number of others who, while perhaps "not-so-extraordinary," are still talented. But we anticipate.

---

*From the initial proposal by the AAAS for its ongoing project on future science education. Originally funded in 1985 by the Carnegie Foundation and first called *Project 2061: Understanding science and technology for living in a changing world,* the project continues as this volume goes to press as *Education for a changing future* (Washington, DC: Author). See also NSTA's 1982 position statement on science, technology, and society in the 1983 Yearbook: *Science teaching: A profession speaks* (pp. 109–112) (Washington, DC: Author).

We should grasp the critical meaning of our responsibility; we should not misplace our efforts and energy. As teachers of science, as curriculum workers, as administrators, *we are not engaged in genetic engineering, in the manipulation of DNA. Are we not, instead, engaged in the critical acts of engineering of the environment?* In so doing, we are required to catalyze an all-important dependence: That is, we are not to isolate nature from nurture; we are to enhance their symbiotic powers: *They interact.* In effect, for data to support our practices, we shall be looking to the findings of both natural and social scientists.

A Thesis: In Concept and in Fact

We are then required to establish a central thesis, a paradigm, if you will. Once we have done so, the problem we have set ourselves flows toward useful solutions, some *verifiable,* some *adequate* for the time, all subject to further probe. The problem: *How do we design an environment in which talent (note, not giftedness) in science expresses itself?* Where do we begin? We reject the thesis that colleges and the universities are, in the end, the sole agents educating our scientists, our engineers, even our science teachers. We know enough, we have experience enough, we have research enough to press the point that the ecologies that encompass the random 16 to 18 years of education at home, at school, and in the community, and the planned environments of 12 years of schooling have a profound effect in seeding the destinies of the young. Willem H. Vanderburg, an engineer and social scientist from the University of Toronto, shares this view and has taken upon himself the description of how technology as a whole evolves and influences culture. He plans three volumes. The first begins with *children.** His reason, as one reviewer puts it: Children's socialization is what gives them, as future adults, *"the intellectual constructs to interpret reality—their culture"* (italics ours) (1987, April, p. 399).‡ That is one way to put it. Another is to say that the young begin to see themselves as part of society as they strive to develop themselves as individual entities within durable personhoods. One such entity may affirm itself in lifelong work in one of the many fields of study in science. We who teach know of this striving of the young. Teachers who probe this book will find us attending to this striving in the service of a common purpose: to fulfill the powers of the young in their undiminished pursuit of personal worth and excellence.

---

*(1985) *The growth of minds and cultures: A unified theory of the structure of human experience* (Toronto: University of Toronto Press).

‡From a review by Kathleen H. Ochs in *Technology and Culture, 28*(2).

Current literature on genetics is filled with cases demonstrating the inter-action of genetic with environmental complements: An individual is *not* the product solely of his or her heredity or solely of the environment. And, as we shall see, originative work or discovery in science, a sign of "talent" in science, must be strongly assumed to be a product of the interaction of heredity and environment at every point in the development of the young.

We consider, then, the variety of school environments that fulfill the powers of all children. Note, *all* children! For how do we know whether the young are talented in science unless they have had access to equal opportuni-ties to participate in fruitful schooling and education *early on in childhood and in youth?* That is, to *think,* to *do,* to *perform* in science. Put another way, in an open society, we are bound to assure access to equality of opportunity to all the young, so that they may try their ability and talent in whatever field and in whatever ways schooling, complemented by education, can devise. So, finally, in the course of development, they may enter science in whatever capacity they choose.

A Thesis: In Intent and Content for This Book

We are convinced that schooling and education should make the widest opportunities available to all; it is, after all, eminently clear that it is the architecture of the environment of schooling—and not the components of the helix of DNA—that is open to redesign by the citizens of the community and by those who staff the schools.

It is further clear that society requires of the schools fundamental partici-pation in any change in its direction—in the evolution of the culture. Thus, the present demand is not simply for rote changes in administration, curricu-lum, or instruction to inform the young of the nature of postindustrial society. We are, instead, obliged to redesign the school environment to allow the young, equipped at various levels of excellence, to find their places in society not only or mainly as *experts* in the incredible variety of work and works possible in an open technocratic society but also as widely literate and numerate citizens within it. Clearly, experts are needed; they will certainly collaborate to invent the new knowledges, skills, even dispositions, to help us to maintain a respectable place in the scientific, geopolitical, social, and economic life of the "global village." But what is also clear is that our schooling and education should fit all citizens—whether or not they are in science—to accommodate newer modes of life and living in this country and over the world. As René Dubos, scientist and conservationist, exhorted, "Think globally, act locally."

However we define the new world forming before our mind's eye, as teachers we have no choice but to confront a serious situation: Our survival as a nation depends not only on equipping agriculture and industry with the requisite new science and technology but also on the redesign of our schooling and education.

What course to follow? The authors represented in this volume have taken on the obligation of proposing—insofar as the redesign of certain aspects of the school environment is concerned—how we may provide for the young so they may in good time find places suitable to the privileges of an open society, but a society rushing into the postindustrial era. We incline to the free exercise of a mode of explanation sanctioned in science. We offer a variety of *works* and *views* suggesting how science talent may be fostered; to these, readers will providentially add their experiences and analyses and so construct explanations idiosyncratically individual.

Early on, Percy W. Bridgman offered a firm view of the architecture of scientific explanation in the service of designing a construct. He wrote: "Explanation consists merely in analyzing our complicated systems into simpler systems in such a way that we recognize in the complicated system the interplay of elements already so familiar to us that we accept them as not needing explanation" (1936, p. 3).*

After this spirit, the various authors writing here probe the phenomenon of talent in science in sufficient analyses that we may in turn sort the "complicated system" into the "simpler systems" undergirding constructs or models, or designs of a special school environment: *one that enables the young with science talent to select the opportunities for performance in science.* The construct in this book lends itself to concept seeking and concept forming along these lines:

• Is there a viable concept of talent in science?
• If so, how do we seek the talent as expressed in the period of schooling?
• Further, how do we foster it in the activities we call schooling and education?

It is becoming ever more clear that the environments invented for the schooling of those we have variously called "gifted in science," or "science talented," or "science prone," or "students with high-level ability in science" are now recognized and have achieved a certain commonality. In this account, we follow a certain route, charting in three interrelated parts of a whole the commonalities that may be posited with some assurance. The contributors to this volume embody not only the disciplines and values of scholars and scientists but also those of teachers. Thus, here and there the reader will find a conscience marching to the cadence of Thoreau's different drummer. Nonetheless, the striving of all the authors is to design *models* or *paradigms* (in the scientific sense of the terms) of "talent in science": that is, modes with which to think and arenas in which to share ideas with others

---

*The nature of physical theory (New York: Dover Publications).

probing similar fields. A caution: The models come out of a science of practice—not yet one of the laboratory. That is, within the authors' experiences and even experiments, the models have worked; the talented in science have prospered.

Purpose and Principles

First, out of his long experience, A. Harry Passow traverses a large terrain to probe the variety of environments designed for the gifted, commonly called "programs" or "curriculums." As he generalizes their purposes and principles, it becomes evident that, within the geography of curriculums for students who have been called "gifted and talented" in science, there are at least two purposes: first, a curriculum that serves all those who will grow up to be citizens of a society in which science and technology have enormous impact; and second, a curriculum that attends to those who intend a lifework in science or technology. Clearly the work of the "gifted" in science is untenable without support. Leaders of research teams in the various fields of science depend on a population of administrators and assisting scientists, as well as a professional managerial and clerical staff that understands and assists the purposes and practices of scientific research. In addition, it is essential that the populations who make up an open society understand in general the aims and the works of scientists; these citizens may thus be empowered to aid such political processes as are required to support scientific work with the resources and funds needed to carry on research. And, what is more, in understanding science such an informed populace can affect the uses of the sciences and technologies; demonstrably the latter have increasing impact on that complex of life and living we call our culture.

Abraham J. Tannenbaum, a psychologist and an experienced teacher of the gifted young, clarifies and systematizes progress in the areas of "giftedness" and "creativity," fields in which he has pursued definitive studies. Because meaningful work cannot be done without the systematic assertions on talent required for further study, his findings, his definitions and conceptualizations are essential. In this volume, he probes what he calls a "tantalizing phenomenon"—creativity—and distinguishes between the promise of talent and its fulfillment. Tannenbaum investigates the singular personal attributes that undergird the visible encounters of learners within those environments that foster the emergence of talent. He identifies five psychological factors and their social linkages between promise and fulfillment, all of which "interweave delicately as if to form a subtle filigree" we call talent. And he addresses some of the paradigms that have accompanied designs of the environment intended to advance the causes of the gifted.

But, recall once again, ours stubbornly remains a science of practice. And, thus, the works of wise, persistent scientists, and those of dedicated teachers

11

who know science and the practice of scientists, and those of the careful creators of appropriate designs for schooling are informed not only by theory but also by practice. We turn to Calvin W. Taylor and Robert L. Ellison, who review certain aspects of their many researches in the field of science talent. They present a broad view of the researches of psychologists, scientists, and scholars engaged in analyzing the "nature and nurture" of aspiring and mature scientists. They posit two ways of nurturing science talent: Knowledge Outcome-Based Education and Talent Outcome-Based Education, melding the two into a schema for the effective nurturing of the talented. Thus, they probe the "multiple intelligences" that stand behind the complex we have called "science talent."

Passow, Tannenbaum, and Taylor were among those who engaged in the early work that laid the foundations of the movement to innovate programs designed to find and educate the gifted in the 1940s, 1950s, and 1960s. It was my good fortune to press on with my probes at about the same time. I* had begun as a research scientist only to find myself more interested in the phenomenon of scientists at work than in their phenomenal work. Later, in the public schools where I taught, my colleagues and I were able to develop a channeling environment that permitted the young to identify their various interests so that they were in turn able—without the hurdle of an entrance examination—to choose (or not choose) to immerse themselves in an augmenting environment. There, sustained by both individual and cooperative effort in the learning of science through observation, experiment, and investigation, students had the opportunity to demonstrate talent in science through *originative work, through performance*. These studies, in turn, led to the development of two models: one, under the rubric of "ecologies of achievement," embraces environments that affect the expression of talent in science; the other, "self-selection," describes a model for students' entry into the kinds of originative work within the capacities of the school.

And then there is Gunther S. Stent, a scientist who sees scientist and artist within a composite sphere, both concerned with the organization of "new and meaningful statements." He probes the hidden likenesses and differences among artists and scientists and, in so doing, informs our sensibilities. He asks, in effect, whether the thrusts of our yet undeciphered minds are similar enough to outweigh the differences of certain obvious habits of scholarship. Is there not "uniqueness" as well as "commonality" in the ways of scientists and artists? Stent's meaningful statements condition our thinking and require us to look to our explanations of the act of origination and discovery. He suggests that, while probing differences between the acts of artists and

---

*The occasional personal reference is to the editor who drafted this section: P. F. B.

scientists, we sometimes inadvertently set aside the hidden likenesses of all those who originate and discover.

The thrust, then, of this opening section of the book is clear. It concerns the schooling and education of those individuals who turn their potential as students into that singular performance of scholarship and personality: the perdurable, verifiable, self-correcting mode and manner of the *performing scientist*.

Programs: Certain Bases for Planning Curriculum and Instruction

We turn now to works of teachers and administrators who have not only designed schools for the gifted but also have devised programs and practices (models, if you will) for those who will turn their gifts to science. We emphasize a paradigm, based in solid studies in the genetics and the biological development of organisms, demonstrable in all the plans that have been proposed by the authors to bring forth talent in science. *The DNA of the individual interacts with environment at every point in the development of the organism.*[3]

What, then, is lost if there is little or no work in science in the elementary school? Or middle school? After all, it is surely a safe notion that experience with science at an early level of development—whether in education at home, or in kindergarten, surely in primary and middle school, and of course the high school—furnishes valuable lessons in both general learning and specific knowledges, attitudes, and skills. Also, the universality of the paradigm leads us to postulate that those modes of instruction that have been found to call forth the kinds of attitudes and intellectual activities characteristic of science should be available at all levels of instruction and utilized where appropriate in the variety of curriculums adopted by different schools.

Indeed, practically every author who has discussed instruction in this volume has adapted the variety of approaches in instruction in free application to early childhood, primary, middle school, and high school programs. Thus, useful among other practices to all learners at many levels of schooling are open classrooms and a variety of inquiry modes, whether practiced in cooperative learning, in individualized and team instruction, and/or investigation in field and laboratory; whether employed in problem solving and problem doing, "hands-on" and "brains-on" activity, or lecture; whether through the utilization of audiovisual aids, computers, libraries, or mentorship in all its varieties. It is to emphasize that the modes of learning that inform all the studies in this book (titles aside) are thus not to be segregated to the elementary or high school levels; it is taken for granted that each reader will select what is necessary to create the most useful of environments for the young within the sphere of his or her responsibility.

Thus, Annemarie Roeper, who, with her husband George, invented the Roeper School for Gifted Children, offers us the richness of experience as she explores the ways of the young in their early growth and development and applies her wisdom to children as they "science." (Some of us view science as a verb masquerading as a noun; therefore, we conjugate it.) Roeper notes not only the "ways" children learn but also the "whys"; she regards each child as preciously idiosyncratic. Thus, while she sees the child within the group, she emphasizes that he or she is seeking individual entity. In a sense, Roeper maintains that experience in sensing and ordering the natural environment furnishes the baby, the preschooler, the kindergartner, the elementary schoolchild with those elements that feed his or her "instinctive craving for the world of understanding." [4] Roeper explains, yes, children can probe; they can perform small acts of origination; they can, and do, science.

Next, Steven J. Rakow proceeds to the architecture of the environment of the middle school, too often a dread land of random experience uninformed by the nexus that frequently shapes the world of the primary school. Taking from his experience as a teacher of the young and as a teacher of teachers, he offers a plan for the young within heterogeneous middle schools—the latter, recent comers to the architecture of public school systems. As yet, the middle school is still seeking its special place in the design of curriculum and instruction particularly applicable to the young in their years of rapid physical, mental, and emotional growth. Rakow takes a simple position: The middle school is to tend the young both in their essential individual development and in the fulfillment of their powers. And, in so doing, to aid their entry into the postindustrial 21st century. For all the young about whom Roeper and Rakow write are precious resources who will advance the society burgeoning beyond the year 2000.

Most of the writers in this volume dwell on students who fall easily into the category of learners. Somehow, such students are able to "accumulate advantage" in the kinds of environments we are recommending. Bill G. Aldridge and Deborah C. Fort would remind us that this good fortune is not shared by all students, in all schooling, in all education, indeed in all of life. Aldridge and Fort insist, rightly, that not all environments in schooling and in life provide equal access to opportunity. Nor do all environments offer equity—justice—nor do they guarantee the right to achieve entity to all in an open society. Aldridge and Fort demonstrate that there can be an "accumulation of disadvantage" but that it can be ameliorated; they describe various stages that could help assure equal access to wide-ranging opportunities in schooling and education. They wonder how an open society can afford to squander the promise in a child. Aldridge and Fort press the case of the single individual who can be lost in the arms of the larger statistic of those measures purported to test "intelligence" or "creativity." Thus, according to the present state of our knowledge: Intelligence and creativity "tests" are not per se measures of talent in science.

Next, Sid Sitkoff calls our attention to an essential organization providing for the science classroom; he explains the role a science center can play in the support of teaching and learning science. He describes and samples for us a most important aspect of a plan for instruction. Science is indeed a significantly "hands-on" activity—not, of course, to set aside the "brains-on" planning involved in doing science or in science teaching. Sitkoff describes a citywide program that supports laboratory and classroom activity by providing equipment, lab materials, learning materials, and a host of aids necessary to "sciencing." The community enters into the support system he has devised. To what end? Science centers help the young learn through all their senses—sight, sound, touch, smell, (rarely taste). Their nervous systems vibrate to many kinds of feeling, kinaesthetic and emotional.

Sigmund Abeles who, as a science supervisor, has guided Connecticut's teachers in their development of appropriate curricular, instructional, and administrative devices in many aspects of science, directs our attention to the variety of programs meeting the needs of citizens and scientists-to-be in urban, suburban, and rural communities. He posits that what is missing is a stable year-in, year-out plan for the instruction of the gifted. Thus, he calls our attention to the dry periods where schooling and education may fail to serve those who have not yet defined their purposes and who seek direction and need guidance and opportunity. Programs for the nurture of gifted students in science are still young; those programs operating are not organized nationwide and are still in search of a guiding philosophy and principles. But certain of the programs he writes about take on a new flavor: They combine schooling and education. This is a long-awaited unity, only a beginning to be sure, but one with great promise.

Then we hear from the specialized schools, which accept their students in numerous ways—for example, on the basis of prior achievement, or admission by tests, and/or recommendations of teachers. Charles R. Eilber and Stephen J. Warshaw describe the environment of the North Carolina School for Science and Mathematics (a state residential school) where they teach and work. Established in 1977, it is already serving as a model for programs both within and without the state. And Milton Kopelman, Vincent G. Galasso, and Madeline Schmuckler describe the architecture of the programs in the Bronx High School of Science, which recently celebrated 50 years of nurturing the gifted who turn their talent to science. Bronx Science's rich program of studies includes opportunity for originative work in science, calling for active performance as well as high general ability. Not only do we gain insight into the personality and scholarship of students and teachers at Bronx Science, but also we may infer from the description of the program the problems of designing so rich an environment for so demanding a population of commuters. Here too we find students using the resources of the community's laboratories and its scientists—a powerful blend of schooling and education. Kopelman and his splendid teaching staff not only carry on the tradition of Morris Meister, the designer and first principal of the Bronx High School of Science, but also advance the learning of the young in another way that would gratify the school's founder. At Bronx Science, large and small groups as well as individuals plan and carry on investigations and, in the time-honored way of the scientist, share with the group in and out of school the results of their research, a practice that is, as it should be, central in that environment we call the ecology of achievement centered in the community. Meister would marvel at the steady high road the school has taken.

"Magnet schools" in the public sector and private "independent schools" for the gifted tend to emulate this model: Performance in the arts of investigation in science acts as an indicator of promise of talent in science. Recall, however, that some heterogeneous schools have devised similarly effective models. But, generally speaking, whatever the structure of schools for the "gifted," if they incline to offer an optimum environment for the "gifted in science," they tend to follow the model given over to a combination of channeling and augmenting environments that demand an earnest of potential: rigorous scholarship in science and a demonstration of ability to perform in the art of scientific investigation.

Irving S. Sato concerns himself with the bread-and-butter task of developing a curriculum for all gifted learners. Then, he stresses the nature of the modifications such a curriculum should manage not only for those "gifted/talented"* in science, but also for those gifted learners who are not especially disposed to science. Sato describes the disciplines required to develop a curriculum. In so doing, he reminds us of the tremendous surge in the 1950s and 1960s in curriculum building for all students in science as well as those "gifted/talented." That wave is spent; now we see another crest of development of new plans for instruction. The thrust and rubric of the current curriculum: Science-Technology-Society. It aims, at last, at a meld of scientist, social scientist, and citizen.

Passow returns to remind us that help for the gifted in science and mathematics is not limited to schooling's offerings. The general community of scholars now makes available educational programs for aspiring scientists and mathematicians; thus, once again the community makes available its educational resources to the young. The programs Passow details offer to the young wide-ranging yet incisive openings into the philosophies and practices of the scientist.

---

*The terminology acceptable to most legislative provisions.

Then, Pinchas Tamir has taken upon himself a difficult, disturbing task, and his striving rewards us with a significant problem. He describes—all too briefly, for the editors presented him with their request too late for the kind of scholarship he would have chosen—programs nurturing the gifted provided by a variety of countries worldwide. Putting aside personal predispositions that governments should offer access to equal opportunity, Tamir sets the offerings of closed and open societies side by side. But the lesson is clear: All societies do not serve all children.

We turn now to examine certain practices that focus on instruction—on teaching the young in large and small groups—and, where desirable, that assist the young in their plans for individual performance in research. In fact, schooling ought never embrace an organization and administration that shut out the "constructive affection" (Margaret Mead's phase) necessary to free all the young in pursuit of their powers; of course, diverse abilities grow in such an environment. That is to say, the young, seeking expression of their gifts and talents, tend to flourish in a beneficent environment within the ecologies of achievement melding school and community.

To reemphasize, these environments, these ecologies of achievement, are to be open to *all* citizens in an open society, a society designed to elicit a balance of opinion without coercion. It is necessary, then, that teachers develop the personal inventions, art-sciences, or crafts we call *teaching*—wide-reaching in purpose and practice, in effort and effect—all-embracing in the new world we are just beginning to apprehend. Here, thus, a number of teachers share their attempts to devise particular practices in schooling and education that have been fruitful in developing environments that call forth the potential within gifted students.

Evelyn Morholt and Sigmund Abeles join me in dissecting out two kinds of environments within science as it is taught in schools generally. Examined first is the work expected in the weekly "experiment" usually finished in a single or double class period. These so-called "experiments" are really exercises in problem doing, for the "discoveries" made in the solutions are known and, indeed, planned for, in advance. Most students with high-level ability as measured by IQ scores do these exercises with good conscience and good results. Examined more fully is another kind of environment, one that exposes in the young their potential to be scientists. This environment offers opportunities to perform in research and appeals to students whose persistent dissatisfaction with present explanations of reality leads them to put their energies and scholarship in a quest for a new expression. For these young, one useful strategy presents itself: Let them try originative work in science. Let them engage in problem solving by wrestling with a "true" unknown over a year or so; of course, there is always the mentor at hand. Thus, we detail the forms of instruction of an early apprenticeship for those who wish to do science. After James B. Conant, we call these strategies and tactics in instruction, modes of *well-ordered empiricism*. This process involves that complex activity in thought and action once called "scientific method," now considered an "art of investigation" (W. I. B. Beveridge's phrase).

So, too, Joseph D. Novak examines critically the methods of teaching that call for rote learning as well as passivity of our students. He approves of neither and proposes a philosophy and practice that is "constructionist," which engages a newer interpretation of how scientists learn and how they investigate. Novak presses for a recognition not only of the dispassionate mode but one that understands underlying feelings. He is mindful of Michael Polanyi's recognition of "passion" in science. And so Novak paints a picture of inquiry—and gives examples of methodology that set aside rote and call for active striving in thought and deed: The young are to think critically and do what they plan constructively in problem solving.

Roger T. Johnson and David W. Johnson also propose a continuation of an effective practice in instruction. Probing failures in instruction, the Johnsons find classrooms inadequate and confining when they fail to offer the "cooperative learning" that serves the young so well. We have, the Johnsons posit, forgotten the one-room schoolhouse and its practice. They also emphasize proficient performance in learning in groups, with students face to face, rather than passively regarding a teacher engaged in a lecture. The Johnsons also remind us of certain kinds of cooperative learning in science: Scientists perforce learn from each other.

Finally, Robert A. Day assists us with a lucid account of why scientists and for that matter all scholars *must* write. He offers perspective into a device too many have neglected: clear exposition of purposes, or plans, or devices for schooling and education, or advances and discoveries, so that others who work in the variety of fields in science may conserve, transmit, correct, and expand their fields. In his dialogue, he reminds us that scientists are part of a society: They interact; they communicate; they inform in speech—and in the more lasting written word. Are we not, as writers and readers of this volume, even as scientists, a gathered people drawing on each other's knowledge, attitudes, and skills? Day develops his thesis so that, in effect, his paper becomes a primer for students at many levels on the planning, drafting, and writing a summary of research—or for that matter, any report requiring the sequential mustering of evidence in science, schooling, or education.

Personal Reflections

Are there indeed any "hidden likenesses" or "hidden differences" in the "crystallizing experiences" (Howard Gardner's term) of scientists and science teachers as they reflect briefly on events that turned them to their life's work? Perhaps we shall find their crystallizing experiences in

- a particular life-affirming event?
- a particular teacher or mentor, a writer perhaps, a key person who has become a permanent and cherished resident in mind and memory?
- a particular experience that strikes home and remains there?
- a period of study or the teacher or teachers who guided it?
- none of these?
- all of these?

Under the contributing editorship of Gerald Skoog, a science teacher and teacher of teachers as well as an administrator of research and academic programs, a constellation of scientists and teachers, who have hollowed out their niches through their distinctive ways of work and thought and being, offer their reflections. All are exemplars of creative thought and action in their work. All are scientists, or teachers of science, or both.

Isaac Asimov
writer

Stephen Jay Gould
scientist and writer

Joshua Lederberg
geneticist (Nobelist)

Lorraine J. Daston
historian of science

Francis J. Heyden, S. J.
geologist and astronomer

Lynn Margulis
biologist and writer

Robert A. Rice
administrator and teacher
of teachers

Gerald Skoog
administrator and teacher
of teachers

Glenn T. Seaborg
chemist (Nobelist)

A Beginning Library: Gifts, Talents, Science

Then Morholt returns with Linda Crow to face this question: What do teachers, in their attempts to design a program for the talented, need to have at hand in order to fashion an appealing and solid program? From the multitude of publications now available, they have gleaned a small but useful library. Such a library was not available in the early years of study of the talented in science—some 50 years ago. But Morholt and Crow have fashioned an annotated bibliography that furnishes a foundation for those who would develop programs and practices for the gifted, and, in so doing, give room to development of the wide swath of excellences, allowing expression of many talents, large and small.

A Promissory Note

Knowing that we stand at the beginning of our understanding of how the human creates, it must be that, at present, work on modes of evoking creativity must remain a promissory note, conclusive only in the fact that its theses continue the necessary search. But note that, even as we state our doubts, the host of scientists of past and present are joined in creating a foundation of newer knowledge, newer skills, and newer attitudes, on which we may manage to create our future. In effect, a new world is being created whose design even now is not fully within the grasp of brain and mind. Yet we know that, in science, the creation of newer knowledge lies not only within the capacities of scientists now at work; it lies now as well in the young and their teachers in our classrooms.

We think of the future and note that we are required to plan for our young, who will be obliged to conduct their lives in a future just barely visible. And we face a discomforting proposition: Society without a future does not need to plan for one. Is it not becoming increasingly true that the young—like us—will find their futures bound not only inextricably with the lives of others in this our land but also with those elsewhere? All our futures are interwoven and fixed on so small a planet whose course is fixed in so vast a universe. It is plain then, is it not, that to teach is to prepare the young for life and living in present and in future? Who is to say that the future may not be bright? Who is to say that the young may not live the lives we—all of us—dream for ourselves?

* * * * * *

Acknowledgments

Clearly, a work is the invention of an individual or of a group, acting within an ecology of achievement—that is, within a singular environment. Also, that environment is almost never the invention of a single individual but comes out of the endeavors of the uncounted. And it is the sharing of the creation of that environment that has been the willing work of the authors and editors of this volume.

Thus, if I may be permitted a personal comment that I believe is to the point, my uses of "we" and my declining in general the use of the personal pronoun is not a matter of style, nor is it sophistry. It is an obligatory acknowledgment of all those who contributed to the thought that nurtures this volume: thought that migrated into the bedrock of our discussion without explicit knowledge of the contributor. Who knows the precise authorship of, or where, or whence, or how a thought expressed in this paper came into being? Hence "we."

In a sense, this book is an affirmation by 34 authors and editors that the intellectual and emotional commitment to science as a lifework may well begin in the schools. The significance of this commitment is no longer couched in the obligatory "aims of education:" It has become emergent. The society that ignores science as part of its cultural imperatives may well be at hazard. So it is that the contributions that are the heart of this book are not in the nature of prolegomena but are statements of the mission by the authors to whom we, as editors, state our deep indebtedness and abiding thanks.

Theirs is not only a vision of science as a root source in advancing an open society launched into a postindustrial era but also is an unmistakable assertion that free and full and equal access to opportunity in science for all the young is a guarantee of our freedom. Further, denial of equal opportunity to demonstrate a talent in science puts us at hazard. A mature schooling in science has not only or mainly become a cultural imperative affecting science, technology, and society, but also in its own right is based on a certain discipline of responsible consent and in its twin, dissent. Thus, science may well teach a certain integrity and the privileges of commonality and diversity in perception and thought about the way the world works.

I, personally, am deeply obligated to A. Harry Passow, my co-editor, and to Deborah C. Fort, association editor for the National Science Teachers Association; their gifts of intellection and hard-won experience are to be found in the development of the substance, structure, and style of the work. Harry Passow brought not only his experience as teacher of teachers and of science but also his profound understanding of the nature and nurture of giftedness as a doyen in the field—so acknowledged worldwide. Deborah Fort, a teacher of writing within and without the university, brought her extraordinary, finely tuned knowledge and skills as a writer and editor to achieving continuity within the diversity of a richness of texts, thus molding the work to its particular purpose: sustaining the young in the pursuit of their special and different excellences as they seek to fulfill their powers. And former NSTA president Gerald Skoog, a teacher of science and of teachers of science, brought as contributing editor his insights into the ways of scientists and scholars to augment the work with vignettes of the reflections of scientists, scholars, and teachers. In all this, we gratefully acknowledge the selfless assistance of Diana J. Holmes, who gave generously of her time to commit her editorial wisdom and superb skills to the completion of the work.

But it is also to the great and good authority of the National Science Teachers Association that all of us who bear responsibility for this work give our thanks. The Association's Yearbook Advisory Committee chaired by John E. Penick and the Advisory Board for Special Education formulated general principles for the volume. It was Phyllis R. Marcuccio, director of publications, who gave origin and impetus to the course of the work. Shirley L. Watt, managing editor of special publications, with a sure hand guided the work through the uncertainties that a first project in any field endures. Both have borne patiently, wisely, and graciously with our demands. And to Bill G. Aldridge, executive director, for his strong faith in the work, go our high regard and admiration not only for his leadership in the schooling and education of the young but also for his acumen in bringing the Association to its present dominion.

We *are* aware that we are at the beginning of the study of creativity in science. This is surely so.

Gregory Bateson (1979)[5] taught me to beware of finality, and I'm grateful for his wisdom. He cautioned: "What remains true longer does indeed remain true longer than that which does not remain true as long."

P. F. B.
August, 1988

Endnotes

1. Daniel Bell summarizes a great deal of history in this fine volume and probes basic knowledge in cultural change. We have drastically redacted his extensive development.

2. Bernard Bailyn presses this distinction (which is still not accepted in common speech) extensively in *Education in the forming of American society* (1960) (Chapel Hill: University of North Carolina Press); and Lawrence Cremin extends it in several works, but mainly in *American education: The national experience, 1783–1876* (1980) (New York: Harper and Row). In our view, the confusion causes mischief when the terms are used interchangeably, especially in the political arena.

3. The model we present is an old one, but honored in misunderstanding. Philosophers spoke of "nature and nurture" in the 17th century; farmers talked of "seed and feed" a century later. Thomas Hunt Morgan, in his laboratories at Columbia University, *demonstrated* it in his seminal genetic studies of fruit flies (*Drosophila*) in the early 20th century.

4. Gardner Murphy's phrase (1961) from *Freeing intelligence through teaching* (New York: Harper and Brothers), p. 32. It is increasingly thought that the child's apparently random groping may be "gene-driven." In effect, there is mounting evidence to consider that an "active" environment (including active parents), rich in learning opportunities, stimulates the child to probe and learn. A passive environment seems to reduce exploration. On this, see Sandra Scarr and Kathleen McCartney (1983) How people make their own environments: A theory of genotype-environment effects, *Child Development, 54,* 424–435.

5. *Mind and nature: A necessary unity.* (New York: Bantam Books). For a more extensive view of the nature of scientific "knowledge," "truth," even "accuracy," see Thomas S. Kuhn (1970) *The structure of scientific revolutions* (2nd ed.), (Chicago: University of Chicago Press), or W. I. B. Beveridge (1957) *The art of scientific investigation* (New York: W. W. Norton).

Part II
Purpose and Principles

The Educating and Schooling of the Community of Artisans in Science

A. Harry Passow

More than three decades ago, Dael Wolfle, writing for the Commission on Human Resources and Advanced Training, observed,

> The nation as a whole profits from the fact that some people possess the ability to design a dam, to plan an automobile production line, to isolate an antibiotic, to conceive an atomic power plant, to develop high-yield hybrid corn, to compose a symphony, to settle a labor dispute.
>
> Creative and imaginative people neither work alone nor in a vacuum: Each builds upon the work of his predecessors; each is dependent upon industrial workers, maintenance crews, farmers, clerks, and technicians; each is nourished by a favorable climate, rich natural resources, and a tradition of freedom and initial initiative. But after all the other credits are properly assigned, the nation still owes much to a small group of able men [and women], men [and women] who play a critical role in scientific, industrial, technological, and cultural developments. (1954, p. 1)

Two overarching purposes of education and schooling in a democracy are the fullest development of individual potential and the nurturance of specialized talents to fill society's need for creative, imaginative, productive individuals. The two purposes are, of course, closely interrelated. The impact of science and technology on society's needs for specialized talent are so well known that no elaboration is required here. As specialization breeds new

27

specialization, society's need for educated, creative scientists and technologists grows exponentially.

To fulfill these two purposes, the task schools and other educating institutions face is to create the environments and the conditions that will enable individuals to identify and understand their potential for outstanding performance in an area of specialized talent and to develop the knowledge, skills, insights, understandings, and values which will enable individuals to recognize and realize that promise—for personal self-actualization as well as to fill society's needs for specialized talents. Talent undeveloped has a significant impact on both self and society in terms of personal self-actualization and service to society. The dual consequences of unrealized potential are a loss in personal fulfillment with accompanying frustrations and a loss to society of the high-level performance that comes from developed and utilized talents.

"Finished" scientists, mathematicians, and technologists are not produced in elementary and secondary schools since postsecondary and even graduate study is essential for the development of most specialized talents. Rather, what such programs do is to activate and motivate the commitment and the development of knowledge, skills, and affective behaviors that contribute to nurturing such talents. In a sense, such programs start the individual from potential talent toward the development of talented performance.

Each area of specialized talent contains the knowledge, content, and substance of a discipline; its special methodologies and processes; and its modes of problem definition and inquiry. All of these, along with the exercise of creativity, innovation, and originality, are essential ingredients of talented performance. It is in the creation of a total educational environment that a community of artisans in science emerges as a source for identifying and nurturing talent potential.

Four Curriculums and the Learning Environment

A curriculum is an instructional plan, a set of intentions to provide opportunities for learning engagements and to create an environment within which such learning will occur. All children and youth participate in what can be thought of as four curriculums: (a) a general education curriculum, (b) a specialized curriculum, (c) a subliminal/covert curriculum, and (d) a curriculum of nonschool educative settings. Together, these four curriculums constitute a learning environment. They are not discrete, separate, or unconnected. Rather, they are interactive and interconnected.

The general education curriculum, which provides the base on which specialized talents can be nourished and nurtured, is aimed at developing the knowledge, skills, insights, and attitudes needed by all individuals if they are to participate effectively in society. At the most basic level, this curriculum fosters the skills of literacy, numeracy, and citizenship and contributes to the development of higher-order thinking and learning-how-to-learn skills.

This general education curriculum provides the substantive cognitive and affective content and processes needed to help a person function effectively in a complex society and builds the foundation for nurturing specialized capacities. No matter how great an individual's potential is for superior performance in some area(s), he/she must acquire the basic knowledge, skills, and attitudes needed for further learning if that potential is to be realized. The general education curriculum provides the common thread of humaneness which ties together the potential poet, scientist, mathematician, musician, and others with specialized talents. Planning for this curriculum takes into account those individual characteristics that affect rate, breadth, depth, nature, and quality of learning.

The specialized curriculum is aimed at providing learning engagements and opportunities that will enable the individual to acquire the skills, knowledge, insights, values, and understandings needed to recognize and to nurture his/her potential. David C. McClelland, Alfred L. Baldwin, Urie Bronfenbrenner, and Fred L. Strodtbeck (1958) observed that "we still know far too little to be confident about how to develop talented performance out of talent potential" (p. 24). Nevertheless, we can be reasonably confident that an individual is not very likely to become a productive scientist without studying science, acquiring appropriate laboratory and computer skills, understanding the basic nature of scientific inquiry, developing competence in related disciplines, committing one's self to science, and practicing the behaviors of scientists by "doing science." The specialized curriculum is the one in which the individual has opportunities for the depth and breadth of study in an area of talent, hopefully where he/she will be introduced to the advanced training and inspirational teaching or role models needed to develop potential into performance.

The covert/subliminal curriculum is that which emerges from the climate or environment created in the learning setting. Self-concepts, values, attitudes, ideas about excellence, willingness to pursue particular lines of inquiry, task commitment and perseverance, and other affective and cognitive behaviors are some of the things which students learn from each other and from the classroom and school environment as well as the larger community of learning. The classroom and school climate, the interpersonal relations, the transactions, formal and informal, which take place are all powerful mediators of affective and cognitive growth. Moreover, this climate has a strong impact on student engagement with more formal or structured learning. Whether, to what extent, in what ways, and how committed students will be in engaging in or designing learning opportunities depends on the climate created for the pursuit of excellence. The classroom, school, and community environments have a profound effect on student motivation, involvement, and performance.

Education and socialization take place in many different settings with student cognitive and affective growth acted upon by the formal and informal, structured and unstructured learning, which makes up the fourth nonschool curriculum and which takes place in many extra-academic institutions and agencies—instruction that goes on in the many nonschool educative institutions and agencies. These learnings are brought into the school and influence school curriculums—or should do so. The family's and community's values affect which talents are valued and supported. And there are many human and material resources in the community which can be used to enrich learning opportunities. Community-centered, experience-based learning extends the classroom, making the entire community a learning environment.

Thus, nurturing individual potential is a process not restricted to a set of courses provided in a classroom but involves the creation of a total learning environment in classroom and school as well as in nonschool settings, beginning with the family and extending into the community and beyond. It is in such a context that a community of artisans in science is created, schooled, and educated.

Creating a Total Learning Environment

The task for educators is the creation of an environment which will help students engage in learning activities to identify their talent potential, stimulate their pursuit of activities that will nurture this potential, and provide opportunities to practice behaviors associated with superior performance in those areas of talent. Basically, what is needed are the channeling or augmenting environments described by Paul F. Brandwein (1987) which "excite the young to pursue an interest" (p. 34). Such environments will be rich enough and sufficiently engaging that youngsters, by their products and their performances, will identify their potential talent.

There are a variety of programs and procedures used to identify "gifted students"; processes that go beyond testing for general intelligence and even specific aptitudes by providing opportunities for identification through performance are probably most productive. Environments that link enriched learning opportunities with identification procedures are most likely both to discover and nurture talents.

Developments in science and technology make an understanding of their nature and impact on society an imperative for all citizens, but especially for those whom we believe are gifted or talented.* Not all gifted students have

---

*I am here using these adjectives virtually synonymously.

the potential, interest, and/or motivation to become scientists or technologists, but all need an understanding of the nature, meaning, and methods of science, if they are to function in a postindustrial society. All gifted students, whether or not they eventually become productive scientists, should have a basic grounding in science as an integral component of a sound general, liberal education.

All need to acquire basic concepts, ideas, and relationships which constitute the structure of science as a discipline. All need to understand the methods of inquiry appropriate to science. All need to understand the nature of scientific knowledge, to learn how new knowledge is generated, and to recognize the sources and bases for extending such knowledge. All need to understand the interactive relationships of science with other aspects of culture and society. All need to recognize and appreciate the aesthetics of science, its form and beauty, as well as the elegance of its solution of problems. All need to comprehend the significance of values and morality (or lack thereof), which affect and are affected by science. All need to develop inquiring minds and learn how to raise questions and define problems worth pursuing.

In sum, the knowledge, insights, understandings, attitudes, and values of science and technology, which should be an integral component of the general education of all children and youth, have a particular significance for that of those with the characteristics and behaviors that lead us to believe they have potential for outstanding performance in socially valuable areas— that is, those we would label as *gifted* or *talented.*

Those students who manifest the special interest, motivation, creativity, critical judgment, high intelligence, and other indicators of potential for becoming productive, practicing scientists need opportunities to develop their areas of specialized talent. These opportunities should go beyond the science-technology education experiences provided all gifted students. Students with special talents in science—some latent and some manifest—need differentiated, appropriate educational experiences in particular contexts or settings.

While we do not have an absolutely precise fix on scientific potential and how to nurture it, studies of scientists and their development over a period of years suggest some common characteristics. Some years ago, Charles C. Cole, Jr., summed up our insights concerning scientific ability as follows:

What makes for good scientific ability? Here again we are dealing with a complex subject. First, there is reason to believe that scientific ability is a function of high general intelligence. Second, whether it is a special ability that is inherited or one acquired and nurtured through opportunities has never been clearly demonstrated. Third, it is easier to identify what accompanies scientific ability than to describe it itself. Scientific ability is linked with unusually good ability to handle spatial concepts, sound judgment, strong powers of inductive reasoning, a certain

31

fluency of ideas, good memory, and a quality that might be called a special kind of stick-to-it-tiveness. The ability to be intellectually and emotionally self-reliant, or "inner-directed," to use Riesman's term, seems to accompany good scientific ability. Other than that, few people seem really to know what scientific ability is. (1956, p. 25)

Cole's statement about "good scientific ability" was based, in part, on Brandwein's earlier working hypothesis: *"High-level ability in science is based on the interaction of several factors—Genetic, Predisposing, and Activating. All factors are generally necessary to the development of high-level ability in science; no one of the factors is sufficient in itself"* (1955/ 1981, p. 12). Included among Brandwein's genetic factors were high verbal and mathematical ability and adequate sensory and neuromuscular control. His predisposing factors included traits he called *persistence*—a willingness "to labor beyond a prescribed time ... to withstand discomfort ... to face failure"—and *questing*—"a dissatisfaction with present explanations of aspects of reality" (p. 10). Brandwein's activating factor included "opportunities for advanced training and contact with an inspirational teacher" (p. 11). Most of these "factors" are not ones that lend themselves to precise measurement by paper-and-pencil tests and, consequently, Brandwein proposed an operational approach—one which created an environment that would facilitate self-identification by offering opportunities for students to demonstrate their abilities, persistence, and questing through their performance and products, and one that would generate activating factors. Brandwein's analogy puts this operational approach in clear perspective:

> The Operational Approach is more like a training camp for future baseball players. All those who *love* the game and think they can play it are admitted. Then they are given a chance *to learn how to play well* under the best "coaching" or guidance (however inadequate) which is available in the situation operating. No one is rejected who wants to play. There are those who learn to play very well. Whether they can play well (in the big leagues) is then determined after they have been accepted into the big leagues (college and graduate school). (p. 23)

In a study focusing on the nature and nurture of exceptionally gifted persons who eventually excelled in their careers, Sidney L. Pressey (1955) observed that all such individuals seemed to have had (a) excellent and early opportunities for abilities to emerge and to be encouraged by families and friends, (b) access to early and continuing individual instruction and guidance, (c) opportunities for regular and sustained practice in their areas of giftedness together with opportunities to progress and develop at an individualized pace, (d) participation with other performers (not necessarily of the same ability) which provided a basis for continuing and stimulating relationships, and (e) stimulation by frequent and increasingly strong opportunities for success (pp. 123–129). Pressey's observations, which underscore the importance of environmental influences and climate for nurturing talent, also have implications for the design of a learning community for students with potential for becoming scientists and technologists.

Creating a Learning Community for Potential Scientists

The elements of a design of a learning community include opportunities for
- acquiring the basic skills and tools for learning
- systematic study of both basic and advanced knowledge in science
- developing the skills needed for designing inquiries into real problems of science
- studying current developments and frontier areas in science and technology
- understanding the nature of science and scientific inquiry
- designing and conducting independent study projects
- interacting with peers in seminar and laboratory situations and developing the concept of an inquiry team
- making meaningful contacts with specialists and settings where science and technology are practiced
- experimenting with ideas and things
- reflecting on the often conflicting values of the impact of science on society
- developing the motivation and commitment to engage in the involved and complex tasks of science inquiry
- becoming connoisseurs regarding the nature of excellent performance in science

The design of a learning community is concerned, then, with access to knowledge, insights, and understandings; with the development of skills and processes; with the evolution of values and attitudes—all within a climate or environment that nurtures and rewards excellence, that is, superior performance and products. Such an environment will become increasingly advanced and complex as students mature and grow. It will be a continuous, articulated program. As part of their general educational experience, the potential specialists in science and technology will participate in many of the learning engagements with other youngsters. Other environments, however, will be designed to enable the potential scientists to go as fast, as far, as deep as they are capable of and are willing to invest their energies and time. This learning community will call for a demonstration of commitment to science in the area which students wish to pursue, but evidence of that commitment will not be sought prematurely, nor will students be precluded from demonstrating it at any point in their development.

A program for identifying and nurturing those youngsters who are potential scientists and technologists must begin early and continue through the secondary school years. Such a program must aim at creating a teaching-learning environment in which children and youth will, by their performance and their products, have opportunities to demonstrate those traits and characteristics which studies and experience have taught us to associate with giftedness or the potential for outstanding performance. Such an environment provides opportunities for self-identification and, at the same time, nurtures aptitudes and abilities: Identification, enrichment, and nurturance are interacting processes, each contributing to the other.

Paper-and-pencil tests, instruments which give youngsters some chance to demonstrate their problem-solving abilities, have generally recognized basic limitations. Providing students with opportunities to interact with concepts and ideas, materials and problems, gives them opportunities to demonstrate their potential by actual performance. In this way, they contribute to self-identification and indicate their needs for more complex, more advanced, more challenging experiences. They are provided with one of Pressey's conditions, namely "excellent and early opportunities for abilities to emerge and to be encouraged" by teachers, peers, families, and others.

Different Goals for Different Ages

At the elementary school level, the aim is not to determine which youngsters will become scientists and which ones will not. Rather, such programs should provide the kinds of experiences which will enhance student interest and stimulate understanding of the meaning and importance of science. Such programs should embody the excitement, challenge, and discovery of modern science, creating a setting in which students will begin to understand their own abilities and aptitudes and will begin to manifest such potential through their performances and their products.

Science programs at the presecondary level should begin to provide access to the substantive content and processes of "real" science in a way that challenges children and begins to lay the groundwork for more complex future understanding. As Lorraine Daston puts it elsewhere in this volume, "World views begin with in-the-fingers knowledge." Young students are ready to explore basic concepts and principles which constitute the structure of science disciplines in an intellectually honest fashion. With access to television programs, toy laboratories, books, and all types of do-it-yourself kits, today's youngsters are ready to begin systematic, serious study of science at earlier ages than ever before. Students need guidance in terms of what they read and what they view so that they come to appreciate what constitutes the challenges for science and technology in the world which surrounds them. Early on, students need to learn that they can satisfy their curiosity by their own efforts. Access to knowledge should not be restricted by artificial barriers, particularly those of grade levels. Students must be encouraged and enabled to study and explore areas and problems of interest to them and must be helped to understand the difference between depth and superficiality, between science and glitter.

Students need opportunities to engage in activities and undergo experiences which will help them to understand the nature of science—its dynamic characteristics ("Its laws, its findings, and the raw materials of its investigations are ever accumulating."); its human element ("Science is not an impersonal subject consisting of accumulated facts with no reference to the human and his[/her] intellect or values ... "); its increasingly cooperative aspects ("The single scientist working alone is being replaced by interdisciplinary research teams and group efforts."); and its inevitable expansion and movement in new directions (Cole, 1956, pp. 4–6). This recognition about the nature of science calls for revision of most current curricular designs and instructional strategies which, while questionably appropriate for transmitting knowledge and facts, are totally inappropriate to developing insights and understandings concerning the dynamic disciplines which constitute science and technology.

Schools for Those Who Might Do Science

As students identifying themselves as having a special interest, motivation, and potential in science mature and move through the program, they should have access to the disciplines and the discipline of science at increasingly more complex, more abstract, more advanced levels. They should be provided with opportunities to examine the relationships among sciences and interrelationships between science and other disciplines. Their study opportunities should offer both acceleration and enrichment as appropriate. At times, they will be ready for experiences at an earlier age or at a more rapid pace than is average. At other times, the instructional opportunities will enable such students to engage in learning experiences of greater breadth and depth than those of their peers. Acceleration and enrichment, both necessary parts of adequate and appropriate experiences for able learners, are complementary, not conflicting instructional strategies.

Students should learn how to explore topics using interdisciplinary approaches in order to understand how science may often be both part of the problem and of the solution. They should become sensitive to the values and moral issues posed by developments in science and technology. All of this requires making available basic courses to provide the foundation for knowledge and understanding, followed by advanced courses in science and technology, which deal with what Jerome Bruner (1961) called the "structure of the discipline" while constantly raising issues of values and meaning and placing the ideas and principles in a broad interdisciplinary context.

A science program for identifying and nurturing potential scientists and technologists will provide a variety of structures—courses, laboratories, seminars, etc. For some, instruction through seminars and colloquia is better than the traditional lecture and laboratory structure. An appropriate balance needs to be developed between two basic approaches to learning: individual and independent study, on the one hand, and cooperative learning situations and group activities, on the other. Independent study has long been a staple in programs for gifted students, providing opportunities for individuals to

- learn how to phrase questions and define problems
- use past experiences in the solution of new problems
- identify resources needed for problem solution
- test possible solutions and to acquire meanings from problem solving

But gifted students also need to learn how to participate in a team or cooperative effort at problem definition and solution, acquiring the skills needed to function effectively in a research team. These experiences also contribute to a sense of community for a community of artisans.

The Hidden Curriculum

Such programs must not neglect the hidden or subliminal curriculum. Within the school itself and beyond the school into the larger society, the subliminal curriculum is the climate, environment, and "feeling tone" that nurtures those affective and cognitive behaviors that contribute to engagement in and commitment to science in society. One does not have to look too far back to recall times when science and technology were viewed largely with suspicion. Abraham J. Tannenbaum (1979) has illustrated this paradoxical situation by describing Barry Commoner's acknowledgment in *Science and Survival* (1969) of the need for brainpower to enrich scientific thinking along with the warning that "no scientific principle can tell us how to make the choice, which may sometimes be forced upon us by the insecticide problem, between the shade of the elm tree and the song of the robin" (p. 21). Tannenbaum noted that scientists can become "tarnished" if they fail "to take account of their human consequences" (p. 21); one result being, perhaps, that fewer able students study science and consider careers in science. Science educators need to deal with and help potential scientists to study and grow in understanding of the continuing dilemmas posed by scientific and technological developments in almost every area of society so that the students will become sensitive and ready to tackle these issues.

Alexander Taffel (1987) has written of "a school ambience in which the mutual stimulation of the students themselves is a major ingredient" (p. 22). He recounts that a Nobel Prize winner who graduated from the Bronx High School of Science reported that he had been influenced there by his fellow students who "were always into something new and interesting" as much as by the formal curriculum. The alumnus reported: "Before long, the rest of us were into the same things. The place was a beehive of activities and ideas" (p. 22). It is the conscious development of this climate, which stimulates and nurtures a "beehive of activities and ideas," that is critical in the development of a community of artisans in science.

Taffel observes that "Clearly a vital element . . . is the ambience created by the interaction of students with each other, and every opportunity to facilitate and encourage should be exploited" (p. 22). Student interactions are important in creating this essential ambience or feeling tone, but there are other elements within both the school and the community—opportunities for interaction regarding science ideas and developments with teachers and other adults; access to laboratory resources and places "to do" science; availability of library resources with advanced-level materials; opportunities to participate in seminars and competitions (individual and group); and avenues for sharing, communicating, and critiquing individual and group investigations.

Perhaps the most critical element in the creation of a learning environment which encourages and nurtures future scientists and technologists is the teacher—that is, the teacher broadly conceived as adult role model. As I wrote over 30 years ago,

> . . . the key to an effective program lies in the quality of the teacher. If the teacher is inspired and inspiring; if he/she understands the meanings of science and the relationships of science to the world in which we live; if he/she is flexible and makes possible the flexibility needed for adequate programming; if he/she encourages individual excellence and devotes the time and effort required to guide the student to locate necessary resources; if he/she is sympathetic to rapid learners and their particular needs; if he/she knows science and its techniques; if he/she is willing to adapt his/her teaching methods to stimulate problem solving— then the teacher has the attitudes and competence which constitute "good quality." (1957, p. 111)

At about the same time, Cole (1956) stressed the importance of the teacher in encouraging the shaping of future scientists and technologists: "The teacher . . . who can ignite that spark of interest in the talented, who can make his[/her] subject live in the minds of his[/her] pupils, and who . . . can stimulate, fascinate and inspire; he[/she], in the final analysis, is the one who can most effectively encourage scientific talent" (p. 21).

In Sum

The creation of a total-learning environment—one which provides students with the opportunities and resources that will stimulate their engagement in activities leading to the acquisition of knowledge, understanding, skills, processes, attitudes, values, and other behaviors at increasingly advanced and complex levels—is at the center of the task of educating and schooling of the community of artisans in science and technology. In a real sense, it is the learning environment that constitutes the "community" in which students are self-actualized. Here, they come to understand their own potential for becoming scientists and technologists by involvement with their peers and mentors in a variety of increasingly complex engagements with the knowledge, substance, and content of science and technology; here, they learn the processes and methods of science and technology; here, they experiment with the modes of problem definition and inquiry. These engagements occur not just in the formal course structure (which is, of course, very important) but in the climate of the classroom, school, and community and in the formal and informal transactions—powerful mediators of cognitive and affective growth—which take place therein.

References

Brandwein, Paul F. (1955). *The gifted student as future scientist: The high school student and his commitment to science.* New York: Harcourt Brace. (1981 reprint, with a new preface [Los Angeles: National/State Leadership Training Institute on the Gifted and Talented])

Brandwein, Paul F. (1987, September). On avenues to kindling wide interests in the elementary school: Knowledges and values. *Roeper Review, 10* (l), 32–40.

Bruner, Jerome. (1961). *The process of education.* Cambridge, MA: Harvard University Press.

Cole, Charles C., Jr. (1956). *Encouraging scientific talent.* New York: College Entrance Examination Board.

Commoner, Barry. (1969). *Science and survival.* New York: Viking.

McClelland, David C., Baldwin, Alfred L., Bronfenbrenner, Urie, and Strodtbeck, Fred L. (1958). *Talent and society: New perspectives in the identification of talent.* Princeton: D. Van Nostrand.

Passow, A. Harry. (1957, March). Developing a science program for rapid learners. *Science Education, 41*(2), 104–112.

Pressey, Sidney L. (1955). Concerning the nature and nurture of genius. *Scientific Monthly, 81,* 123–129.

Taffel, Alexander. (1987, September). Fifty years of developing the gifted in science and mathematics. *Roeper Review, 10* (l), 11–24.

Tannenbaum, Abraham J. (1979). Pre-Sputnik to post-Watergate concern about the gifted. In A. Harry Passow (Ed.), *The gifted and the talented: Their education and development, Part I* (pp. 5–27). (Seventy-eighth Yearbook of the National Society for the Study of Education.) Chicago: University of Chicago Press.

Wolfle, Dael. (1954). *America's resources of specialized talent.* New York: Harper and Brothers.

Probing Giftedness/Talent/ Creativity:* Promise and Fulfillment

Abraham J. Tannenbaum

The power center for giftedness is the brain, which controls both the magnitude and diversity of individual potential. It can transport an Einstein into heights of abstraction and a da Vinci into flights of creativity that are so far beyond ordinary accomplishments as to seem truly wondrous. It can also generate nearly endless traces of genius, ranging from the esoterica of plasma physics through the aesthetic and engineering beauty of the Taj Mahal, to the magical cadences of a Miltonic sonnet, the sublime sounds of an *Eroica* Symphony, the gustatory delights of gourmet cooking, the intricacies of Oriental knot designs, and on and on into every possible domain of individual activity.

But whereas the psyche, that is, a person's mental or physical structure, especially as motive force, determines the *existence* of high potential, society decides on the *direction* toward its fulfillment by rewarding some kinds of achievement while ignoring or even discouraging others. Rare brainpower has to fit into its own *Zeitgeist* (i.e., spirit of the age) in order to be recognized and appreciated. There also has to be a match between a person's particular talent and the readiness of society to recognize it. Otherwise, genius may remain stillborn or mature to serve an unappreciative public that may regard it either as passé, if it is a throwback to an earlier period, or as too

---

*I am using these terms virtually synonymously unless otherwise indicated.

avant-garde, if the times are not ready for it. There is always room to speculate whether Einstein would have been able to make a contribution to theoretical physics, or whether the scientific world would have been ready for his kind of contribution, if he had been born only half a century earlier. Would he have been capable of creating any spectacular theories at all if he had been born in 1950, and others had formulated the theory of relativity ahead of him? This is pure conjecture, of course, but clearly evident is the fact that gifted individuals who achieve great breakthroughs in the world of ideas do not operate simply as free spirits detached from the temper of their times.

By the same token, scientists affect the world in general as well as their own professional domains. For example, Robert Goddard's pioneering experiments with rocketry in the early part of this century eventually revolutionized ordinary people's long-distance travel habits, scientists' methods of space exploration, and nations' ways of making war. These developments, in turn, helped create a political climate in which U.S. space scientists are once again being pressured to catch up with their Soviet counterparts, who are far ahead in establishing space stations as bases for exploring distant planets (Wilford, 1988). Shades of the post-Sputnik fever in American science and society more than 30 years ago! It is quite evident, then, that scientific creativity changes the world and that the changed world pressures or becomes a fertile ground for new kinds of scientific creativity. This reciprocity between the distinguished human mind and the general human environment exists not only in science but probably in every discipline. As for the "chicken/egg" question, who cares what set the cycle in motion, as long as seminal thinkers and the societies in which they live are mutually enriching?

Early Promise

Despite differences in orientation as to the meaning of giftedness, psychologists do not hesitate to look for signs of it in children because of the assumption that precocity among the young is a fairly valid forerunner of their future success. Still, a distinction has to be made between *promise* and *fulfillment*.

Work accomplished during a person's maturity can be evaluated by objective standards if its aim is very specific: For example, if one's goal is to prevent rejection of transplanted human organs, one either succeeds or fails. Or if one creates a product, it can be subjected to critical review, as in the case of poetic composition, to determine whether it deserves to be disseminated and treasured. Not so with children's achievements. Children identified as potentially gifted would mostly fail to qualify for renown if they were judged on the basis of universal criteria (except in the rare instances of a young Mozart or Mendelssohn). Instead, the young have to be compared to others of their age for early signs of talent that is amenable to nurturance and that promises to grow in the future.

Since there can never be any assurance that precocious children will fulfill their potential, defining giftedness among them is necessarily risky. One set of criteria may be *ineffective* because it excludes too many children who may grow up to be gifted; other qualifying characteristics may prove *inefficient* by including too many who turn out to be average in most respects. There is inevitably a trade-off between effectiveness and efficiency, and educators invariably should opt for a definition that enables them to cast the widest possible net at the outset to be sure not to neglect children whose high potential may be all but hidden from view.

Keeping in mind that a child's brilliance usually cannot compete with that of an adult but often foreshadows later stages of excellence, giftedness in children is therefore a shorthand way of denoting *potential* giftedness. Potential for what? For becoming critically acclaimed in one of two ways, either as
• *performers* who render valuable helping services to society or who display stage artistry before appreciative audiences or
• *producers* of ideas in spheres of activity that enhance the moral, physical, emotional, social, intellectual, or aesthetic life of humanity

In one sense, this definition is broadly inclusive, embracing a wide range of talents; from another perspective it is restrictive, since there is no place in it for rapid learners of existing ideas (i.e., consumers of knowledge) or for admirers of great performance (i.e., appreciators of what others can do). True, producers and performers are often consumers and appreciators, but not *necessarily* so.

And Fulfillment

Those who have the potential for succeeding as gifted adults require not only the personal attributes that lodge mostly in the psyche, but also some special encounters with the environment to facilitate the emergence of talent. Altogether, there are five psychological and social linkages between promise and fulfillment, all of which interweave delicately as if to form a subtle, complex filigree. They include (a) superior general intelligence, (b) exceptional special aptitudes, (c) nonintellective facilitators, (d) environmental influences, and (e) chance, or luck (Tannenbaum, 1983).

The five factors combine in a rare blend to produce great performance or productivity. Each of them has a fixed threshold that represents the minimum essential for giftedness in any publicly valued activity. Whoever achieves some measure of eminence has to qualify in all of these standards; the person who is unable to measure up to just one of them cannot become truly outstanding. In other words, success depends upon a combination of facilitators, whereas failure can result from even a single deficit. Furthermore, for each of the five intellective, personal, and social-situational factors connecting potential with high-level accomplishment, there is also a threshold level that varies according to specific areas of excellence.

Thus, for example, the talented artists of Jack Getzels' (1979) sample were able to demonstrate their exceptional talent in various art forms even though their general academic abilities were no better than most other college students'. On the other hand, without high academic promise, no college students could become as distinguished as the creative scientists Anne Roe studied (1953). It is reasonable to speculate that the IQs, along with spatial and scientific aptitude thresholds, are different for artists and for scientists. Those who fail to measure up to any of these minimum essentials for their respective fields of endeavor never compare with the populations observed by Getzels and Roe. By virtue of its "veto" power, then, every one of the qualifiers is a necessary requisite for high achievement.

General Ability

General ability can be defined roughly as the g factor, which is itself defined roughly as some kind of mysterious intellectual power common to a variety of specific competencies. The g factor, reflected in tested general intelligence, figures on a sliding scale in all high-level talent areas. This means that different threshold IQs are required for various kinds of accomplishment, probably higher in academic subjects than, for example, in the performing arts (cf. Getzels, 1959). There is no basis for making extreme assertions about the IQ, either discounting its relevance to giftedness entirely or claiming that all those destined to become great producers or performers in any area of human activity need to score at the 99th percentile or better. Instead, positions along this continuum should be adjusted according to the talent area, which means taking a stance closer to one extreme for some kinds of giftedness and nearer the opposite extreme for others.

There is reason to suspect that, as scientific knowledge accumulates and inventions become more and more sophisticated, the threshold IQ has to be high in order to qualify a child as a potential producer of certain kinds of ideas. In a large study of intergenerational gains in IQ in countries where such data are available, J. R. Flynn (1987) found that present-generation 20-year-olds in Holland scored about 20 points higher in IQ than did their counterparts some 30 years earlier.

Both groups were compared on the same test, using the same norms. An increase of such dramatic magnitude means that those with IQs of 150 and above have increased proportionately by a factor of almost 60 from the previous to the current generation. Yet, the number of patents granted has actually diminished, with the 1980s showing only 60 to 65 percent of the yearly rate for the 1960s. *To the extent,* therefore, that IQ figures in the inventive mind, it appears that the abstract-thinking abilities measured by such instruments have to be even more sophisticated to qualify a child as potentially gifted today than a generation ago.

The relative importance of IQ in the sciences, as compared to other occupations, is also noteworthy. In sobering essays on the use of tests to forecast various kinds of performance, David C. McClelland (1973) and Michael A. Wallach (1976) argued that measures of human potential have been validated on success at school but tell us little about accomplishment after graduation. The usual supportive data provided by test manuals pertain to classroom-type performance, and the evidence is often impressive. But it means little from a lifetime perspective, except that those who excel scholastically have a better-than-average chance of achieving well in some unspecified high-level work in adulthood. In other words, no existing battery of measures can predict with much confidence (1) *who* will qualify for *what* occupation and (2) *how* those who manage to qualify will perform at their jobs. Yet, in Roe's study of creative scientists, she found median IQ scores, or their equivalents, well beyond the 99th percentile. This does not mean, of course, that every child, or even most children, with a high IQ is destined to excel in science or enter a science field. But as Jonathan R. Cole (1979) points out, measured intelligence figures in the careers of scientists in specific ways. Its association with the status of a person's occupation is greater than father's background, father's education, or family size. Although it is not a vital factor in forecasting the initial placement of a scientist in the prestige hierarchy of the profession, it is of critical importance later on. After the scientist's first job, IQ correlates with positional recognition, regardless of first-job prestige, educational background, and scholarly performance.

Special Aptitudes

Some educators assume that specialized abilities in science and in other fields do not begin to develop until adolescence. Therefore, elementary education should consist of a smorgasbord of subjects, taught by generalists in self-contained classrooms, and no content area should be emphasized more than others. Unfortunately, behavioral scientists have rarely investigated the cognitive structures or processes of children who excel in a single area of study.

A notable exception has been the work of Gerald S. Lesser, F. B. Davis, and L. Nahemow (1962), who constructed a test to assess children's mastery of previously learned science material and also their ability to combine elements into principles and to apply "the scientific method." The instrument addressed itself to the following behavioral objectives:

1. Ability to recall information: knowledge of
   a. scientific vocabulary
   b. scientific principles
   c. tools and scientific instruments
   d. the natural environment
2. Ability to assign meanings to observations
   a. formulation or verbalization of a principle to explain an effect described
   b. identification of crucial elements of a problem
3. Ability to apply new principles in making predictions
   a. utilization of available information in novel situations
   b. utilization of a scientific principle in a familiar situation
   c. analysis of the factors influencing predictions
4. Ability to use "the scientific method": planning steps leading to a solution

Lesser and his associates administered the test to 58 third graders at the Hunter College Elementary School. The children ranged in age from 6 years, 9 months, to 7 years, 9 months, and their Stanford-Binet IQs ranged from 136 to 171, with a mean score of 151.4. The predictive validity of the Science Aptitude Test proved to be extremely high when correlated with a battery of seven science achievement tests. In fact, the aptitude scores were much better predictors of the children's accomplishments in science than were the IQs. It would therefore seem that at some point—perhaps somewhere in the upper 2 percent of the IQ distribution—there is a threshold beyond which IQs are no longer predictive of achievement in science, at least in the elementary grades. Much more has to be known about special aptitudes and the different branches of science before determining to what extent these tested abilities reinforce the predictive validity of IQ.

There are also auxiliary aptitudes that somehow seem to relate to high achievement in science, not only in the school years but also beyond. Camilla Persson Benbow and Lola L. Minor (1986) conducted a follow-up study of 1,996 seventh and eighth graders who qualified for special acceleration programs in mathematics on the basis of their scores on Scholastic Aptitude Tests (SATs) (verbal and mathematics portions). At the time of retesting, over 90 percent of the students were freshmen in college and 67 percent had completed the high school sequence of biology, chemistry, and physics (74 percent of the boys and 56 percent of the girls). The researchers found that the SAT mathematics scores obtained in junior high school were strongly associated with science achievement in senior high school for both sexes.

Robert S. Root-Bernstein (1987a, 16 March; 1987b) suggests that creativity in the sciences is enhanced by creativity in the arts; Gunther S. Stent also notes a "commonality" in this volume. As evidence, Root-Bernstein cites a long list of famous scientists, such as Louis Pasteur and Lord Lister, who were also graphic artists, Johannes Kepler and Max Planck, who were musicians, and J. Robert Oppenheimer and George Washington Carver, who wrote poetry. Furthermore, Root-Bernstein is convinced that creativity in the arts and sciences is so highly correlated that genius in science cannot develop in the absence of high-level proficiency in the arts. He therefore posits a number of "tools of thought" that are common to both, including

> perceptual acuity, pattern recognition, pattern forming, analogizing, abstracting, and imagining . . . . also transforming skills: learning to "see" what a mathematical equation looks like, or to "hear" what data sounds like; or learning to "feel" like the molecule or "act" like the machine that one would like to comprehend. All of these skills are absolutely essential to scientific and technological curricula. All of them, intrinsically at least, already exist in the fine arts. (1987a, p. 18)

Nonintellective Facilitators

Conceptually, it is easy to distinguish between intellective and nonintellective factors in human functioning. One denotes the mental powers and processes needed to master or create ideas; the other refers to the social, emotional, and behavioral characteristics that can release or inhibit the full use of a person's abilities. Problems in separating the psychological domains do arise, however, when they have to be assessed. Mental measurement, for example, is accomplished inferentially, through tests of performance, which are always refracted by nonintellective factors. These traits are integral to the achieving personality regardless of the areas in which talents manifest themselves. However, it is impossible to tell which of the nonintellective attributes are *responsible* for creative achievement, which are merely *associated* with it, and which are *by-products* of it.

Of all personality traits, none has drawn more attention than motivation to achieve. Joseph S. Renzulli (1978) counts task commitment as one of only three major factors that characterize giftedness, the other two being above-average ability and creativity. A strong distinction is made between intrinsic and extrinsic motivation by Teresa M. Amabile (1983), who shows evidence that children and adults perform more creatively when the urge to excel comes from within rather than from without. It is possible that Thomas Edison was exaggerating when he claimed that genius is 1 percent inspiration and 99 percent perspiration, but nobody would argue that either ingredient should be excluded from the mix.

There are other nonintellective ingredients, too. In studies made in the 1940s and 1950s of high school students who showed promise in science, Paul F. Brandwein (1955/1981) found a "predisposing factor" to comprise *persistence* and *questing*. (See also Brandwein's papers in this volume.) In one of Louis N. Terman's (1954a with Melita H. Oden; 1954b) follow-up studies of high-IQ children, he attempted to discover why scientists and politicians could not work harmoniously in government by exploring personality constructs in the two groups that might have led to mutual distrust and conflict. His method was to compare high-IQ children who grew up to be scientists with those comparably high in IQ who eventually elected to major in law, the humanities, and the social sciences. Since a great deal of data had been accumulated on these subjects over the years, it was possible to search for whatever developmental differences may have existed between the two subsamples.

Of the many variables on which comparisons were made, 108 yielded significant differences. The first had to do with an early and persistent interest in science. As 11-year-old children in 1922, those who were later to fall in the science group were already showing a far higher tendency toward interest in science than were those who were to go on to other careers. Moreover, the inclinations were still evident 18 years later in 1940, when the subjects were already launched on their careers. Even in later years, the interest patterns were surprisingly constant. Of the 250 men Terman studied who took the Strong Vocational Interest Test as college freshmen and again 20 years later, few showed appreciable changes in their scores, and such changes as did occur bore little relation to the kind or amount of involvement in their respective fields in the interim period.

Furthermore, comparisons between the scientists and politicians on their interests in business occupations showed sharp differences, with the political group favoring business careers far more strongly than did their scientific counterparts. Also, politicians scored higher than did scientists in the need to develop informal, wide-ranging social relationships. There were various degrees of social adjustment and social understanding within each group, but overall, the scientists Terman studied tended to resemble Roe's subjects in showing relatively little dependency on strong interpersonal ties in order to accomplish their work successfully. Commenting on these findings, Terman pointed out that physical scientists, engineers, and biologists are the ones who do most of the secret research for the government and are compelled to operate under rules laid down by a Congress composed mainly of lawyers and business executives. Although it would be oversimplifying the matter to assume that the difficulties of these contrasting groups in trying to understand each other are explained merely by their different interests per se, personality factors can be critical in such circumstances.

Similar findings were reported by Robert MacCurdy (1954) in his survey of 600 men and women who had earned honorable mentions or better in the Westinghouse Science Talent Search of 1952 and 1953. In this study, comparisons were made between superior science students and their contemporaries in general education who had outstanding scholastic ability. MacCurdy found that, with respect to personality, the science students had a strong curiosity about causality in relation to natural phenomena. Their thoughts often took the form of daydreaming or mental puzzle solving, and they often enjoyed symbolic art and classical music. They tended to choose activities that were solitary or nearly so; yet they did not enjoy the roles of spectators entertained by others' accomplishments. Preferences for independent activities led them to read science, take nature walks, tinker with gadgets, practice photography, and build radios. Typically, their experiences in high school were scholarly, with much emphasis on scientific problem solving. There was also an important place for an inspiring science teacher who helped them to determine the course of their professional lives.

In sum, there may be more than a grain of truth in the way scientists have been stereotyped. David E. Super and P. B. Bachrach (1957) found considerable research evidence to show that people in the engineering, mathematics, and natural science professions tend often to be lonely, socially awkward, slightly withdrawn, curious, self-disciplined, unemotional, tolerant of others, and intensely devoted to their work. Perhaps these traits make sense for people engaged in independent scientific inquiry, with all the complex abstractions and distancing from intrapersonal preoccupations that it entails.

Environmental Influences

Human potential cannot flourish in an arid cultural climate; it needs nurturance, urging, encouragement, even pressure, from a world that cares. The child lives in several worlds, the closest being those of the family, peer group, school, and community, the more remote including various economic, social, legal, and political institutions. These environments all help to determine the *kinds* of talent that society is willing to honor as well as the *amount* of investment it is prepared to make in cultivating them. Societal conditions are therefore critical in stimulating the gifted child's pursuit of excellence.

Conventional wisdom suggests that giftedness thrives best in an atmosphere of love and support. But some studies (cf. Goertzel and Goertzel, 1962; and Goertzel, Goertzel, and Goertzel, 1978) suggest that there are some children who seem to respond to pressure or even adversity that galvanizes them to fulfill their potential. They succeed, apparently, *because* of the pressures, not *despite* them. Obstacles that discourage most people from achieving somehow challenge a few to "beat the odds" and "make it big." Their drive toward excellence may be basically an act of defiance against what they consider hostile, inhibiting forces in the world. For most children, however, a nurturing and stimulating home environment are indispensable not only for maximizing potentialities but also in helping specify the directions they take.

A supportive peer culture is also important, especially for children who need to be accepted by their age-mates. And teachers make a difference, despite the belief of many people that talent is irrepressible in some children and impossible to nurture in others. There is evidence to show that elementary and high school students in American schools actually improved somewhat in science achievement from 1970 to 1983 (Jacobson and Doran, 1985). Such gains should be credited, at least in part, to better teaching over those years. However, to prepare the best young scientific minds for careers in the early part of the 21st century, there will probably have to be some radical revisions in teacher preparation and in the design of enriched curricula for the gifted.

Chance, or Luck

Generally overlooked in studies of the fulfillment of talent are the entirely unpredictable events in a person's life critical both to the realization of promise and to the demonstration of developed talents. It is not only a matter of being in the right place at the right time, although that is important, too. Many unforeseen circumstances in the opportunity structure and in the prevailing lifestyle can make a difference in the outlets for gifted performers. For example, negatively, a budding medical researcher who is ready to achieve a breakthrough in disease control may suddenly and unpredictably be distracted by a personal crisis or by the lure of a social issue that is considered more immediately relevant to human concerns. Chance factors can also determine the choice of career, as in the case of a gifted young biology student who happens to meet and study with a brilliant, inspiring teacher and makes the most of the opportunity.

Obviously, chance factors defy systematic characterization, but this should not obscure their powerful influence on achievement. In his study of personal opportunity in America, Christopher Jencks (1972) suspected that luck has at least as much influence on income levels as does competence. John W. Atkinson (1978) seems to ascribe all of human behavior and accomplishment to

> two crucial rolls of the dice over which no individual exerts any personal control. These are the accidents of birth and background. One roll of the dice determines an individual's heredity; the other, his formative environment. Race, gender, time and place of birth in human history, a rich cultural heritage or not, the more intimate details of affluence or poverty, sensitive and loving parents and peers, or not, all of them beyond one's own control, have yielded the basic personality: a perspective on the life experience, a set of talents, some capacities for enjoyment and suffering, the potential or not of even making a productive contribution to a community that could be a realistic basis for self-esteem. (p. 221)

Every biochemist knows how chance factors figured prominently in Sir Alexander Fleming's discovery of penicillin. Roughly speaking, the sequence of events was as follows: First, he noticed that moldy bread crumbs had accidentally fallen into his culture dish; he then saw that staphylococcal colonies residing near it stopped growing; he realized that the mold must have secreted something that destroyed the bacteria; this brought to mind a similar experience he once had; and he finally postulated that perhaps this mysterious ingredient, the mold, could be used to destroy staphylococci that cause human infections. It is interesting to note that the "similar experience" Fleming remembered was his suffering from a cold some nine years earlier, when his own nasal drippings accidentally fell on a culture dish, killing the bacteria around the mucus. He followed this lead with further experimentation but got nowhere until the accident with the bread reminded him of the nasal drippings.

This time he was on to a most celebrated medical discovery. The finding was truly serendipitous, but lest it be forgotten, it took the brainpower, motivation, training, and hard work of a Sir Alexander Fleming to make the most of the lucky sequence of events. One shudders to wonder how long we would have waited for penicillin if Fleming had worked in a neater laboratory or if he had been more diligent about wiping his runny nose!

Implications for Identification

From the foregoing discussion, it seems as if the cause of extraordinary accomplishment can be described best as resembling some kind of not-so-clear, complex, moving target. The number and variety of antecedent variables preclude any easy designation of a child as gifted on the basis of a few performance measures. And the causes are not the same for all kinds of "giftedness." Every area of excellence has its own mix of requisite characteristics, even though general ability, special aptitudes, and the nonintellective, environmental, and chance factors under which they are subsumed apply to all kinds of talent. These factors should be viewed as common denominators always associated with giftedness, no matter how it manifests itself. Yet, within each of them, the threshold levels, below which outstanding achievement is impossible, must adjust to fit every talent domain. That adds to the difficulty of making predictions about the fulfillment of a promise.

Potential in children means different things to different people. But no matter what the definition, identifying childhood promise has to be counted among the inexact sciences, partly because the methods and instruments available for the purpose are imprecise. Besides, childhood is usually too early in life for talent to be full-blown, so it is necessary to settle for dealing with talent-in-the-making and to keep in mind the uncertainties of the future. In creating a pool of "hopefuls," it is best to admit any child who stands a ghost of a chance of someday making it to the top of the world of ideas. Of course, many of those "hopefuls" may be really "doubtfuls," but nobody can know for sure in advance which are which. Bringing into the pool under liberal admission criteria the widest possible group uncovers as much hidden talent as possible.

Because there is no foolproof, formal test procedure to identify budding research scientists, the best alternative is to engage seemingly qualified children in laboratory activity, including some kinds of experiences left out of the regular science curriculum. Those with potential in science would then be those who respond most successfully to the special challenges. Such straightforward procedures, practiced frequently and with gratifying results in the arts and in sports, are unaccountably ignored in many other talent domains that are of interest to educators.

Schools should depart radically from the usual two-step, diagnosis-and-then-treat process advocated in medicine and in the education of the handicapped. The approach recommended here is an oscillating one (not vacillating!) between diagnosis and treatment. Not only should the gifted be identified and then educated, they should also be identified *through* education. In other words, *prescribed enrichment becomes a vehicle for identification as much as identification facilitates enrichment, the relationship now being reciprocal.*

~ Identification should begin as early as possible in the child's life, and it should go on for as long as possible, because there are always opportunities for developing new insights and correcting old errors of judgment. The stages are (1) *screening,* (2) *selection,* and (3) *differentiation.* This sequence goes on continuously for children not yet screened and also for those who have previously not "made it" into the first stage.

Screening

At the beginning, the criteria for inclusion are liberal, and many of the instruments used at this stage assess remote and sometimes farfetched indicators of potential, not just actual performance at school. The purpose here is to include all children who show vague hints of giftedness/talent/creativity in order to determine later if they possess real potential. To obtain the proper kinds of initial information, it is necessary to consult multiple sources, including, but not limited to, the following:

Evidence of General Ability. I believe that it is more appropriate to rely on an IQ-type test in the search for scientific talent than in locating other kinds of giftedness. At least one standard deviation above the mean should be a minimum essential, except when the search is among disadvantaged children, whose performance on such measures is often deficient. For these children, it is best to engage in dynamic assessment of potential, in which the examiner serves as a participant observer rather than an objective monitor of success at test taking (Feuerstein, 1979).

Evidence of Special Aptitudes and Achievement. Tests of special skills are useful in assessing children's progress in a few content areas, more so in the sciences than in the arts. Among widely used examples in research and in the schools: The Primary Mental Abilities Test, at the elementary level; the Differential Aptitude Test, at the high school level; and the Scholastic Aptitude Test, at the college entrance level. A serious problem can arise when schools play mostly to the strengths of these kinds of tests and offer enrichment programs only to children who can perform well on them. In such cases, the instrument exercises inordinate power over the program rather than serving its intended purpose as a tool to help implement the program.

Evidence from Noncognitive Traits. Limitations in the predictive validity of performance measures should encourage educators to correct the under-emphasis on personality variables and behaviors, including self-direction, pride in accomplishment, persistence, dedication, efficient work habits, and other traits associated with scientific achievement. In the questionnaire developed recently by the Bureau of Educational Research and Service at the University of Kansas for aspiring Merit Scholarship winners, answers to the following items distinguished most consistently between the finalists and the others:

• How would you rate yourself in terms of willingness to withstand discomfort (a cold, illness, etc.) in completion of a school task?

• How would you rate yourself in terms of willingness to spend time *beyond the ordinary schedule* in completion of a given school task?

• How would you rate yourself in terms of *questioning* the absolute truth of statements from textbooks, newspapers, and magazines, or of statements made by persons in positions of authority such as teachers, lecturers, and professors? (personal communication, 1988)

This kind of information can be obtained not only from the children but also from their peers, parents, and teachers. A large number of trait lists now exist, and although they are not all fully validated for a wide range of talents, even the "soft signs" they reveal in the screening stage can be enlightening.

Evidence of Productivity or Performance. It is important for teachers to keep constant records of children's accomplishments in and out of school. A cumulative file that shows samples and other evidence of such projects may reveal unusual potential in a valuable area of work that is not emphasized in the classroom. One source of such information is often the school, usually the home, but not always so. Parents and peers can certainly help to keep a child's record up-to-date, and teachers should be eager to obtain and record whatever information can help build a case for high potential in an individual child.

Selection for Science

After the screening stage, to reduce the proportion of average children in the pool requires shifting from remote indicators to those more clearly in the context of the science curriculum. All children in the pool are then given a chance to prove themselves in real and simulated enrichment activity and to show how well they respond to the challenge. (This process is an example of the "self-selection" Brandwein recommended [1955/1981].) For example, if a unit on meteorology is part of a program for the gifted in science, and no existing pretest of skills in that subject exists, the only way in which to make a proper identification is to allow children in the pool to "try out" for the unit to prove whether or not they can engage in it productively. This is what is meant by an oscillating process between identification and enrichment. The special curriculum should not be a privilege for a predetermined group of children labeled as "gifted"; it should be initially a testing ground in which the gifted in science sort themselves out from children with other interests, pretty much as do aspiring athletes in scrimmages in football or basketball or prospective actors in auditions. The quality of identification, therefore, depends to a great extent on the quality of the program as a vehicle for observing hopefuls under "game-like" conditions in the talent search.

In the course of exposure to enrichment experiences, a child can reveal potential giftedness by a variety of behaviors that can alert teachers and parents to monitor the child more closely. These behaviors include the student's sophisticated use of language; the quality of the student's questions; the quality of illustrations or elaborations that the student uses in communicating an idea; the student's ability to adopt a systematic strategy for finding or solving problems and to change the strategy if it does not work; the student's innovative use of materials found in or out of the classroom; the student's breadth or depth of information relevant to a particular learning experience; the extensiveness of the student's exploratory behaviors; the student's preferences for complexity, difficulty, and novelty in learning tasks; and the student's critical view of his or her own performance. These selection criteria are more demanding than those used in the screening stage, but even here mistakes can be made if the procedures are adhered to too rigidly. There is still the danger of accepting some who do not qualify and of rejecting others who do qualify as potentially gifted. It is therefore necessary to refine the process further in the next stage, which is the longest lasting of the three.

Differentiation

The final step in identification is to move to separate the gifted from the gifted as well as the gifted from the average. This process should continue indefinitely, with several sifting and sorting educational activities to help along the way. The objective is to begin distinguishing potential mathematicians from computer programmers, scientists from technicians, physicists from engineers, biologists from chemists, and so on. Much depends on the breadth and inspirational quality of the enrichment program because the gifted need exposure to a variety of opportunities to avoid being locked into an area of specialization too early in life.

Thus, progress is made from the initial screening stage, with its heavy reliance on measures that are indirectly related to life in the classroom, to the final differentiation stage where identification is mainly through the curriculum itself. If enrichment is continuous throughout the children's schooling, differentiation should never really end as long as they are in the program. In the last analysis, identification of the gifted is related not only to systematic observation and intelligent interpretation of test data but to the development of the right kinds of educational enrichment to facilitate self-identification. But what are "the right kinds of enrichment"? This question, which is addressed by many papers in this volume, still needs full and varied answers based on theory and research in the best scientific tradition.

One Scientist's Glimpse into the Future

In his speculation about forthcoming trends in the sciences, Lewis M. Branscomb (1986), formerly chief scientist at IBM and then professor and director of the Science, Technology, and Public Policy Program at the John F. Kennedy School of Government at Harvard University, observed that, by the year 2006, several dramatic changes will have taken place in various fields of science. These include at least five major reintegrations of the sciences:

- neurophysiology, brain studies, and cognitive and behavioral science
- cosmology, high-energy physics, astrophysics, and mathematics
- biochemistry, medical sciences, and molecular, cellular, and developmental biology
- geophysics, meteorology, oceanography, and paleontology
- geography, a resurgence of which synthesizes studies of human habitats, ecology, geomorphology, social anthropology, and economics

Branscomb also expects the social and natural sciences to work more collaboratively than ever before, sees a growing association of science and the arts, and envisions a resurgence of pure mathematics as the handmaiden of science. To hasten these revolutions, learning centers around the world will have access to each other's theories and research through highly sophisticated networks that enable students, staff, and faculties to communicate with a minimum of inconvenience. If any of Branscomb's crystal-ball gazing has merit, the implications for enriched education for the child gifted in science, whose career will blossom in the 21st century, are nothing less than revolutionary.

References

Amabile, Teresa M. (1982). *The social psychology of creativity.* New York: Springer-Verlag.

Atkinson, John W. (1978). Motivational determinants of intellective performance and cumulative achievement. In John W. Atkinson and Joel O. Raynor (Eds.), *Personality, motivation, and achievement* (pp. 221–242). New York: John Wiley and Sons.

Benbow, Camilla Persson, and Minor, Lola L. (1986). Mathematically talented males and females and achievement in the high school sciences. *American Educational Research Journal, 23*(3), 425–436.

Brandwein, Paul F. (1955). *The gifted child as future scientist: The high school student and his commitment to science.* New York: Harcourt Brace. (1981 reprint, with a new preface [Los Angeles: National State Leadership Training Institute on the Gifted and Talented])

Branscomb, Lewis M. (1986). Science in 2006. *American Scientist, 74*, 649–657.

Cole, Jonathan R. (1979). *Fair science.* New York: Free Press.

Feuerstein, R. (1979). *The dynamic assessment of retarded performers.* Baltimore: University Park Press.

Flynn, J. R. (1987). Massive IQ gains in 14 nations: What IQ tests really measure. *Psychological Bulletin, 101*, 171–191.

Getzels, Jack. (1979). From art student to fine artist: Potential problem finding and performance. In A. Harry Passow (Ed.), *The gifted and the talented: Their education and development, Part I* (pp. 5–27). (Seventy-eighth Yearbook of the National Society for the Study of Education.) Chicago: University of Chicago Press.

Goertzel, V., and Goertzel, M. G. (1962). *Cradles of eminence.* Boston: Little, Brown.

Goertzel, M. G., Goertzel, V., and Goertzel, T. G. (1978). *300 eminent personalities.* San Francisco: Jossey-Bass.

Jacobson, Willard J., and Doran, Rodney L. (1985, February). The second international science study: U.S. results. *Phi Delta Kappan, 66*(6), 414–417.

Jencks, Christopher. (1972). *Inequality: Reassessment of the effect of family and schooling in America.* New York: Basic Books.

Lesser, Gerald S., Davis, F. B., and Nahemow, L. (1962). The identification of gifted elementary school children with exceptional scientific talent. *Educational and Psychological Measurement, 22*, 349–364.

MacCurdy, Robert D. (1954). *Characteristics of superior students and some factors that were found in their background.* Unpublished doctoral dissertation, Boston University, Boston.

McClelland, David C. (1973). Testing for competence rather than for intelligence. *American Psychologist, 28*, 1–14.

Renzulli, Joseph S. (1978, November). What makes giftedness? Re-examining a definition. *Phi Delta Kappan, 60*(3), 180–184.

Roe, Anne. (1953). *The making of a scientist.* New York: Dodd, Mead.

Root-Bernstein, Robert S. (1987a, March 16). Education and the fine arts from a scientist's perspective: A challenge. (A "White Paper" written for the College of Fine Arts, University of California, Los Angeles)

Root-Bernstein, Robert S. (1987b). Tools of thought: Designing an integrated curriculum for lifelong learners. *Roeper Review, 10*, 17–21.

Super, David E., and Bachrach, P. B. (1957). *Scientific careers and vocational development theory.* New York: Teachers College, Columbia University.

Tannenbaum, Abraham J. (1983). *Gifted children: Psychological perspectives*. New York: Macmillan.

Terman, Lewis M. (1954a). Scientists and nonscientists in a group of 800 gifted men. *Psychological Monographs, 68* (7, No. 378)

Terman, Lewis M., and Oden, Melita H. (1954b). *The gifted group at mid-life*. Stanford, CA: Stanford University Press.

Wallach, Michael A. (1976). Tests tell us little about talent. *American Scientist, 64,* 57–63.

Wilford, J. N. (1988). Shades of Sputnik: Who's ahead in space? *New York Times,* Jan. 3, Section E, p. 7.

Recognizing and Fostering Multiple Talents—Reflections on Yesterday's Achievements and Suggestions for Tomorrow

Calvin W. Taylor
Robert L. Ellison

In the history of science, stars, remote from human beings on earth, were among the first things studied. Almost the last—but by no means the least—phenomena on which science focused were human beings. In education, the traditional emphasis has been upon knowledge. But it is now scientifically feasible to make human beings and their potential resources, especially their multiple creative talents, become education's primary and most important concern.

What we present here is some of the evidence about methods of teaching and developing students which bring out their highest potentials for functioning effectively, not only in their schooling but also in their careers and throughout all other aspects of their lives.

Over a period of years, our Utah-based team has done research on the selection, education, and training of many types of scientists. This work follows earlier studies and research at the national level—most notably Taylor's two years of service as director of research of the Office of Scientific Personnel at the National Academy of Sciences' National Research

Council (1953–1955). There, selection procedures were developed to identify future scientists for the National Science Foundation's graduate and postdoctoral fellowship programs. These organizations also helped to stimulate, nationally and internationally, research and conferences on creativity.

All our studies considered creative scientists of many sorts—basic researchers, engineers, and others, as well as career scientists and science teachers. Together, we have had many years of experience in a wide variety of organizations doing research on personnel selection and training of scientists and the measurement of their career performances and accomplishments (see Thurstone, 1964). Our findings and experiences demonstrate that both the discovery and the schooling of students for scientific careers can be greatly improved. (The same is essentially true for practically all types of high-level careers where traditional academic preparation and retention practices are followed.) Our educational systems could provide better scientists for research, better practicing scientists for the professions, and better teachers of science, who better develop students into more effective scientists of all types.

Ross L. Mooney (1963) reported four ways to study creativity: through creative persons, processes, products, and places or environments. Our team has done research in all these areas. We are convinced that, if findings from this research were implemented for students and scientists in the schools and in workplaces, the selection and functioning of scientists would improve. Thus, we primarily consider in this paper what scientists of the future could be like if we used in their behalf what we know about the abilities and characteristics of the best scientists of the past. Educational systems could begin to identify, develop, and utilize human resources of science students and of career scientists in a variety of scientific and professional programs.

The quarter-century-long series of nine Utah-sponsored conferences among leading researchers on creativity are among many major U.S. investigations of creativity. We have done more research in the field than any other nation. Nonetheless, we think that more mainstream scientists and technicians have been produced than truly creative scientists on the frontiers of knowledge. And, even when the latter do emerge, they are not always recognized. In this context, the case of Barbara McClintock's creativity is noteworthy. Her latest and greatest research—a truly revolutionary genetic breakthrough—was treated with silence and hostility for a couple of decades, especially by workers in her own field. However, eventually—without knowing she had been nominated—she found herself rewarded, first, with a John D. and Catherine T. MacArthur Foundation prize which freed her financially (with a no-strings-attached $60,000 annual stipend) and, later, with the Nobel Prize.

During our first major research project, we studied scientists in action at a basic scientific research center in the Air Force. The scientists said that they would give us time to undertake and complete our required studies if we would incorporate their perceptions about the problems about the environments and organizations that deterred them from doing their best scientific work.*

The first phase of the Air Force project was to study the performances, products, creations, and other accomplishments of the scientists involved, who urged us *not* to obtain measures from their written products and inventions alone. While we analyzed the latter, we also included many other measurements to indicate the scientists' other accomplishments, ending up with over 50 criterion measures obtainable throughout their careers. A multiple-factor analysis then reduced the number to 14 different criterion factors (Thurstone, 1947).

The second phase of the study of Air Force scientists was to develop and select potential predictor test scores related to 1 or more of the 14 dimensions of performances and accomplishments of the scientists, as discovered in the first phase. Eight different types of predictor tests administered to all the scientists yielded 130 scores. The scores from four of these kinds of tests produced validities below the chance level; that is, the scores were of no essential value in predicting the performance criterion targets of scientists. Three other test types predicted barely above chance; thus, their validities were also low. The *Biographical Inventory* scores were the only ones with high validity.

Multiple Talent Teaching

These projects and subsequent studies have provided us with the basis for and insights into why Multiple Talent Teaching works:

> All the results on 10 projects plus several replications scattered across the nation have almost uniformly been in favor of Multiple Talent Teaching over traditional academic-only type of teaching . . . . The probability would be essentially infinitesimal that these strings of differences, practically all in one direction and across ten or more projects, could ever occur by chance. *It is suspected that no new educational approach has ever attained such powerfully significant results on measured student performance across such a wide range of relevant classroom activities.* ([italics ours] Taylor, 1980, pp. S12407–12411)

For two decades we have searched for procedures, using a career criterion-oriented measurement approach, to identify persons who will become effective scientists. Early on we discovered that undergraduate grades, even in

---

*The result was three "by-product" reports on the need to improve working environments and overcome such deterrents. However, we neither saw nor heard of substantial changes in the workplaces of Air Force scientists.

science courses, though expensively obtained through four college years, were of little or no value in spotting productive scientists, engineers, teachers, or physicians. In contrast, we have found that in the entertainment fields, particularly in professional athletics, a high correlation exists between one's chosen undergraduate activities and later performances in one's professional career.

Our classroom approach in Multiple Talent Teaching could alternately be described as "asset-focused teaching." Students generally will live up to the high concept teachers have of them. The opposite is also often true. If teachers see their students as liabilities, the students may come to meet that low expectation.

Students could be *talent tested* to determine their potentials and then *talent taught* to develop potentials into actual behaviors, first, through scientific development of talents during schooling and, second, through self-actualization of talent potentials accomplished by the students themselves.

Talent Searches

The Institute for Behavioral Research in Creativity has worked for over two decades on a series of research projects to develop valid biographical inventories (Taylor and Ellison, 1967, 1983) that could predict science talent. One outcome has been the *Biographical Inventory,* which measures high-level brainpower talent potentials in four areas, including (1) academics, (2) creativity, (3) leadership, and (4) the arts, yet analyses of school grades show that (2), (3), and (4) are practically unrelated to school transcripts. The test generates additional scores in (5) vocational maturity (manifested in thinking through possible careers) and (6) educational orientation (manifested in anticipated length of schooling to be completed).

It is now possible to do a systematic and scientific talent search among students in an individual school, school district, state, or nation. The *Form U Biographical Inventory* is our special technique for searches and tests of upper elementary school-level students; *Form CC,* of beginning college students.

Striking samples of talent profiles include one person with an academic percentile rank score of 05 and a leadership score in the high 90s. Another person had a 90th percentile rank in academic talent, with all other talent scores below 50. Still another person, with a below-average academic score, had creativity and leadership scores above the 90th percentile. Together with other information, we determined that his current potential for dropping out of school was very high. Offering him some creativity and some leadership training and experiences might have kept him in school.

A recent talent search survey has tested students in two high school vocational centers, both "alternative" high schools. We suspected that some of these students would earn higher percentile rank scores than colleges and universities would predict; our belief that brainpower talents can be found in high school technical programs was borne out by our research. In one school, we found that 30 percent of 155 vocational center students scored 4 pluses (above average on all 4 talent scores); 20 percent, 3 pluses; 20 percent, 2 pluses; and a last group over 15 percent, 1 plus. Over 85 percent were above average in at least 1 talent area—academics, creativity, leadership, or the arts. In general, each student was above average on approximately 2 of the 4 talent scores. Detailed examinations of the 16 patterns showed that, if the school were to enroll specific individuals in courses emphasizing creativity, leadership, or the arts, the tendencies of some students to drop out could be reduced or even overcome.

Through our testing, we found rich veins of precious different kinds of brainpower in students in whom many traditional academically oriented educators would never conceive that such high potentials could exist. Without the discoveries from our talent tests, these students might have hidden within themselves treasures of high-brainpower talents, untapped and unknown.

With a challenge to awaken higher education to an increased role in recognizing and developing creative potentials in students with many kinds of talents, we have also started talent searches and tests in two technical colleges. The results of our work in these colleges are similar to those found in typical high schools; consequently, we believe that educators in the two-year colleges should teach in ways that develop multiple talents. We strongly anticipate that, when two-year college graduates have been talent tested and talent trained to activate deliberately and systematically the eight creative talents in our talent totem pole picture (see below), they will become both more employable and more promotable than traditionally taught university sophomores.

Double Curriculum Teaching

What we call a *double curriculum* simultaneously develops two radically different types of human resources, namely, brainpower talents and knowledge. Education should be concerned with identifying and cultivating *all* known human resources, inner and outer.

In contrast to inner talent resources, much knowledge begins as an outside "library resource," which can potentially become, in the long or short range, an inner resource. Knowledge only becomes effective when a person fully possesses it. But it must have and retain a sufficiently full form that can be retrieved, worked on, added to (if necessary), and transformed into shapes that can be applied appropriately.

If knowledge has only a temporary effect at the time of taking a test and is largely forgotten thereafter, it has failed to become a permanent, useful, transferable inner human resource. Such knowledge has mainly served the short-range purpose of passing a test and getting a record on a transcript and, in general, does not become a potential resource for the future. Yet short-term knowledge has always been the main focus of teaching and testing for students at all levels of schooling. However, as we will argue below, knowledge can serve in a supportive role as the input that can activate and develop natural brainpower resources.

Intellectual talents are a second major human resource. The inborn brainpower talents can be stimulated to produce multitalented people who function effectively. Multiple intellectual talents, stimulated by knowledge, become tools (mental processes) for acquiring additional knowledge. Talents differ widely in the degree to which they actively process incoming knowledge by associating it with what already exists within one's intellectual storehouse. But the greater links to association between incoming knowledge and this "inner library," the more educated and learned a person can be. In fact, such knowledge will be more retrievable and applicable and, therefore, more transferable and usable in later situations than knowledge only passively learned.

Basic research findings have convinced us that persons become more effective professionally and in general by having a rich functioning inner library of knowledge and multiple talent processes. These talent processes are needed for input reception processing, for inner central processing, and, later, for both retrieval and output expressional processing. The more effective the multiple input channels, the multiple central brain channels, and the multiple output channels of expression, the better. The great need lies in weaving knowledge and talents together as functioning teammates.

Talent Totems Produce Talent-Taught Students

We have chosen to describe the highest-level powers of the brain as *talents,* rather than as skills. Our theory of education conceptualized two broad categories of resources inside students, namely, intellectual and nonintellectual. Because of the availability of many studies on the intellectual brainpower talents, we chose to focus initially on implementing them (Taylor, 1963), postponing the study of nonintellectual resources until later. We have gradually formulated several intellectual talent categories (see below) but have remained open to additional categories by concluding our list with "other."

The project remains open-ended: We continue to expand the total number of talent totem poles, just as other analysts have continued to find new factors (talents) in the mind (e.g., *The Vectors of Mind* [Thurstone, 1935], *Multiple Factor Analysis* [Thurstone, 1947], and *Way Beyond the IQ,* [Guilford, 1977]). We continue to cross frontiers into additional high-level brainpower talents, which are relatively independent of each other.

Over the years, we have expanded the number of totem poles three times. We started teaching for talents typically encouraged by schooling, calling our first pole academic talent, the kind of broad, traditional skill used most of the time in classrooms. In general, academic talent calls for learning-and-returning knowledge and passing tests on it. Presently, attention to this talent almost swamps classroom time; we believe the focus should shrink back to allow other talents time to grow and develop. Then, we injected the broad, complex human resource called "creativity" into both talent testing and talent teaching. This combination yielded the two quite separate and different totem poles of traditional academic and creative talents, which lead to their separate excellences.

The creative talents involve inner processes of thinking and producing and can create at the higher levels of brainpower. These types of talents, which can have positive lifelong impact, need only minimal supervision, with assignments generated largely by the student. Research analyses of talents of the mind led to the next expansion—the splitting of creative talents into five categories, initially in the following sequence: productive thinking, planning, communicating, forecasting, and decision making. We have retained academic talent, adding the five new ones to make the set of six talents used for a decade and a half to develop students. More recently, we rearranged some of the first six and we gradually added three more: implementing, human relations, and discerning opportunities, making a current total of nine talents.

The multiple talent totem poles show that no one is found at the top, or at the bottom, or in the middle across the nine different talents. Instead,

Figure 1
Taylor's Talent Totem Poles*

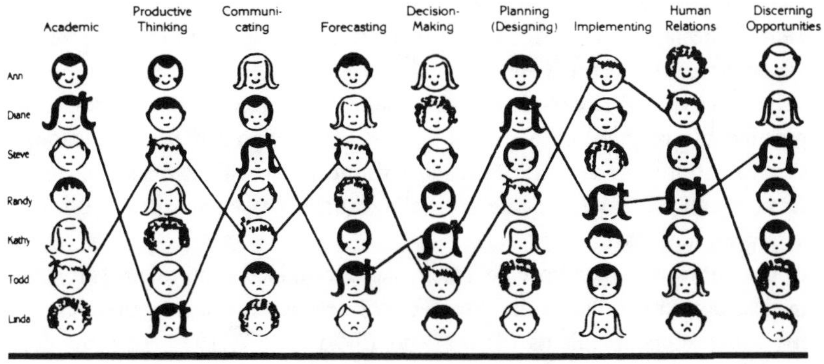

*1984 extended version.

everyone has strengths and weaknesses. The greater the number of different brainpower talents that all students learn to use, the more "equal," on the average, the scores across the nine talents students will earn. Finally, almost all students are potentially high in at least one of an abundant variety of important talent resources.

Academic grades tend to be almost uncorrelated with extracurricular activities, but surprisingly, extracurricular activities not only tend to bring out more whole-brainpower and whole-person performances but also have considerable success in predicting who will be the most effective in later careers. In striking contrast, traditional classroom activities have little if any predictive validity for later career and total life performances and accomplishments (National Federation of State High School Associations, 1987).

Beverly Lloyd (1984), having talent-trained her second-grade class throughout the year and placed all the children on the first six totem poles, studied these students through high school. She found that their high school grades were largely the result of their academic talent. In secondary classrooms, these students were largely functioning one-dimensionally, showing only a small amount of a second, weak talent. However, the students were functioning with other talents in their chosen extracurricular activities. This observation argues that multitalent training, implemented successfully into regular classrooms, can lead to multitalented individuals' later participation in extracurricular activities in high school.

After amplifying the number of talents, we saw a need for simplification and synthesized the nine talents back into two. We obtained the eight nonacademic talent scores in the classroom and applied best-estimate appropriate weights for each talent to derive creativity composite talent scores separate from academic talent scores. Our pair of academic and creative composite talents thus derived illustrates a major issue: We compared, on the one hand, the academic talents demonstrating excellence traditionally to learn the "library of the past" to, on the other, the creative talents British historian Arnold Toynbee sees as "the history making talents" (1964/1967, 1968). The latter produce creative excellence that leads to improvements which "make the future."

Starting with the pair of academic and creativity composite scores, it is possible, by applying another set of best-estimate weights, to produce leadership composite scores for all students. This process would yield a set of three broad totem pole talents, namely, academic, creativity composites, and leadership composites.

This set of three Totem Pole Talent scores obtained from talent teaching in classrooms is similar to the three *Form U* test scores—academic, creativity, and leadership talents (Taylor and Ellison, 1967, 1983; Institute for Behavioral Research in Creativity, 1968, 1978). (The combination of creative leadership measures can thereby be obtained separately, both from talent-tested and from talent-taught approaches.) Once teachers are properly trained to teach in this manner, our experience in research and implementation convinces us that it will cost little more for a simultaneous double curriculum to function in the classrooms than to continue the existing system, which overuses narrow academic mindpower and results in less efficient learning and retaining of knowledge. In fact, the greatest potential resources in the classrooms are the students whose multitalents function effectively and who use more than one of their high-level talents.

Students in our undergraduate university classes have previously used, for most of their 13 or 14 years of schooling, only their academic talents. It takes these students several days or even a few weeks to find out that they can and should be using the more active talent powers of their brain. They can learn quickly, academically, about the talent totem poles. But they are not ready to learn to function that way nearly as quickly as they were when they were younger. Nonetheless, our 400 university students in 10 classes of 25 nominated all but 1 percent of their peers as being in one of the two "best" (top 8 percent) of the 7 talent areas. We have recently added inventive talents as a component to our list of the collective set of creative talents.

Combining Knowledge Outcome-Based Education (KOBE) with Talent Outcome-Based Education (TOBE)

Having repeatedly stated that knowledge and brainpower talents are two radically different types of human resources, we have urged that one should cultivate both knowledge and talents—not one without the other. Each type of Knowledge (K), at one time or another, can be processed by every type of Talent (T). In this way, there is no need to debate "knowledge versus talents" or "learning knowledge versus learning to function effectively."

This is a challenge, since teachers and other educators often have been trained only how to teach knowledge and have developed a curriculum for it alone—perhaps the easiest approach in the first place. However, we now see knowledge as one among other talents—perhaps a less lasting human resource than some of the others, less likely to be relearned, reused, retained, retrieved, and less necessary as the main focus in schooling. We can now develop many talents and thus move toward fulfilling the whole person. Schooling's outcomes can now be based on knowledge in combination with talent. The interaction strengthens knowledge acquisition and lengthens the durability, transferability, and usability of both knowledge and talents. We can now combine *K*nowledge *O*utcome-*B*ased *E*ducation (KOBE) and *T*alent *O*utcome-*B*ased *E*ducation (TOBE) to form the powerful and lasting TOBE-KOBE combinations.

Students esteem creative teachers who use creativity after school to plan tomorrow's lesson and use it as they teach. But most important, their creative teaching must develop creativity in their students.

We see no shortage of teachers for the classroom in which students are trained to use talents and simultaneously to acquire knowledge. Instead of being knowledge dispensers ("sages on the stage"), teachers will develop students who can learn to participate more actively and more fully than they do in and from traditional classes. Such teachers for talents will acquire self-esteem from direct observation of their students' growth. The students will progress through working with knowledge and through stretching their minds.

Students are not nearly as bottled up in talent-focused classrooms as they sometimes are in traditional ones. Instead, they become more spontaneous and function more freely than children who are overcontrolled in strict classroom climates. In the latter situations, the process of grade getting is almost through nonthinking ways of learning.

Teachers for talents use classroom processes to prepare students so that they can reflect on the past, tackle unknowns, and try to eventually create a better world, while revering and perpetuating what is good about the past.

Resurgences Through Education

We have developed a list of what we call "resurgences," proposed or begun, to improve nations, states, their schools, and therefore their students. Toynbee (1964/1967) proposed U.S. reform, hoping it would lead to a worldwide resurgence. He suggested that creativity has been among "history-making talents" and asked, "Is America Neglecting Her Creative Talents?" He believed that America has been great because of citizens active in three major creative pioneering movements: the first settlers on the Atlantic seaboard; the founders of the Republic; and the pioneers who settled the West. Following these earlier movements, Toynbee thought, the U.S. was still the hope of the future. He forecast that this nation can for the fourth time become a creative history-making one if Americans

> return to those original ideals that have been the source of her greatness .... The pioneers were not primarily concerned with money-making; if they had been, they could never have achieved what they did. *America's need, and the world's need, today, is a new burst of American pioneering, and this time not just within the confines of a single continent, but all around the globe.* (1964/1967, p. 29)

To Toynbee, the next chapter of America's history is to help the indigent majority of humankind toward a better life. This task requires unlocking and activating as much creativity as is possible. Americans must cooperate to foster and treasure all the potential creativity they have within them and also encourage other nations to join in this worldwide creative resurgence.

If the recommendations made by a 1968 Presidential Talent Task Force Report on the topic of *Talent Development: An Investment in the Nation's Future* had been implemented, American education would have begun to develop techniques for talent teaching of scientists.* Still, many of its recommendations have reached the international arena a decade later. Then, in 1979, Luis Alberto Machado, a former chief of staff to his president, was selected for a five-year cabinet term as Venezuela's "Minister for the Development of Intelligence." Machado defined intelligence as total brainpower, not narrowly as IQ. In his cabinet role, Machado undertook a mission within Venezuela and to nations around the world. He believed that—for the good of themselves, their society, and the world—people's basic human right was to have their total brainpower activated and developed; therefore, he charged all societies to educate everyone as fully as is scientifically possible.

Machado observed that Venezuela's potentially greatest wealth is not in black gold (oil) but in grey gold (brains). He thought only 10 percent or slightly more of the intellectual potential of typical Venezuelans was active. But, by applying the latest scientifically based methods of teaching, he hoped that average persons in the future would be able to develop 60 or 70 percent of their potentials.

In 1980, parts of our program were recommended by Ernest Boyer, president of the Carnegie Foundation for the Advancement of Teaching and former U.S. Commissioner of Education. Twenty-three scholars and 40 college and university presidents gathered in a symposium which included a day spent discussing creativity. Boyer reported that many college presidents told him that they were no longer leaders. They saw themselves not as decision makers who moved from ideas to constructive action but as cogs in a machine. They believed they had lost their time for imagining; they felt powerless. If universities become lifeless, uncreative institutions, we wonder how anyone can expect students to be taught to think creatively.

But just five years later, in 1985, Frank Newman produced a report for the same Carnegie Foundation calling for an American resurgence in and through higher education. Later that year, Newman became the president of the Education Commission of the States, under whose auspices he recommends policies to the 50 U.S. governors. His call for resurgence through education now applies to all educational levels—preschool, elementary, secondary, undergraduate, graduate, and professional. The themes in his report suggest providing activities to develop all students' creativity, risk taking, independence of functioning, and shouldering of responsibilities in society.

---

*Unfortunately, due to the political and social turmoil of the late 1960s, the report was neither released nor disseminated.

Newman's report, unlike *A Nation at Risk* (1983), focuses on creative excellence as well as traditional excellence.

Another resurgence occurred after John Goodlad's 10-year study titled *A Place Called School* (1983). That report was disconcerting though accurate in its findings. The problems it posed led John Raven (1987), a Scottish educator, researcher, and reformer, to warn that, if the U.S. keeps running the kind of "educational systems" described by Goodlad without making major changes, our nation as well as others is truly at risk.

Many of the typical classrooms in Goodlad's 10-year study were *classrooms without laughter*—which is no laughing matter. Where life is joyless in school, and students feel that they have been assigned to hard labor in an emotionally flat classroom climate, they are not going to grow. In contrast, classroom programs can foster serious playfulness of the mind, serious toying with ideas, and serious fun. Some students found such a classroom climate delightful, commenting that they felt they had paid their dues in humorless classrooms for years. They wanted to move into more natural toying with ideas and more exciting experiences where they learned through thinking. Such programs can help stem the crisis in schooling in our nation. In recent years, fortunately, there has been attention given to the concept of whole brain development or whole person development. We recommend both traditional and creative excellence in schooling—not aiming at one without also nourishing the other. No longer should we teach for academic talent alone.

Traditional excellence tries to perpetuate what is best about the past, whereas creative excellence aims to produce a new and often better future. Both are necessary. Misused and mistaught, traditional academic learning can be primarily rote, can proceed by cramming, can be a nonthinking (or unthinking) way of gaining knowledge. Such rigid methods do not develop and use students' nonacademic brainpower talents as ways of acquiring (learning) knowledge. Such students are not becoming self-educators. But learning should not stop with school. Academic and talent learning should be *lifelong.* Self-education could keep the postschool population up to date with new knowledge and new techniques that will improve their society and their physical and mental health. Learning that calls for thinking should underlie every successful resurgence proposal. *A mind (or any large part of a mind) is a terrible thing to waste,* according to Sam D. Proctor (1974). Many have agreed with him. We are convinced that practically everyone is talented in something, and if we let each person go through enough somethings, the talent(s) will emerge.

Better Uses of Multiple Talents—at All Levels

In another study, we correlated the 6 *Form U* scores of fifth graders with their 14 scores in the Iowa Test of Basic (Academic) Skills. We found that our academic talent score almost typically produced correlations from .50 to the mid-.60s with the 14 basic skills scores. The educational orientation score also produced correlations between .35 and .50 with all the basic skill scores. The striking finding, however, was that the crucial scores of creativity, leadership, and artistic talents had no significant correlations whatsoever with any of the Iowa Basic Skills scores. We fear that creativity, leadership, and artistic talents are not being emphasized fully enough in official school curriculums, and grades for these talents do not appear on transcripts. The field is wide open for teachers and educators in science to have all the *Form U* talents tested and taught within their official curriculums. The multiple creative talent teaching approach has been and can be successfully inserted into classrooms to accompany the first curriculum of knowledge (typically learned mainly by the academic talent). Even if educators and teachers resist new paths, multitalented students who discern such opportunities can later create new paths themselves. Parents, teachers, and supervisors should become talent developers as well as knowledge dispensers, focusing on both talents and academic knowledge. Computers and other media approaches can provide and dispense some of the knowledge, relieving educators so they may concentrate on talents.

Out of school, the same processes can apply. An organization is often created when a person can't do or accomplish something alone. Organizations can enable good things to happen, things that could not occur if the organization didn't exist. A study at one large research center found that the many people in its organization whose duty it was to provide support and service to the research scientists had been classified as *can doers* and *can't doers*. The latter excelled in generating explanations as to why it was impossible for them to provide a particular service. In contrast, the can doers practically never failed to provide successful solutions. The same can be true of scientists, at any or all levels and types of functions in which they engage. If selected and trained to be able to see and believe success is achievable, they should become can doers, then will doers, and then doers.

One rule in science is that hindsight is as important as foresight and can lead to new insight. If a study gains no new insight, then it is really merely a verification study. In one investigation of a project in military research, it was found that about two-thirds of the knowledge necessary to complete the project was available in advance; the remaining third, new ideas found during the course of the study, was essential in yielding a fruitful result. Similarly, a major chemical company found that many of its projects as originally proposed and funded did not yield a successful *end product*. But if there were *new insights* gained during these and other ongoing projects, scientists often moved quickly (and often informally) to another project, which did produce a needed product.

If masters' and doctoral research has as an unstated emphasis a wish to foresee the positive results of all studies, such work may produce persons inadequately trained to gain new insights. Many studies for graduate degrees are almost nonrisk verification projects leading to no new ideas and excitements; such degrees do not prepare graduates to gain new insights in their research careers.

All is not well in professional schools, even in the health fields. Since professional schools, which tend to focus on academic talent alone, do not look for other important talents, they often overlook highly qualified applicants. This predicament occurs because the selection process assumes too strongly that the academic talent is an extremely high predictor of the implementation talent—a false assumption, in our view. After their professional training, the graduates often diversify widely in their practice on the job. Requiring that students be *talent taught and tested at the preprofessional level* and taking these factors into account, admissions programs could better select applicants for both graduate and professional study. The creative brainpower talents can make history. Yet they are still largely neglected at all levels of schooling. We emphasize the development of multiple creative talents in all students. In brief: When multiple talents function more creatively, people function more effectively, and when multiple talents function more effectively, people function more creatively.

References

Boyer [Ernest] assails lack of creativity by "tired" universities. (1980, November 21). *Higher Education and National Affairs, 29* (38), 1, 4.

Goodlad, John. (1983). *A Place Called School.* New York: McGraw-Hill.

Guilford, J. Paul. (1977). *Way Beyond the IQ.* Buffalo, NY: Creative Education Foundation.

Institute for Behavioral Research in Creativity. (1968). *Manual for the development of the Alpha Biographical Inventory.* Salt Lake City: Author.

Institute for Behavioral Research in Creativity. (1978). *Preliminary administration and research manual: Biographical Inventory—Form U.* Salt Lake City: Author.

Lloyd, Beverly. (1984). *The longitudinal effects of multiple talent training on 28 second-grade students: The totem pole kids.* Unpublished doctoral dissertation. University of Utah, Salt Lake City.

Mooney, Ross L. (1963). A conceptual model for integrating four approaches to the identification of creative talent. In Calvin W. Taylor and Frank X. Barron (Eds.), *Scientific creativity: Its recognition and development* (pp. 331–340). New York: Wiley.

National Commission on Excellence. (1983). *A nation at risk.* Washington, DC: U.S. Department of Education.

National Federation of State High School Associations. (1987). *The case for high school extracurricular activities.* Kansas City, MO: Author.

Newman, Frank. (1985). *Higher education and the American resurgence.* Lawrenceville, NJ: Princeton University Press. (The Carnegie Foundation for the Advancement of Teaching)

Proctor, Sam D. (1978, November). A mind is a terrible thing to waste. *Phi Delta Kappan, 60*(3), 203–210.

Raven, John. (1987). *Developing the talents (career and life) and competencies of all our students.* Paper presented at the 7th World Conference for the Gifted, Salt Lake City.

Taylor, Calvin W. (April, 1963). Many-sided intelligence. *Childhood Education, 39*(8), 364–366.

Taylor, Calvin W. (1980). Multiple talent teaching results. In *The Congressional Record* (pp. S12407–12411). Washington, DC: U.S. Government Printing Office.

Taylor, Calvin W., and Ellison, Robert L. (1967). Biographical predictors of scientific performance. *Science, 155,* 1075–1080.

Taylor, Calvin W., and Ellison, Robert L. (1983). Searching for student talent resources relevant to all USDE [U.S. Department of Education] types of giftedness. *Gifted Child Quarterly, 3,* 99–106.

Thurstone, Louis L. (1935). *The vectors of mind.* Chicago: University of Chicago Press.

Thurstone, Louis L. (1947). *Multiple factor analysis: A development and expansion of the vectors of mind.* Chicago: University of Chicago Press.

Thurstone, Louis L. (1964). Criterion of scientific success and the selection of scientific talent. In Calvin W. Taylor (Ed.), *Widening horizons in creativity* (pp. 10–16). New York: Wiley.

Toynbee, Arnold. (1964). Is America neglecting her creative minority? In Calvin W. Taylor (Ed.), *Widening horizons in creativity* (pp. 3–9). New York: Wiley. (Paper reprinted 1967 as Is America neglecting her creative talents? [Salt Lake City: University of Utah Press])

Toynbee, Arnold. (1968, April). Creativity in our schools. *The Instructor, 77,* 21, 132. (An interview with Margaret Mason)

Science Talent: In an Ecology of Achievement

Paul F. Brandwein

The purposes of this analysis are fivefold. First, we intend to devise an environment in schooling in which the young seek and find opportunities to think and do science. The environment, structured in an ecology of achievement, may thus affect the course of change in the young as they fulfill their diverse powers in the pursuit of their special excellences. Second, we are obligated to devise environments that channel the interests of the young so that they learn the arts of investigation and, in a sense, become performing scientists akin to performing artists; that is, they demonstrate their talent in a work or performance.

Third, given the opportunities so devised, the young who participate discover that there is a need for individuals of different gifts, different talents, and different levels of achievement. Fourth, we shall describe an environment in which the young serve their apprenticeship to various kinds of well-ordered empiricism, a significant methodology of the performing scientist. Thus, we will probe essential questions: Does an individual's talent in science demonstrate a private gift? Or is an effective channeling and augmenting environment integral in evoking talent? Finally, is it possible to turn our experience and investigation into some form of generalizable knowledge? We think it is. But I anticipate.

A First Thesis: An Ecology of Achievement

The early environment of the child—home, family, community, and school—forms an extraordinary interrelationship of environments, ecologies that contribute to shaping the individual. David T. Suzuki, Anthony J. Griffiths, Jeffrey H. Miller, and Richard C. Lewontin (1986) furnish us with an intriguing model and a masterful analysis of the relationships of genes interacting with the environment. Consider, they say, two monozygotic ("identical") twins, the product of a single fertilized egg that divided and produced two sisters with identical genes, that is, with identical complements of DNA. Say the two were born in England but separated at birth. Suppose one were raised in China, by Chinese-speaking foster parents. The other, in Hungary. The former will speak Chinese, the latter, Hungarian. Each will behave in accordance with the customs and values of her environment. But consider: The twins began life with identical genetic properties (equivalent and equal DNA, and identical genomes), but in the end the different cultural environments produce great differences not only between the sisters but also from their parents. Clearly, Suzuki and his colleagues maintain that "differences in this case are due to the environment and the genetic effects are of little importance" (p. 5).

It boggles the mind to consider the effects of the multitude of non-DNA differences in the environment that determine the actual course of change in the individual. Since the early work on genetics in 1900–1940, there has been no study in genetics generalizing findings of research on an organism's structure or function that does not base its hypotheses on the generality that an organism is the product of the interaction of its genes and its environment.

For example, the color of the fat of certain rabbits is changed from yellow to white, or vice versa, by the color of the mash they are fed; the arteries of the human are clogged or remain clear depending on the diet (an environmental effect), obesity aside. Inherited pink lungs turn gray-black from the smoke of tobacco; light-pigmented skin darkens in the sun; obviously, we learn and do not genetically inherit a knowledge of history. Further, certain inherited disorders can be treated as easily, or with as much difficulty, as those arising from environmental difficulties and accidents, before and after birth. By way of example, Wilson's disease is characterized by the steady degeneration of the nervous system and liver because the body cannot synthesize normal amounts of a certain copper-containing blood protein. Instead, copper atoms from food are deposited in the brain and other tissues. However, an available drug (an environmental factor) removes the copper atoms and prevents degeneration of nerve tissue.

Another example: The absence of a specific gene results in *Phenylketon-uria* (PKU), a condition that pushes a child into mental retardation. The amino acid phenylalanine accumulates in the body, resulting in brain damage. A child may be spared the effect of its genetic defect when given a phenylalanine-free diet. In this case, as in Wilson's disease, the intervention of a changed environment reduces the deleterious effects of the genetic condition. Thus, in these two examples, we see an explicit interaction of gene and environment through medical intervention. The notion that some inherited traits are unchangeable or inevitable is no longer acceptable.

Clearly, the ability to do originative work or discovery in science, a sign of talent in science, must be strongly assumed to be a product of heredity and environment. As Liam Hudson (1966) suggests, the intellectual operations of the scientist depend on a huge accumulation of experience as well as on vast accumulation of knowledge. Neither can be ascribed to genes formed long before that knowledge was available. We seek then an environment (an ecology) which will give all our young the multiples of opportunity for development, allowing them to benefit fully from the consequences of the interaction of their genes and their environment.

Recall now that the term *ecology* describes a relationship, an interdependence, among organisms and their environment. Although an ecology seems a loose relationship, it is in effect a structure built on strong interdependencies among organisms in a *particular environment.* Together, organisms and environment form an ecosystem; if we alter the environment, the organisms may not survive. The interdependence or ecology we seek is that which leads to performance; we try to relate the manner in which the design of the ecology in schooling and education translates or metamorphoses potential in science into a performance: a *discovery* (the term I should substitute for "creativity").

As we shall see, the sole use of intelligence tests can result in a single reified judgment of ability. IQ, too often a dominant criterion in the selection of the "science-gifted" in isolation from the powerful environments that affect talent, is thus not central to our probe.

Note, however, that the transformation of an organism from one stage of life to another is a result of the unique interaction of its genes and its environment at each moment of life (Suzuki et al., p. 5). But this significant statement may be inadequate unless one considers two types of heredity—one *genetic,* the other *cultural.* We are aware that biological heredity (or inheritance) consists of the transmission and transmutation of DNA. Cultural heredity (or inheritance) consists of the transmission and transmutation of learned elements: knowledge, values, and skills. Put another way, the transformation of the human from one stage of life to another is a result of the unique interaction of its biological and cultural heredities.

In other words, a child is not the result solely of gene-driven factors. The very young child is already a complex of gene- and environment-driven factors interacting to form structure and function. Further, subsequent development stems from a newer base: the result of the up-to-this-point interaction of the two heredities, genetic and cultural.

Surely cultural heredity is within our control. Yes, but not historically—or certainly not in present cultural history. The life span of a child born in an underdeveloped country is an average 40 years; the inner uterine environment is as critical as the outer. In the United States, at present, the life span for those with optimum health is 75 years. What was the cultural history of a South Korean child in 1910, as compared and contrasted with one in 1988? Surely, the former did not enjoy a civilization centered in science and technology. Of a Japanese child, before the Japanese industrial and cultural revolution? Of a woman born to mature to her majority before, say, women were allowed to vote or readily accepted, generally, into the scientific professions? The word "environment" hides too much and is often too bland. For example, the brilliant young British physicist, H. G. Moseley, was killed at Gallipoli during a failed attack in World War I. Are not wars and their effects environment driven? You will know of other instances—dismaying ones, or those made glad—some by human and humane intent. You will find we have survived countless terrors.

For this reason, we shall be stressing throughout a diptych in human development consisting of two universes of factors or traits that are gene driven and environment driven. The simplest example of the former is height, which is mainly gene driven, but even here affected by diet and, perhaps, exercise. The most complex, perhaps, is intelligence, which is dependent upon many genes, or, as the term is used, is multifactorial, to an extent as yet unknown and surely affected by a host of environments also not fully known.

Environment-driven factors come out of life's experiences. Schooling is precisely that construct of the environment that is intended to nurture all. It is the environment that is intended to offer varieties of experience for multiple intelligences. In brief, schooling is the social construct designed to offer multiple channeling and augmenting environments for the varieties of young who make up a school—and by extension, the varieties of young who make up a democracy.

Richard H. de Lone (1979) in his *Small Futures*, a study for the Carnegie Foundation, furnishes impressive evidence of the effect of the socioeconomic environment on the futures of children in minority groups. The data he has gathered demonstrate powerfully how environmental disadvantages and injustices limit achievement. Kenneth Keniston's foreword to the study states this view clearly:

> For well over a century, we Americans have believed that a crucial way to make our society more just was by improving our children. We propose instead that the best way to ensure more ample futures for our children is to start with the difficult task of building a more just society. (p. xiv)

Simplistic consideration will yield the truth that, even if undertaken immediately, reconstructing society (and particularly the family) demands careful and incredibly intricate and elaborate plans over the definable short and long terms. Yet daily, each and every year, without halt, the young keep coming to our schools. The school environment, its effect on intellection, on personhood, on knowledges, attitudes, and skills—if you will, the ecology of achievement—is the heart of a school supported by a community that can stand some rebuilding. The impact of schooling is notable with regard to the opportunities for both the gifted (the seemingly advantaged) and those disadvantaged by the many social and economic ecologies that affect them. The term *environment* is thus best placed within the widest ecology of achievement, which includes not only development of the organism in its biological, psychological, social, and educational components but also in its full political and economic range.

Schooling per se, to be effective in our present technocratic society, with its techno-electronic technology—a society hastening into a postindustrial era—must accommodate an ecology of achievement. But a school is only a part, certainly not all, of this ecology. That is, schooling is reasonably a part of the society and the culture that mother it, just as the separate communities of the forest and the sea are parts and not the whole of the ecosystem. As we shall note again and again, a community furnishes a part of the ecology of achievement in which a school succeeds or fails. The ecology of achievement that nourishes a successful school is characteristic of a successful community. And conversely, it is difficult to find a successful school within a community that does not support the kind of schooling required in an open society, especially one entering the global economy of the postindustrial period (Brandwein, 1981).

The community of scientists within a given field also establishes an ecology of achievement. Scientific knowledge is cumulative knowledge; scientists cannot ignore precedent work: Indeed, they build upon it. The library and the computer's data base precede and endure during the scientist's work in the laboratory. Simplistically stated, "brains on" before and during "hands on." And, if you will, as the young mature and rush to learn, to do on their own, a certain dose of "hands off" is desirable.

It seems also that ecologies of achievement affect the kinds of problems accessible to the problem-solving activities of science. James J. Gallagher (1964) remarks in the 1970 preface to *Teaching the Gifted Child* that much of the material contained within was "nonexistent" five years ago. That is to say, studies on giftedness seem to be grouped in periods—in the early 1950s to 1960s and again in the early 1970s. Possibly studies of the nature of the cyclical change in interest in giftedness generally, and in science specifically, parallel closely the crises of society—particularly those in economic, social, and political ecologies. So Abraham J. Tannenbaum (1979), both a teacher of the gifted and a student of giftedness, remarks: "The cyclical nature of interest in the gifted is probably unique in American education. No other special group of children has been alternately embraced and repelled with so much vigor by educators and laymen alike" (p. 5). And again, Tannenbaum, quoting Spaulding (undated), writes, "A review of the state of research for the years 1969 to 1974 reveals a fairly bleak picture; only 39 reports on the gifted had been published in that period" (1983, p. 33). On this oscillation by the culture or "change in signals" by society, see also Harry S. Broudy (1972) and Brandwein (1981).

Thus, there appear to be concerted efforts in a given field at a given circumscribed period, ecologies of achievement apparently responding to periods with similar problems. On this, Robert K. Merton (1961) remarks:

> I should like now to develop the hypothesis that, far from being odd or curious or remarkable, the pattern of independent multiple discoveries in science is in principle the dominant pattern, rather than a subsidiary one . . . . Put even more sharply, the hypothesis states that all scientific discoveries are *in principle* multiples, including those that on the surface appear to be singletons. (p. 306)

Merton's account is, thus, a reference in our terms to an interrelationship in effort of different individuals and groups of individuals, concentrating on achieving solutions to the problems of society (an ecology of achievement).

Derek de Solla Price (1961/1975) in *Science Since Babylon*, his thesis on the nature of "scientific civilization," refers to discoveries by foremost scientists: "Probably it follows that to double the population of workers in the few highest categories, there must be added eight times their number of lesser individuals" (p. 120). That is, lesser individuals who prepare the ground, or assist in the investigation, or add to the field. Among them are individuals who, given a certain intellection and personality (neither yet fully understood) as well as the necessary opportunity and luck (also not yet fully understood), may become giants in their own right.

Yakon M. Rabkin's (1987) analogy of the contributions of those scientists in the "lesser" categories compared to those in the "higher" categories is found in his discourse in *ISIS* on "Technological Innovation in Science: Adoption of Infrared Spectroscopy by Chemists." There, he states,

> Among the most important new methods was infrared spectroscopy, which acquired remarkable popularity during the 1950s and 1960s. The number of infrared instruments, a handful before the war, rose to 700 in 1947, to 3,000 in 1958, and to 20,000 in 1969. The technique's use in scientific research, as recorded in a 1965 report issued by the National Academy of Sciences in Washington, D.C., skyrocketed correspondingly. (p. 31)

Obviously the corresponding rise in numbers of scientists working in infrared spectroscopy included those in both the "lesser" and "higher" categories, presumably the greatest number in the "lesser" categories. Question: Could those in the "higher" categories have done their work without the efforts of those in the "lesser" categories? May we not say all were involved in similar ecologies of achievement? On this aspect, please see Glenn T. Seaborg's "Letter to a Young Scientist" in this volume.

I refer here only to a very few studies that offer firm cognizance of the notion that science discovery (synonym: "creativity") does not come out of a single "Eureka!" or "Aha!" but from a network, a seamless fabric of the effort and work of many individuals—of all levels of ability and temperaments—over time. It is almost with regret that I must redact Newton's fabled statement that even a dwarf can see farther when seated on the shoulders of giants.*

In essence, the figure of dwarf seated on giant is one aspect of the methods of intelligence of the scientist, namely: Science as a field, and a scientist working in an area of special knowledge, depends on cumulative knowledge. Thus, I must revise the Lucan-Burton-Newton model (Virgil probably had an earlier hand in the aphorism) of dwarf on giant. That is to say, a dwarf may stand on a gigantic pyramid of bricks composed of the clay and straw of prior massive effort, and so even a dwarf may see farther. But if a dwarf may see farther, so may a giant. This is not to reject the truth that, in an ecology of achievement, one or two individuals may have knowledge, skills, and attitudes that make them clear leaders in a field. There surely are seminal thinkers, or giants, in the general attributes of intellection (say, Aristotle, Galileo, Newton, Kant, Einstein, the Curies, Mead) or in a special talent (say, Mozart, Beethoven, Stravinsky, Rembrandt, Monet, Van Gogh, Shakespeare, Dickens, Cassatt—select your own) who grasp a field entire and set it into a new context. However, they too have learned from others before them. Further, children and adults are not gifted, or talented, in *all* things, are they? They are talented in *some* things—even many, but certainly not all. The towering genius does not tower in all fields but leaves some towers to others. That is to say, the processes of work in any field encumber us with the methods of thought and the ways of work, as well as the knowledges and skills gained in time known and unknown, but *probably* prior to the time of discovery and *possibly* within the same period.

It seems, then, we cannot escape the ecology of achievement no matter how we examine it, any more than the talented scientist can escape the past. We may thus safely take a first look at the devices that channel the interests of the young so that they may undertake significant role exploration in the

---

*Even here there is an ecology of achievement. Surely among others, it occurred to Lucan (39–65 A.D.) to note: "Pigmei gigantum humeris impositi plusquam ipsi gigantes vident." (Pygmies on the shoulders of a giant see farther than the giant.) Robert Burton (1577–1640) referred to Lucan's *Didacus Stella*— and it would not be surprising that Newton read Burton. I am indebted to Merton for bringing this to light in his *On the shoulders of giants* (1965/1985).

acts of discovery in science as coming out of a *dyad: genome interacting with environment*. Yet, we as teachers do not tamper with the genome, we tamper with its environment, a precious yet dangerous opportunity. Precious, because we are designing an environment for the young, who deserve an optimum environment in which to fulfill themselves. And dangerous, because in the sense of the Hippocratic oath we must "first, do no harm."

A Second Thesis:
The School Environment, as Dyad

Are there any environments that would approach the condition that would permit us to use the term "identical environment" even as we use the term "identical genomes" for identical twins? We should expect to find such a home environment for twins or siblings reared in the same family in one household. Yet the consensus of recent findings is that, in a sense, each child within the family (a macroenvironment) is in an environment of her or his own (a kind of microenvironment). For example, siblings reared in the same home environment are generally different in significant behaviors. That is, the siblings are reared within different ecologies of achievement—their genotypes and initial environments interact to furnish them with a singular, not similar, ecology of achievement. Indeed, the parents are likely to react differently to children of different temperaments, behavior, and ability. Parents furnish offspring with different microenvironments, thus creating different ecologies. This should be expected if the siblings have different genotypes, and further, if, as is thought, as much as 30–40 percent of the traits of temperament are inherited, are gene driven.

Is there then a hypothesis that may guide our efforts to determine whether there is a trait we may call "giftedness" or "talent" in science? Forty years ago (1947) I tried to define one, based on my early work in scientific research and thus on my observations of scientists at work in their laboratories. As a participant in scientists' researches for six years, I learned their "methods of intelligence" (a term Percy Bridgman so often used). I used what I had learned in my own research. When I turned to teaching, I applied what I had learned from my observations of the laboratory environment into environments nurturing those whose wishes and intentions were to become scientists and who came forward (i.e., they selected themselves) to undertake the work open to them. The program was described in its initial plan in *The Gifted Student as Future Scientist* (Brandwein, 1955/1981) and more recently and briefly in "A Portrait of Gifted Young with Science Talent" (Brandwein, 1986).

The titles are significant. First I had noted that all the scientists (some 26) I had worked with were gifted students; that is, they had mastered their fields, could use the tools of "uncovery" in library and laboratory in order to come to their aim: a *discovery*, a new and meaningful work. What seemed to characterize the scientists who made new knowledge through that definitive, creative act, discovery?

First, being *gifted students*, they obviously had a certain level of intelligence. Recall, intelligence is a gene-driven factor; it is multifactorial (Guilford, 1968; Gardner, 1983) and thus is not to be reified in a single measure, IQ. But multifactorial intelligence should not be confused with *achievement*. While the former is strongly gene driven and, of course, highly factored by environment, the latter depends strongly on an environment channeled in schooling and education.

Not only were the mature scientists gifted students in their particular field in science (and apparently also in general intelligence as indicated in their verbal and mathematical skills), but they also were persistent in pursuing the solution of problems. Generally, their observed behavior may be expressed thus: Theirs was a quest, a search for verifiable knowledge. Embraced in the quest was their notable and observable dissatisfaction with present explanations of the way the world works—particularly in their chosen field of study. They often spoke warmly of certain experiences in school and university—and of their mentors in research. In short, their channeled interest was augmented by personal and activating attention of a memorable sort.

On the basis of these observations made in a laboratory setting, and the basic interpretations that sprang from them, early on I developed a "model" to guide our work in preparing an environment that would

• *Channel* the energies of those who wished to carry on the study of science (a channeling environment)

• *Augment* this channeling environment with opportunities to do such research (an augmenting environment is specially designed to furnish situations in which "original" problems in science would be found and an attempt made to engage in the kind of discovery characteristic of the scientist: Students would engage in problem solving, not the usual problem doing, of the scheduled lab)

• *Transform* potential into performance (once the young had performed with success and personal satisfaction, it seemed as if they had transformed themselves)

Our test of "creativity" in science was thus to be a test of ability to do an experiment or an investigation leading to a "new" bit of knowledge, a *work*. In order to describe in detail a plan to test ability to do an investigation, with Evelyn Morholt and Sigmund Abeles I have prepared a companion paper, "Apprenticeship to Well-Ordered Empiricism," in this volume. This paper is concerned with the channeling and augmenting environments open to the young seeking to determine whether their talent might be expressed in science.

Our first concern in the design of the "model" was to focus on the nurturing face of the environment. We would create a novel environment, and thus we could attempt to specify the competencies of our students in terms of performance. Ours was a kind of psychology or "unpsychology" (Michael A. Wallach's term) based on discerning and discovering conditions that stimulated field-specific, real-world performance in a particular area such as science. We would substitute the students' *work* for their scores on tests purporting to measure general "giftedness," "talent," or "creativity."

As my colleagues and I observed youngsters at work over the years, it became apparent to us that *questing* included a reaching out, perhaps an avid search for experiences that were stimulating, perhaps "experience in search of meaning," as Einstein would say. These students seemed to want an environment fitting certain personal tendencies toward autonomy; they were bent on hollowing out a kind of capsule of freedom. Further, our interviews with a number of parents of these young who used novel approaches in problem solving led us to conclude that these youngsters had the privileges of early "independence training." (Anne Roe's phrase in Roe and Marvin Siegelman, 1964, p. 5). And, in their active search for experience, these young exercised a certain autonomy. Indeed, Lois-Ellen G. Datta and Morris B. Parloff (1967) suggest the possibility that the "main influence on early scientific creativity was autonomy versus parental control" (quoted in Tannenbaum, 1983, p. 295).

Our observations of mature scientists at work and of young aspirants to scientific careers at Forest Hills High School (New York City) led to the development of the working hypothesis summarized as follows: *"High-level ability in science is based on the interaction of several factors: Genetic, Predisposing, and Activating. All factors are generally necessary to the development of high-level ability in science; no one of the factors is sufficient in itself"* (Brandwein, 1955/1981, p. 12). This hypothesis guided us in our efforts to develop a channeled and augmented environment for those who *selected themselves* for the program we called OPUS—*Occupational Program Undergirding Science* (Brandwein, 1955/1981; 1986).

Clearly, what appears to be a *triad* rests in an interaction of gene and environment, a *dyad*. Probably the elements of the genetic factor display themselves in the early intellectual and physical development of children, as they interact with the people and things in their early environment. *Possibly* the predisposing factors, involving as they do personality, that is, temperament, are at least in part gene driven but are expressed in interaction with the environment. Thus Sandra W. Scarr and Kathleen McCartney (1983) state,

We all select from the surrounding environment some aspects to which to respond, learn about, or ignore. Our selections are correlated with motivational, personality, and intellectual aspects of our genotypes. The active genotype-environment

82

effect, we argue, is the most powerful connection between people and their environments and the most direct expression of the genotype in experience. (p. 427)

The predisposing factor I postulated comprised two identifiable traits: *persistence* and *questing*. Early on, Catherine M. Cox (1926) observed that high but not the highest intelligence, combined with the greatest degree of persistence, would achieve greater eminence than the highest degree of intelligence with somewhat less persistence. The trait questing (embracing a free-floating curiosity) demonstrated itself in the need to know and in a dissatisfaction with certain explanations of phenomena. Roe (1953), in her study of mature scientists, and Donald W. MacKinnon (1962, July) agree. MacKinnon, for example, states, "Our data suggest, rather, that if a person has the minimum of intelligence required for mastery of a field of knowledge, whether he performs creatively or banally in that field will be crucially determined by nonintellective factors" (p. 493). That is to say, factors of personality, even luck. Hudson, in considering the performance of creative work, states plainly that the work depends not so much on an individual's "intellectual apparatus but the use he sees fit to make of it" (p. 30). Hudson asserts as well that, given a certain level of ability, the personal, not the intellectual, factors are crucial.

From our observations of working scientists as well as from common sense, it seemed clear that genetic and predisposing factors were not all that operated in the making of a scientist. Opportunities for further training and the inspiration of the individual teacher and/or mentor were clearly factors to be considered in reaching a working hypothesis on the nature of high-level ability in science. Robert H. Knapp and Hubert B. Goodrich (1952) have studied the place the college teacher has in stimulating individuals with high-level ability in science. I recall that, without exception, the scientists who gave me my early training stated their indebtedness to one or more teachers and cited the opportunities these teachers made available to them. What we then called the *activating factor* turned out to involve the channeling and augmenting environments mentioned earlier. Indeed, Roe found the scientists she studied recalled those teachers who stimulated them to find things out for themselves.

It is important to emphasize that the hypothesis does *not* postulate talent, giftedness, or creativity in science per se—as rooted in the gene. We postulate that *high ability* (rooted generally in genetic factors) interacting with *predisposing* and *activating factors* (rooted generally in the effects of environment) are necessary to the development of scientists.

Why not use tests of creativity, for example, those of E. Paul Torrance (1966) and Jacob W. Getzels and Philip W. Jackson (1962)? Note that Tannenbaum (1983), in a major review of tests of creativity, states, "It remains to be demonstrated conclusively, however, that divergent thinking and creativity are synonymous and that so-called 'creativity tests' have strong predictive value" (p. 298). Richard S. Mansfield and Thomas W. Busse (1981) have also remarked that the diversity of definitions of creativity has produced a jumble of findings with only dubious applicability to real-life creative performance. On this, also see MacKinnon, who concurs, "Our conception of creativity forced us further to reject as indicators or criteria of creativeness the performance of individuals on so-called tests of creativity" (p. 485).

We found that the young who selected themselves for apprenticeship in research and were thus faced with the tasks of the research scientist exhibited what we know to be the behaviors of the scientist. In selecting themselves for the demanding work, they opened for themselves a period of instruction in problem solving as a central part of an *augmenting environment*: research over a period of 6 to 18 months during and beyond the course. This experience, differentiated from that of students who preferred the accelerating enrichment of the *channeling environment*, may well be what is sometimes described as "differential education for the gifted." The design of the augmenting environment, its strategy and tactics in teaching and learning, is described in "Apprenticeship to Well-Ordered Empiricism" in this volume. For example, those young who sought out the augmented environment were able to

- note discrepant events
- discover a problem situation within the event they wanted to investigate
- uncover the prior literature related to the work on the problem
- propose a hypothesis
- design an investigation involving observation and experiment on the basis of the hypothesis
- record their data (including error)
- design control experiments in an attempt to defeat their hypothesis
- offer a tentative solution
- propose new experiments in an attempt to defeat their solution
- state their solution in a systematic assertion
- present their work in seminars with other apprentice-scientists
- present their assertions and predictions in a paper
- present the paper to their peers in a science congress
- offer their work for critique to their mentors—scientists in the field
- enter their work in the Science Talent Search for further appraisal by scientists (if they wished; on this, see pages 90–92)

True, the congeries of activities called forth by the individual research these young could and did undertake were entirely unlike the creativity tests

Getzels and Jackson and Torrance proposed. However, for us and the two psychologists we consulted, the activities represented a decent test of the ability to *discover*. For us, the research constituted a test of creativity directly related to the lifework these young were contemplating. As such, the research was perhaps similar to the audition of the aspiring musician, the portfolio of the aspiring painter, the tryout of the aspiring athlete, the story or essay of the aspiring writer. Or, as Hudson put it, confirming our view, at least in part,

> When we ask a scientist to complete a verbal analogy for us, or a numerical series, we are asking him to perform a skill insultingly trivial compared with those he uses in his research: when he grasps a theory; reviews the facts for which it is supposed to account; decides whether or not it does so; derives predictions from it; devises experiments to test those predictions; and speculates about alternative theories of his own and other people's. In all these maneuvers he exercises skills of a complexity greater than we can readily comprehend. (p. 109)

We must content ourselves then, at this point, in rooting science talent in a dyad: genes interacting with environment. And the *evidence* of the presence of science talent was to be a *work*, which would necessarily come out of the interaction of heredity and environment undergirding the qualities of thought and action interpreted in our hypothesis stated on page   82 of this paper.

At this point, we may postulate strongly that high-level ability *in science specifically* is not to be conceived of as lodged in DNA. The gifted student's high-level ability in intellection (or critical thinking) and in numerical and verbal skills may be *initially* gene driven. But high-level ability *turned to science*, to the cumulative knowledge and the skills in inquiry that characterize science, are *environment driven*. The triads offered by Brandwein (1955), by Joseph S. Renzulli (1977), and by Robert J. Sternberg (1985) all seek a nexus within genetic and environmental expression as a sign of giftedness. The field-specific hypothesis expressed here for science, embracing high ability as emerging from the interaction of the *triad* of genetic, predisposing, and activating factors, nonetheless is subsumed by the reality of the *dyad*, the highly evidential interaction of an individual's DNA (genome) with the environment. Our thesis proposes that—as part of the ecology of achievement acting on the antecedent development of the young—carefully designed field-specific curriculum and instruction in science that encourage originative laboratory work (an environment) would catalyze the activity of the young as "performing scientists," albeit in the early stages of development. As early as 1957, A. Harry Passow had developed the essentials of a science curriculum undergirding a channeling and augmenting environment. (On curriculum, see also Passow, 1983; and F. James Rutherford, 1985.)

A Third Thesis:
The Student as "Performing Scientist"

How does one "select" those who have a potential for becoming performing scientists? Selection implies acceptance of some and rejection of others. Between 1947 and 1952, when we were engaged in researches to investigate whether there is such a trait as giftedness in science (also called "science talent"), measurement of "science talent" was in its infancy. Still, at Forest Hills High School, examinations were not a prelude to entrance. The school doors were open to all the students in the district. We were both required and glad to make opportunities available to all who wished to do the work; we could not, and indeed, we would not, make an examination requisite to entry to any program in science.

If we use the phrase "potential *to* performance," we seem to imply that first we find a potential and then turn it into performance. The actuality is that in giving young aspirants their opportunity to develop a personal art of investigation through performance, we seek out potential *through* (not *to*) performance. The strategy is to furnish students, their interest perhaps now channeled and augmented, with an environment, a problem-solving situation, in which they further augment interest and fix it in observable behaviors.

In 1947–1948, we gave students the opportunity to enter a program of individual work in science in the second half of the year in their study of biological science. They could select themselves for the opportunity, one that included individualized work and instruction (including mentoring) to solve a problem through modes of research resembling closely those of research scientists. Our model for the test was thus a simulated "real-world" process of testing an ability to solve a problem using the scientists' methods of intelligence. To a high degree, the "apprentices" would reveal how they faced problems that required an invented, that is, a novel approach. And yet the apprentice experience would embrace an unknown, approaching a decent tincture of the complexity found in adult research. Solving the problem would require sustained effort over six months to a year, perhaps even longer. A certain originality in insight and evaluation and, what is more, in overcoming countless failures, small and large, was required. In effect, the apprentice would face a paradigm of persistence in scientific critical thinking and ultimately, in origination, a *discovery*.

Is *performance* evocative of, even a *test* of, *potential* in science? It may well be a crystallizing experience masquerading within a test of talent in science, and it is possible that such an experience may exert powerful, long-term effects on the individual. Further, to quote Joseph Walters and Howard Gardner (1986), crystallizing experiences "are a useful construct for explaining how certain talented individuals may first discover their area of giftedness and then proceed to achieve excellence within the field" (p. 309).

Let me state one aspect of our program plainly: Once students had been accepted as members of the school, they were not to be excluded from participation in any activity devised by the school; we let each one of them have a try at any program with full attention to physical and psychological safety. Some 40 to 60 students were to apply (that is, select themselves for the science work) each year, and all students were accepted, no matter their IQ. The environment was so channeled and augmented to permit their performance to be the test of their ability.

True, the students in the research program tended toward four years of science, three years of formal mathematics plus one special mathematics course of their choice, usually the calculus. But it was clear to us that, in this act of self-selection, the vast majority who applied for individual work had come through a period of self-appraisal based on their achievement. Indeed, the oft-repeated research that, given the opportunity to do so, students can judge themselves by examining their own achievement was validated. In effect, to us, self-selection implied an active seeking of a channeling environment leading to an augmenting environment.

In any event, the students who selected themselves for the work displayed several sets of traits: high interest, generally high ability, persistence, and questing (the predisposing factors). But would they all be able to generate, pursue, and complete an investigation (a work performed) that would test their ability in discovery, in creativity? Similarly, but not in the field-specific context of science, Renzulli, Sally M. Reis, and Linda H. Smith (1981) utilize Renzulli's triad to describe giftedness as

> an interaction among three basic clusters of human traits—these clusters being above-average general abilities, high levels of task commitment, and high levels of creativity. Gifted and talented children are those possessing or capable of developing this composite set of traits and applying them to any potentially valuable area of human performance. (p. 27)

Further, Renzulli and his colleagues, in stressing task commitment, press the point that "whereas motivation is usually defined in terms of a general energizing process that triggers responses in organisms, task commitment represents energy brought to bear on a particular problem (task), a specific performance area" (p. 24). Possibly, by task commitment, Renzulli and his co-workers confirm the elements of the predisposing factor (questing and persistence) stipulated in our hypothesis (see page 82). Nancy E. Jackson and Earl C. Butterfield (1986) are content with the following definition: "Gifted performances are instances of excellent performances on any task that has practical value or theoretical interest. A gifted child is one who demonstrates excellent performance on any task of practical value or theoretical interest" (p. 155). Robert S. Siegler and Kenneth Kotovsky (1986) propose that "... careful observation of the products children produce may prove to be the most practical way to improve on intelligence tests as assessment devices" (p. 432).

We are persuaded, then, by practitioners in research in the field of giftedness that performance in a field that has practical value or theoretical interest is a useful test of promise, or potential, for a given talent. However, one of these indicators is certainly the ability to perform an experiment or to do a theoretical investigation or analysis in an augmented environment in science.

One Search for Science Talent: Effects of Ecologies of Achievement

Why is it that some secondary schools, especially those apparently endowed with all the possibilities for developing a fruitful program for a body of gifted students who might seek out a lifework in science, do not offer an augmented environment in science—that is, one with individualized instruction that fosters and crystallizes experience in the arts of investigation characteristic of scientists? We found that the use of the Science Talent Search as an *instrument* to study various aspects of the ecology of achievement for the putatively talented in science was effective in dissecting out certain aspects of the augmented environment that seemed to be missing in these schools. The "Search" is instrumental in two ways: first, as a possible test of science talent; and second, significantly for our purposes, as an indicator of the nature of certain ecologies of achievement that affect the demonstration of science talent.

Early in my observations I thought I had clues to the answer to this perplexing question. Apart from special schools whose practice was and is to select science-prone students for admission by a series of tests, why do certain schools with heterogeneous populations seem to succeed in developing a rich environment in science and mathematics in which the *potential* for science talent is expressed, while others with a similar student population are relatively devoid of such an environment?

Siegler and Kotovsky (1986), in considering the question "What will be the most fruitful approaches for research on giftedness in the next 5 to 10 years?" suggest,

One useful approach would be to focus on people in the process of becoming productive—creative contributors to a field, for example, high school students who win Westinghouse Science Competition prizes; ... [or students] who publish articles in nationally circulated magazines, or who have their drawings shown in major exhibits. Members of these groups are of special interest for two reasons.

They already have made creative contributions—they have not just learned to perform well on tests—but they are still in the process of becoming eminent. (p. 434)

And so too, Julian C. Stanley, Director of the Study of Mathematically Precocious Youth at Johns Hopkins University, states in the *Phi Delta Kappan* (1987, June),

I firmly believe that a residential state high school of science and mathematics should follow the lead of those prestigious programs [referring to those in the Bronx High School of Science and Stuyvesant High School, both in New York City] by preparing most of its students to compete in the Westinghouse Science Talent Search when they are seniors. To do less is to underdevelop the investigative scientific spirit of highly talented students.* (p. 771)

Further, Robert D. MacCurdy's study (1954) of 600 men and women who had been awarded honorable mentions or were finalists in the Science Talent Search further suggests its validity as a measure of manifest originality in science. E. G. Sherburne, director of Science Service, which administers the Science Talent Search, stated (1987),

The Science Talent Search is unusual among scholarship competitions in that it puts primary emphasis on the quality of a paper reporting an independent research project in some area of science, engineering, or mathematics and only secondary emphasis on academic achievement.

In short, the evaluation is on the basis of the student's ability to "do" science in a way that is analogous, though at a less sophisticated level, to what a professional scientist does [italics ours]. To use a sports analogy, one does not test a student's ability to play tennis by giving a paper-and-pencil test. One puts the student on the tennis court to play so the performance can be observed.

For the purposes of this book, then, it is useful to add to the literature the thrust of the Science Talent Search in respect to the place of the school in the ecology of achievement that results in the emergence of young who are manifestly originative in science and mathematics. As I see it, the ecology of achievement conducive to performance in science, and thus to demonstration of science talent, assumes a certain constellation. This ecology calls for an appropriate curriculum to include an individual investigation; a mode of instruction necessary to pursue independent investigation; and the mentors essential to guide and advise those aspiring scientists, who are willing to enter

---

*Note that Stanley recommends "preparation to compete"—not an insistence on competing. Preparation would mean giving students an opportunity to perform, that is, to do an investigation, the major requirement of the Science Talent Search. Note, too, that, as my investigations disclose, a number of students did not choose to compete (see page 95, note).

the Search, in the problems attending individual effort. (On this, see Brandwein and Morholt, 1986.)

In the period 1952–1962, I had the opportunity to study the science programs of 103 schools. I was able to select 22 of these schools for further study and to visit 17 of these heterogeneous schools two to four times during this period.* These 17 had a somewhat similar socioeconomic group of students and similar high levels of acceptance to colleges and universities; all, confirmed by my observations, had developed effective science and mathematics departments. All had developed channeling and augmenting environments; the latter gave opportunity for performance in an investigation requiring originative work on the high school level. In the period 1944 through 1954, the records of the 22 in the Science Talent Search were as follows (please refer to table 1, page 91).

Recall that participation in the Westinghouse Science Talent Search is voluntary among hundreds of schools in the nation. The 22 schools in my study do not compose a statistical sample but are studied (aside from the select schools) because of their similarity to the school in which my teaching was done and the one that gave me opportunity to carry on work with the talented in science.

Schools 1, 3, and 7 had populations selected by tests specially designed to attract those with "science potential"; schools 9 and 11 were independent schools with selected populations and high socioeconomic levels; the remaining schools had heterogeneous populations, with somewhat similar socioeconomic levels. School 13, which selects its students for ability in performing arts, is especially interesting.

My observations and notes, based on checklists and honed by work for the Board of Examiners of New York City in selecting individuals seeking admission to the posts of science teacher or chair of a science department, show that all of the 22 schools offered exceptional course work. Their success in placing students in "sought after" colleges and universities was considerable.

Within the 103 schools, I also had opportunity to study another sample of 17 schools that served as a control group; these 17 schools did not enter the Search. Explicit statements by 14 heads of departments and/or deans of the 17 schools summarized their belief that the *training of scientists could well be left to the universities.* The explicit policy of their schools was the task of preparation for future study in universities where further extended prepara-

---

*Certain of my observations during this period were part of my functions as chair of the Gifted Student Committee of the Biological Sciences Curriculum Study (BSCS) as well as those of a member of the steering committees of both BSCS and the Physical Science Study Committee (PSSC).

Table 1
Westinghouse Science Talent Search 1944 Through 1954*

School	Finalists	Honorable Mentions
1	17	79
2	17	57
3	17	53
4	8	34
5	8	19
6	8	8
7	7	48
8	6	38
9	4	3
10	3	74
11	3	0
12	1	17
13	2	3
14	2	1
15	2	0
16	1	1
17	0	5
18	0	0
19	0	0
20	0	0
21	0	12
22	0	2

Key: Schools 1, 3, 7—select schools for science (based on entrance examinations)
Schools 9, 11 —independent schools
School 13 —a school of performing arts
School 2 —Forest Hills High School, a school with a heterogeneous population, furnished the students included in this 10-year study and the one summarized in "Apprenticeship to Well-Ordered Empiricism" in this volume.
The remainder are schools with heterogeneous populations.

*Grateful thanks to Dorothy Schriver and Carol Luszcz of Science Service, Washington, D.C., for their aid in supplying me with the data from past Westinghouse Science Talent Searches.

tion would be available for chosen careers, especially science, but also most areas of scholarship involving research activity. This view seemed acceptable and generally was found to be the prevailing one. Indeed, Lloyd G. Humphreys (1985) reports that "differences among chemistry, physics, geology, and engineering measures of attainment are obviously produced during postsecondary education" (p. 344).

In the schools that did not make the effort to enter the Search, there was a demonstrated absence of the individualized instruction and mentors necessary to help students do "an experiment," that is, to express their *potential through performance*. For example, a special sample of 18 of the 103 schools consisted of schools with populations under 600–1,000 students. There were 15 in rural areas that were unable to organize a program of 4 consecutive years of science and did not have the facilities for individual research (in an augmenting environment). However, 3 of the 18 attempted a mentorship program for those 6 students who aspired to careers in science.

Another Aspect of the Ecology

After I had left Forest Hills High School to take the post of Director of Education for the Conservation Foundation (now situated in Washington, D.C.), my colleagues in the school continued their work in the program with somewhat similar success. Then, three of them were promoted to the chairs of science departments in three schools in different areas of New York City, areas where the culture was not disposed toward intensive work in science and mathematics and where the teachers were not specially trained in experimental protocol and techniques. The records of these schools in the Science Talent Search were not noteworthy—three honorable mentions in five years. In spite of the efforts of these highly effective teachers and supervisors, the nature of the population and the environment per se (parts of our ecology of achievement) were not then conducive to the development of science-talented individuals, as measured either by their own standards or those of the Science Talent Search. That is to say, a channeling environment and augmenting environment in science in a given school may not function if the culture (a part of the ecology of achievement) is not disposed to support or interact with it.

Still another factor, seemingly minor but worthy of attention, was the availability of laboratory assistants educated sufficiently to supervise individual laboratory work. Without such staff, it was necessary for a teacher-mentor to be available to ensure safe use of equipment and substances while students were doing individual experimentation. This safety factor, rarely considered, acts as a block to individualized experimentation in science throughout the country. But it seems clear that necessary to place students in the finalist or honorable-mention categories in the Science Talent Search are a student body with a sufficient number possessing the genetic, predisposing, and activating factors and a supportive community cognizant of the significance of science in our culture. One aspect of participation in the Search depends on a certain performance in solving a scientific or mathematical problem and, therefore, on an environment that supports the activities facilitating the performance. It can also be demonstrated that whether or not an ecology of achievement undergirding science talent exists and persists in a school depends at least on these or similar factors:

• the early education of children in the home environment and the behavior necessarily reinforced in the community
• the early schooling and education of the student population within the school and community
• the policies of the school and community as they affect a decision to enter the Search*
• the preference of the school to offer special opportunities in an augmenting environment for the gifted, and, moreover, the ability of the school or community to provide for the *one* or the *very few* who aspire to a career in science
• the decision of students who, for various reasons, apply to enter or decline to enter the competition
• the recognition by the faculty, supported by the community, that it is the performance of the student in a *work* that is significant; its caliber and completion is itself a test, whether or not the student enters the Search
• the judgment that science is a collaborative enterprise, in which a given work reflects the antecedent and present efforts of many with different functions, levels of ability, and skills

All are essential to an ecology of achievement, which must be considered as a whole. Thus, the development of all—of high and modest ability—who can contribute on whatever level plays a part in sustaining the ecology.

On the other hand, in those schools that entered the Search, the belief prevailed that the young aspiring scientist or mathematician might well benefit from the judgment of others, just as writers, musicians, painters, and other artists, and, of course, athletes, submit their work to the review of expert and peer. It was interesting to determine whether schools (table 1) maintained a steady course in the Search over the years, say a quarter of a century later. Below (table 2) are the 1942–1988 distributions of finalists and honorable mentions from the same schools studied earlier (1944 through 1954).

---

*This is not, however, intended to convey the impression that schools need prove themselves through participation in the Science Talent Search.

Table 2
Top High Schools in
Westinghouse Science Talent Search 1942–1988*

School	Location	Winners
Bronx HS Science (1)[a]	New York, NY	106
Stuyvesant HS (3)	New York, NY	63
Forest Hills HS (2)	Forest Hills, NY	42
Erasmus Hall HS (8)	Brooklyn, NY	31
Evanston Township HS (6)	Evanston, IL	26
Benjamin Cardozo HS	Bayside, NY	25
Midwood HS (4)	Brooklyn, NY	20
Jamaica HS (22)	Jamaica, NY	19
Martin Van Buren HS	Queens Village, NY	15
Brooklyn Technical HS (7)	Brooklyn, NY	11
Central HS	Philadelphia, PA	11
Abraham Lincoln HS (5)	Brooklyn, NY	11
Hunter College HS	New York, NY	9
Lyons Township HS	La Grange, IL	9
New Rochelle HS (10)	New Rochelle, NY	9
Coral Gables Senior HS	Coral Gables, FL	9
North Phoenix HS	Phoenix, AZ	9
Phillips Exeter Academy (9)	Exeter, NH	8
Melbourne HS	Melbourne, FL	7
Newton HS (16)	Newtonville, MA	7
Ramaz HS	New York, NY	7
Niles Township HS West	Skokie, IL	7
Columbus HS	Marshfield, WI	7
Stephen Austin HS	Austin, TX	7
Woodrow Wilson HS	Washington, DC	6
Wakefield HS	Arlington, VA	6
Princeton HS	Princeton, NJ	6
Nova HS	Fort Lauderdale, FL	6
James Madison Memorial HS	Madison, WI	6
Alhambra HS	Alhambra, CA	6
McLean HS	McLean, VA	6
Eugene HS	Eugene, OR	6

[a]The numbers in parentheses refer to the schools' coding in table 1.

Note please, while only winners are listed, the order of listing may be interesting. What is significant in the above is the relative stability over 40 years of the ecologies of achievement of certain participating schools. (See below "Select and Heterogeneous Schools.")

*Grateful acknowledgment goes to Dorothy Schriver and Carol Luszcz of Science Service for furnishing the data in table 2.

Select and Heterogeneous Schools

It is clear from the data in tables 1 and 2 that two of the select schools of science maintained their positions; they had maintained a steady ecology of achievement. Essentially, select schools accomplish this not only by the selection of their populations of students, but also by the support of parents (a gathered community), the continued support of the boards of education or boards of advisors, and the selection of teachers, as well as the maintenance of their channeling and augmenting environments. Others with heterogeneous populations retained their presence over the years. Some schools that were not present in the earlier study have now begun to make significant showings. One set of assumptions may explain some of these trends:

• Select schools have developed their own following, a dispersed but likeminded community, a kind of homogeneity. Therefore, they have maintained and sustained their niche in an ecology of achievement.

• Over a period of time, populations in heterogeneous communities may change economically, politically, socially, and, thus, in the objectives of their schooling. It follows that, in a given school where there are strong efforts to maintain an augmenting environment, there may be temporary changes in administration, faculty, student body, or community support. Such changes could affect not only interest in or commitment to science and technology and the school's own ecology of achievement but also the ecology of the community.

Nonetheless, over the years, within a given geographic area the relative total number of finalists and honorable mentions coming from heterogeneous schools (however they shift in identity) compares favorably with that of the select schools. The ratio of achievement of heterogeneous to select schools in a defined geographical area seems to be fairly steady.* And it is clear that the vast majority of students attend heterogeneous schools. As do the select schools, the heterogeneous schools (in communities that have developed notable ecologies) may also serve as models affecting the establishment of channeling and augmenting environments for communities in the surrounding areas. Thus, the Science Talent Search, reflecting as it does the achievement of certain students, is not only a demonstration of individual talent but a demonstration, as well, of the presence of an ecology of achievement in home, school, and community.

---

*Discussions with several administrators of schools indicated that they were loath to enter a national competition in which the number of prizewinners was drastically limited. Is it possible that Science Talent Searches held annually in all the 50 states might serve to attract students at a variety of levels of ambition to seek a lifework in science? The National Science Talent Search might then be extended to draw on the combined populations involved in 50 distinguishable Searches.

However, many entrants have still another opportunity for recognition through State Science Talent Searches (33–43 in number depending on the year of the Search). The number of state talent searches has increased by approximately 25 percent over the 40 years of the Search. The Science Service duplicates the written entries and forwards them to directors of the State Searches. The states then conduct their own competitions, many of which offer numerous awards, including scholarships.

Contrasting Ecologies

Recall that, as we have defined it, an ecology of achievement affecting a child embraces the sustaining environments of the home, peers, and community in the *educational* functions that support the particular construct of schooling in that area. Thus, the performance of a child in schooling per se and, of course, in a particular area of performance—art, science, or athletics—is not solely an outcome of the particular curricular and instructional practice in a given school but is also a test of the family and community support of both the education and schooling of the young. In sum, this support is not only exemplified in the attitudes of parents and in their competence in rearing their children but also is an earnest of the traditions, attitudes, and practices of the community reflected in the school.

It is also necessary to remind ourselves that, even when there is evidence of a child's tendency to excel, superior—or mediocre or failing—performance is still not solely or even mainly an index of the child's potential. Performance on any given occasion may well be a reflection of the effect of an entire complex of prior environments within schooling as well as the educational opportunities (or lack thereof) and influences (good and bad) afforded by home, peers, and community. A case in point: hapless addiction to drugs of able young. We may not disregard, as well, the steady accumulation of disadvantages those already disadvantaged may not be able to set aside, that is, to overcome without constructive affection. (On this, see Bill G. Aldridge and Deborah C. Fort's paper in this volume.)

Recall Tannenbaum's reflection on the cyclical nature of interest in the gifted. In his paper (1979), he documents an easily verifiable observation: that gifted children as a special group have been "alternately embraced and repelled with so much vigor by educators and laymen alike" (p. 5). It is also observable that certain states or communities with a notable dropout rate may, in the same period, also have schools with notable records in admission to elite universities—and, in fact, notable records in the Science Talent Search. Nonetheless, the schools that I have observed promoting the kinds of channeling and augmenting environments that make provision for the gifted (described here, in "Apprenticeship to Well-Ordered Empiricism," and in Passow's papers in this volume) also provide well-planned programs for the disadvantaged. The effectiveness of these programs is reflected by the considerable improvement in the tendency of the disadvantaged young in these schools to continue there (in contrast to the high dropout rates elsewhere). It may well be that the philosophy and practice of most of these schools is to attempt to provide for all their young. In my experience, Tannenbaum's remark about the swings in affection and disaffection for the gifted also applies to the disadvantaged.

On the other hand, throughout the country, we appear to alternate in our attention to the schooling of our gifted and, then, to that of our disadvantaged. These cycles or the recurrent formula of our "crises" in schooling have been well documented. We find too often a periodic rise in concern about the effectiveness of our schools stimulated by the momentary prominence of some event such as rises and falls in test scores, or Sputnik, or the economic success of another nation, or possibly our failures at the Olympics. If serious enough, this concern is followed by a period of vigorous effort mounted, almost *entirely,* through "reform" of schooling—to ameliorate certain symptoms of the decline. In time, this is followed by another "failure" of the schools.

It is fairly easy to demonstrate that what is considered to be a general failure of the public schools is in effect a failure within the ecologies of achievement. To repeat, it is rare to find an effective school within a community that does not support its schools and teachers. However this may be, there remains undeniable evidence that an individual's traits are the result of the interaction of heredity and environment at any point in development.

To recapitulate, we understand the term *environment,* in respect to its various effects on the traits of the young, to mean precisely the effects of the ecologies of achievement that are at the heart of this discourse. At this point, it is observable that the environments conducive to developing our young's learning capacities and personal growth have not achieved the stability that reduces oscillations in the effectiveness of the communities—whatever their precursors in political, economic, or social events—which support schooling and education. In time, these oscillations may become intolerable. It remains to be seen whether an open society can maintain the requisite stability of a schooling system that can respond equitably to its supreme responsibility of attending to the future of *all* its young in their various capacities. It is obvious that in so doing we attend to our own future, as is required of a society that acquits itself so nobly in a document beginning with that epiphanous phrase, "We, the people. . . ."

The Need for the Talented in Science
However this may be, the world is in dynamic change. Daniel Bell (1973), among other observers, made the case early on that essential individuals in industry in the decades to come will be those centered in science and mathematics. The postindustrial society is here; it has global consequences; we now accept that the scientist and science-trained individual are necessary to the well-being of present and future societies. However, it is not the scientist of earlier years whom we seek; we need individuals educated in societal as well as scientific aspects. We need not only biologists, chemists, physicists, geologists per se; they are required, in a certain measure, to be ethicists as well. For scientists, through their discoveries, have indeed made the planet a global village.

Perhaps it will now be necessary to consider whether it is not a general function of schooling to design the channeling and augmenting environments that evoke the potential of those whose driving interest is to think and do science and technology. If so, it will become clear that these environments *cannot* all be sequestered in a limited number of select science schools. The young who are to seek their lifework in science must come out of the schools that exist, and these, for the vast majority, are part of some 15,000 school districts across the country serving mainly heterogeneous populations. Should not all the young have available a mature ecology of achievement in which they can fulfill their powers in pursuit of their personal swath of excellence?

It has been shown in earlier periods of work on the gifted; and again in the most recent spurt of activity in research on the gifted young; and further in this paper and in this volume in the accounts of a number of scholars; and it has been and is being shown through the validating data of the Science Talent Search that it is feasible to develop programs that bring forth the young who may become our future scientists at various levels of originative work. We do not, in my view, find the science talented through paper-and-pencil tests of creativity or necessarily in those whose abilities are reified in the IQ. The confluence of traits and competencies of the young secured in antecedent environments and reflected in a performance, that is, in a work, more nearly reflects the scientist's processes. The process is central to our thesis that the completion of such a work is a sign of talent in science.

A strong hypothesis may thus be advanced: The demonstrated opportunity of the *school* to develop the curriculum and instruction devised as the channeling and augmenting environment, enabling individualized work in science, and the demonstrated ability of the *student* to plan and complete an experiment or theoretical analysis, with the meticulous application required, may well be a valid test of science talent. The hypothesis is in support of Wallach's (1985) suggestion that creativity in a specific field, say science, might be a "by-product of field-specific instruction" (p. 115).

Schools are filters of feasibilities. In the environments they construct lie the seeds of the destinies in which the young and their various but remarkable capacities find perdurable ways of advancing the culture.

Reflections and Conclusions

Recall that Gregory Bateson remarked, when asked to define "scientific truth," that he contented himself with this definition: "What remains true longer does indeed remain true longer than that which does not remain true as long" (1979).

As one who aspired to spend his life in science and once was welcome and worked in various laboratories, I am keenly aware that in the field of study called "giftedness and/or talent," we are not yet in the area we may call science. In this area, we are still in a science of practice and not of the laboratory. We are beginning to know something of the kinds of environments that may bring forth the metamorphosis of potential into performance. We *may* know more of one part of the dyad, the environments that encourage potential, than we know of the other part, the genes that are basic to performance.

There are thus certain clusters of paired reflections and conclusions I am obliged to put before you.

First: We are teachers. We are obliged therefore not to use the young as implements of our particular sociopsychological warfare in measurement or method. Because we cannot change the genetic complexes in the young that come to us, we are obliged to take the only course available to us: to change the environment to fulfill the powers of the young so that they may pursue the best course of development within their capacity. For the facts are these: An individual is the result of the interaction of inherited genes and of environment; this interaction occurs at every moment in the course of development. Yet, although the concepts of the ecology of achievement and the dyad (in numerous diverse phrases) are, and have been, strongly evident, the culture generally and societies specifically continue to attribute to each individual aptitudes, attitudes, and achievements without reference to augmenting or suppressing environments. Indeed, the attempt is to reify an individual's history in a single measure—the IQ. *Not* in works, *not* in achievements, but in a measure of potential that does not describe the individual's advantaged or disadvantaged prior history.

Indeed, our study and numerous others support this hypothesis: *The ability to apprehend the arts of investigation and to complete an empirical study is, in conjunction with other indices, a better predictor of future entry into successful scientific work then are paper-and-pencil tests of creativity.*

In other words, *not only the work but also the doing* summarizes the originative traits found in the self. The painting, the musical score, a building's architecture, the scientific or entrepreneurial act of creation—these furnish pictures of the self in the act of creation. We are thus required to wait until geneticists have analyzed the DNA of a host of gifted and talented grandparents, parents, and young (for genetics is a study of families) to determine whether attributes of talent are inherited, and if so, in which gifts or talents.

I *conclude*, then, that we are obliged—no, required—to develop the most fruitful course of development in the schools: a channeling and augmenting environment fulfilling the most generous of auguries. These environments are to nurture in fruitful curriculum, instruction, and mentorship the aspirations of all young who wish to demonstrate a talent in science *through performance* in the arts of scientific investigation.

Second: We work in an admirable environment, the architecture of a long cultural history: the school. But the school is not an isolate; it is part of an *ecology of achievement* affected by and participating in all the activities accommodating the particular community, society, nation, and culture of which it is a part. Just as a gaunt deer is not generally to be found in a rich, capacious forest, so a poorly supported school is not found in an effective community, state, or nation. Put another way, the human is the result of interaction of a biological and social (cultural) environment of which one important environment is the school.

I am obliged to *conclude* that the environment of the young who will eventually act to conserve, transmit, expand, and correct the culture consists of *all* the advantages the entire school-community-nation can and should offer and afford to offer. We are acutely aware that the talented, whether scientists or not, whether giants or auxiliaries maintaining giants, or the young not yet tall enough do not—and cannot—complete their work in isolation.

We know enough about curriculum and instruction and administration; we are competent and compassionate enough to develop models of apprenticeship in science to fit all manner of competency. If we will it. Further, there is evidence that both select schools and those with heterogeneous populations design and utilize similar models of channeling and augmenting environments.

I *conclude*, then, that *both* select and heterogeneous schools can develop appealing and adequate models of curriculum and instruction that channel and augment the interests and abilities of all who wish to enter the various levels of excellence required in science.

Third: We know enough about the traits of scholars, whether scientists or artists, to afford them sufficient opportunity. Our society cannot speedily be made perfect, although we—in the United States—are fortunate enough in minds and resources to travel that long and tortuous road. If we would. But the pieces of the constructs of most excellent schools are everywhere; while society is not yet perfectible, the school, an instrument of society, is a small enough community to approach the perfectible within an appreciable, even predictable, time.

I am obliged to *conclude* that within a society intending perfectibility, the schools may be exemplars in the design of a perfectible social construct—an ecology of achievement—for developing youth. For in the schools lie our future resources of mind and personhood. It is entirely conceivable, then, if the perfectible school could become a prime objective in the humane use of human beings, then talent would indeed be a by-product of study in a field freely chosen by each individual.

In a sense, our schools are strong signs of our character. They are bulwarks against the enemies that plague us: ignorance and indifference, mindlessness and meaninglessness. They are signs of our persistent, character-rooted passions: It is better to know than not to know. They are signs of our intent—to construct an open society in which all of us, without coercion, can behave as if we could find out what is true.

References

Bateson, Gregory. (1979). *Mind and nature: A necessary unity.* New York: Bantam Books.

Bell, Daniel. (1973). *The coming post-industrial society: A venture in social forecasting.* New York: Basic Books.

Brandwein, Paul F. (1947). The selection and training of future scientists. *Scientific Monthly, 54,* 247–252.

Brandwein, Paul F. (1955). *The gifted student as future scientist: The high school student and his commitment to science.* New York: Harcourt Brace. (1981 reprint, with a new preface [Los Angeles: National/State Leadership Training Institute on the Gifted and Talented])

Brandwein, Paul F. (1981). *Memorandum: On renewing schooling and education.* New York: Harcourt Brace Jovanovich.

Brandwein, Paul F. (1986, May). A portrait of gifted young with science talent. *Roeper Review, 8*(4), 235–243.

Brandwein, Paul F., and Morholt, Evelyn. (1986). *Redefining the gifted: A new paradigm for teachers and mentors.* Los Angeles: National/State Leadership Training Institute on the Gifted and Talented.

Broudy, Harry S. (1972). *The real world of the public schools.* New York: Harcourt Brace Jovanovich.

Cox, Catherine M. (1926). The early mental traits of three hundred geniuses. In Lewis M. Terman (Ed.), *Genetic studies of genius* (Vol. 2, pp. 11–842). Stanford, CA: Stanford University Press.

Datta, Lois-Ellen G., and Parloff, Morris P. (1967). On the relevance of autonomy: Parent-child relationships and early scientific creativity. (Described in Abraham J. Tannenbaum [1983], *Gifted children: Psychological and educational perspectives.* New York: Macmillan)

De Lone, Richard H. (1979). *Small futures: Children, inequality, and the limits of liberal reform.* Foreword by Kenneth Keniston (pp. ix–xiv). New York: Harcourt Brace Jovanovich. (Carnegie Council on Children)

De Solla Price, Derek. (1961/1975). *Science since Babylon* (enlarged ed.). New Haven: Yale University Press.

Gallagher, James J. (1970). *Teaching the gifted child* (2nd ed.). Boston: Allyn and Bacon. (Original work published 1964, with a new preface for this printing)

Gardner, Howard. (1983). *Frames of mind: The theory of multiple intelligences.* New York: Basic Books.

Getzels, Jacob W., and Jackson, Philip W. (1962). *Creativity and intelligence: Explorations with gifted children.* New York: John Wiley and Sons.

Guilford, J. Paul. (1968). *Intelligence, creativity and their educational implications.* San Diego: R. R. Knapp.

Hudson, Liam. (1966). *Contrary imaginations: A psychological study of the young student.* New York: Schocken Books.

Humphreys, Lloyd G. (1985). A conceptualization of intellectual giftedness. In Frances Degen Horowitz and Marion O'Brien (Eds.), *The gifted and talented: Developmental perspectives* (pp. 331–360). Washington, DC: American Psychological Association.

Jackson, Nancy E., and Butterfield, Earl C. (1986). A conception of giftedness designed to promote research. In Robert J. Sternberg and Janet E. Davidson (Eds.), *Conceptions of giftedness* (pp. 151–181). New York: Cambridge University Press.

Knapp, Robert H., and Goodrich, Hubert B. (1952). *Origins of American scientists.* Chicago: University of Chicago Press.

MacCurdy, Robert D. (1954). Characteristics of superior students and some factors that were found in their background. Unpublished doctoral dissertation, Boston University, Boston.

MacKinnon, Donald W. (1962, July). The nature and nurture of creative talent. *American Psychologist, 17*(7), 484–495.

Mansfield, Richard S., and Busse, Thomas V. (1981). *The psychology of creativity and discovery: Scientists and their work.* Chicago: Nelson-Hall.

Merton, Robert K. (1961). The role of genius in scientific advance. *New Scientist, 259,* 306–308.

Merton, Robert K. (1985). *On the shoulders of giants.* New York: Harcourt Brace Jovanovich. (Original work published 1965)

Passow, A. Harry. (1957). Developing a science program for rapid learners. *Science Education, 41*(2), 104–112.

Passow, A. Harry. (1983). The four curricula of the gifted and talented: Toward a total learning environment. In Bruce M. Shore, Françoys Gagné, Serge Larivée, Ronald H. Tali, and Richard E. Tremblay (Eds.), *Face to face with giftedness* (pp. 379–394). Monroe, NY: Trillium Press. (World Council for Gifted and Talented Children)

Rabkin, Yakon M. (1987, March). Technological innovation in science: The adoption of infrared spectroscopy by chemists. *ISIS, 78*(291), 31–54.

Renzulli, Joseph S. (1977). *The enrichment triad model: A guide for developing defensible programs for the gifted and talented.* Mansfield Center, CT: Creative Learning Press.

Renzulli, Joseph S., Reis, Sally M., and Smith, Linda H. (1981). *The revolving door identification model.* Mansfield Center, CT: Creative Learning Press.

Roe, Anne. (1953). *The making of a scientist.* New York: Dodd, Mead.

Roe, Anne, and Siegelman, Marvin. (1964). *The origin of interests.* Washington, DC: American Personnel and Guidance Association.

Rutherford, F. James. (1985). *Education for a changing future.* Washington, DC: American Association for the Advancement of Science. (Originally called *Project 2061: Understanding science and technology for living in a changing world*)

Scarr, Sandra W., and McCartney, Kathleen. (1983, April). How people make their own environments: A theory of genotype-environment effects. *Child Development, 54,* 424–435.

Science Service, Inc., 1719 N Street, NW, Washington, DC 20036. (Agency that administers the Westinghouse Science Talent Search)

Sherburne, E. G. (1987, January). Washington, DC: Science Service, Inc. (Announcement relating to the Westinghouse Science Talent Search)

Siegler, Robert S., and Kotovsky, Kenneth. (1986). Two levels of giftedness: Shall ever the twain meet? In Robert J. Sternberg and Janet E. Davidson (Eds.), *Conceptions of giftedness* (pp. 417–435). New York: Cambridge University Press.

Stanley, Julian C. (1987, June). State residential high schools for mentally talented youth. *Phi Delta Kappan, 68*(10), 770–773.

Sternberg, Robert J. (1985). *Beyond IQ: A triarchic theory of human intelligence.* New York: Cambridge University Press.

Suzuki, David T., Griffiths, Anthony J., Miller, Jeffrey H., and Lewontin, Richard C. (1986). *An introduction to genetic analysis* (3rd ed.). New York: W. H. Freeman.

Tannenbaum, Abraham J. (1979). Pre-Sputnik to post-Watergate concern about the gifted. In A. Harry Passow (Ed.), *The gifted and the talented: Their education and development, Part I* (pp. 5–27). (Seventy-eighth Yearbook of the National Society for the Study of Education.) Chicago: University of Chicago Press.

Tannenbaum, Abraham J. (1983). *Gifted children: Psychological and educational perspectives.* New York: Macmillan.

Torrance, E. Paul. (1966). *Torrance tests of creative thinking.* Princeton: Personnel Press. (Norms technical manual)

Wallach, Michael A. (1985). Creativity testing and giftedness. In Frances Degen Horowitz and Marion O'Brien (Eds.), *The gifted and talented: Developmental perspectives* (pp. 99–123). Washington, DC: American Psychological Association.

Walters, Joseph, and Gardner, Howard. (1986). The crystallizing experience: Discovering an intellectual gift. In Robert J. Sternberg and Janet E. Davidson (Eds.), *Conceptions of giftedness* (pp. 306–331). New York: Cambridge University Press.

The Unique and the Commonplace in Art and in Science

Gunther S. Stent

In the fall of 1974, in the first issue of the new journal *Critical Inquiry,* there appeared a 50-page essay on the relationship of art and science by the University of Chicago musicologist Leonard B. Meyer. Meyer began his essay by pointing out that for the past few decades the relationship between art and science has been the subject of a lot of confusing debate. Much of that confusion Meyer attributes to doubtful analogies made by such people as "Gunther S. Stent, a molecular biologist . . . [who represents] a viewpoint not infrequently espoused by scientists, and occasionally by artists and laymen as well . . . Like a number of other writers, Stent contends that in essential ways science and art are comparable" (p. 163). Although Meyer expresses his sympathy for my attempts to bring the so-called Two Cultures together, he doubts that their viable union can be achieved by ignoring or glossing over important differences. Meyer says that he will argue in his essay that my "union between art and science is a shotgun marriage, not one made in heaven, and that [my] attempt to wed different disciplinary species results not in fertile but barren misconceptions" (p. 163). Meyer then implies that a shotgun marriage between the Two Cultures is bound to fail because artist and scientist can only make a go of what in California we call a significant relationship if they take in also a humanist as a roommate.

I felt honored that a brief popular article on art and science that I had published two years earlier in the *Scientific American* had become the subject of a lengthy scholarly essay by a leading theorist of the arts (Stent, 1972). But I was quite surprised by Meyer's critique, because I had believed all the while that in my article I had presented merely a watered-down version of what I thought were Meyer's very own views. His excellent book *Music, the Arts and Ideas* (1967) had actually been the main source of my own ideas about the nature of art in the first place.

Meanwhile, in these past 13 years I have often wondered why these debates about the relation of art and science are so confusing, why it seems self-evident that art and science are essentially similar, and yet, as Meyer rightly claims, also essentially different. Finally I came to realize that at the root of the difficulty is the unsolved, and possibly insoluble, deep problem of semantics, namely to say what it is that we are saying about a structure when we say that it has "meaning."

My 1972 *Scientific American* article had been inspired by my reading of the many reviews (and preparing a review myself [1968]) of James Watson's autobiography *The Double Helix* (Watson, 1968/1980). These reviews had appeared immediately after that book's publication in the spring of 1968. Probably more than any other book, Watson's personal account of his and Francis Crick's discovery of the structure of DNA contributed to the present decline of the traditional view that science is an autonomous exercise of pure reason carried out by disembodied, selfless spirits, who are inexorably moving toward an objective knowledge of Nature.

The reviews of *The Double Helix*, almost all of them written by scientists, turned out to provide (mainly unwittingly) as much insight into the sociology of science and into the moral psychology of contemporary scientists, as did the book itself. Peter B. Medawar (1968) was one of the few early reviewers who recognized the considerable literary merits of Watson's book. He predicted that it would become a classic, not only in that it will go on being read, but also in that it presents an object lesson on the nature of the creative process in science. Two other scientist-reviewers, both about to become best-selling authors themselves, also identified *The Double Helix* as a future classic, namely Jacob Bronowski (1968) (of the TV series *The Ascent of Man*) and Alex Comfort (1968) (about to become a millionaire with his *The Joy of Sex*).

But the biochemist Erwin Chargaff (1968), who has himself an important role in Watson's story, found as little merit in Watson's literary attainments as he had found in Watson and Crick's discovery of the DNA structure in the first place. Not only did Chargaff not care for Watson's book, but he declared that scientific autobiography is a most awkward literary genre. The reason for this awkwardness is, according to Chargaff, that scientists are dull people who, he says, "lead monotonous and uneventful lives . . . " (p. 1448).

But why *are* the lives of scientists so monotonous and uneventful, in contrast to the exciting lives of, say, artists, which make much less trite biographical subjects? Because, according to Chargaff, there is a profound difference in the uniqueness of the creations of artists and of scientists: *"Timon of Athens* could not have been written, 'Les Demoiselles d'Avignon' could not have been painted, had Shakespeare and Picasso not existed. But of how many scientific achievements can this be claimed? One could almost say that, with very few exceptions, it is not the men that make science, it is science that makes the men. What A does today, B and C and D could surely do tomorrow" (pp. 1448–1449).

On reading this passage, I was surprised to find Chargaff embracing the "great man" view for the history of art, that is to say, regarding the development of art as wholly contingent on the appearance of a particular succession of unique geniuses, while at the same time viewing the development of science from the Hegelian or Marxist perspective of historical determinism, which sees history as shaped by immutable forces rather than by contingent human agency. Since I found it hard to believe that Chargaff would really hold such incoherent ideas, I suspected at first that he had made his point about the irreplaceability of Shakespeare and the *re*placeability of Dr. A only to downgrade the importance of Watson and Crick's discovery. But I soon found that my suspicion was quite mistaken. In the following months, I asked many scientific friends and colleagues whether they too think that the achievements of art are unique whereas the achievements of science are inevitable, and hence commonplace. To my surprise, I found that most of my respondents (including Francis Crick [1974] himself) agreed with Chargaff in believing that we would not have had *Timon of Athens* if Shakespeare had not existed, but if Watson and Crick had not existed, we would have had the DNA double helix anyway. Therefore, the deficiencies of the proposition of differential uniqueness of the creations of art and science do not seem to be as self-evident as I had thought at first. Accordingly, I wrote my little *Scientific American* article to show why this proposition has little philosophical or historical merit.

Semantic Content in Art and Science

In order to examine the proposition of differential uniqueness of creation, I provided an explicit statement of what I understand to be the meaning of the terms "art" and "science." Both art and science, I wrote, are activities that endeavor to discover and communicate truths about the world, about the reality in which we live our lives. Thus, art and science share the central features of discovery and communication, and hence both involve the search for novelty and the encoding into a semantic medium the meaning of what has been discovered. Where art and science differ fundamentally is in the domain of reality to which the semantic contents of their works mainly pertain. The domain addressed by the artist is the inner, subjective reality of the emotions. Artistic communications therefore pertain mainly to the relations between private phenomena of affective significance. The domain of the scientist, by contrast, is the outer, objective reality of physical phenomena. Scientific communications therefore pertain mainly to relations between public events.

This dichotomy of domains does not mean, however, that a work of art is wholly devoid of all outer meaning. For instance, a Canaletto painting communicates something about the public phenomenon that was Venice of the *settecento*. Nor does it mean that a work of science is wholly devoid of all inner meaning. For instance, Freud's *Interpretation of Dreams* is addressed mainly to the private phenomena of the subconscious.

Hence, despite this fundamental difference in their principal foci of interest, art and science actually form a thematic continuum, and there seems to be little point in trying to draw a sharp line of demarcation between them. In any case, the transmission of information and the perception of meaning in that information constitutes the very essence of both the arts and the sciences. In other words, works of art and works of science are not merely there. They have a semantic content; they are meant to mean something.

So I was now ready to ask in my essay whether it is reasonable to claim that only Shakespeare could have formulated the semantic structures represented by *Timon,* whereas people other than Watson and Crick might have made the communication represented by their paper published in *Nature* in April of 1953. Here it is at once evident that the exact word sequence of Watson and Crick's paper would not have been written if the authors had not existed, no more than the exact word sequence of *Timon* would have been written without Shakespeare, at least not until the fabulous monkey typists complete their random work at the British Museum. Thus paper and play are both historically unique semantic structures.

But in assessing the *creative* uniqueness of a linguistic structure, we are not concerned with its exact word sequence; we are concerned with the uniqueness of its semantic content. And so I readily admitted that it was very likely that meanwhile, even without Watson and Crick, other people would have published a satisfactory molecular structure for DNA. Hence the semantic content of their paper would not be unique.

As for the semantic content of Shakespeare's play, however, I pointed out that the story of the trials and tribulations of its main character, *Timon,* not only *might* have been written without Shakespeare but in fact *was* written without him. Shakespeare had merely reworked the story of *Timon* he had read in William Painter's collection of classic tales, *The Palace of Pleasure,* published 40 years earlier, and Painter in turn had used as his sources the ancient authors Plutarch and Lucian.

But then the creative aspect of the play is not Timon's story; what counts is the novelty of the deep insights into the human emotions that Shakespeare communicates with his play. He shows us here how a man may make his response to the injuries of life, how he may turn from lighthearted benevolence to passionate hatred toward his fellow men. Can we be sure that *Timon* is unique with regard to the play's semantic essence? No, because who is to say that had Shakespeare not existed no other dramatist would have communicated very similar insights? Another dramatist would surely have used an entirely different story to treat the same theme (as Shakespeare himself did in his much more successful *King Lear*), and he might have succeeded in pulling it off.

Hence we are finally reduced to asserting that *Timon* is uniquely Shakespeare's (or maybe the Earl of Oxford's) because no other dramatist, although he might have communicated to us more or less the same insights, would have done it in quite the same exquisite way as the Great Bard (of Stratford or Oxford).

But here we must not shortchange Watson and Crick by taking for granted that Doctors B, C, and D, who eventually would have found the structure of DNA, would have found it in just the same way and would have published a paper that produced the same revolutionary effect on contemporary biology. On the basis of my personal acquaintance with the people engaged in trying to uncover the structure of DNA in the early 1950s, I expressed my belief that if Watson and Crick had not existed, the insights they provided in one single package would have come out much more gradually over a period of many months or years. Indeed, as Medawar found in his review of *The Double Helix,* the great thing about Watson and Crick's discovery was "its completeness, its air of finality" (p. 3). Medawar thought that "if Watson and Crick had been seen groping toward an answer, . . . if the solution had come out piecemeal instead of in a blaze of understanding, then it would still have been a great episode in biological history" (p. 3). But it would not have been the dazzling achievement that it, in fact, was.

Creative Processes

Why, then, is it that so many scientists seem to believe in both the uniqueness of artistic creation as well as in the commonplace, inevitable nature of scientific discoveries? In my article, I had put forward a variety of explanations, such as the scientists' lack of familiarity with the working methods of artists. Scientists tend to picture the artist's act of creation in the terms of Hollywood: Cornel Wilde, in the role of the one-and-only Frederic Chopin, is gazing fondly at Merle Oberon, as his muse and mistress George Sand, while he is sitting down at the Pleyel pianoforte, and, one-two-three, he composes his "Preludes." As scientists know full well, science is done quite differently: Dozens of stereotyped and ambitious researchers are slaving away in as many identical laboratories, all trying to make similar discoveries, all using more or less the same knowledge and techniques, some of them succeeding and some not.

Artists, we might note, tend to conceive of the scientific act of creation in equally unrealistic terms: Paul Muni, in the role of the one-and-only Louis Pasteur, is burning the midnight oil in his laboratory at the Institut Pasteur. He has the inspiration to take some bottles from the shelf, he mixes their contents, and, Eureka!, he has discovered the vaccine for rabies. Artists, in turn, know that art is done quite differently: Dozens of stereotyped and ambitious writers, painters, and composers are slaving away in as many identical garrets, all trying to produce similar works, all using more or less the same knowledge and techniques, some succeeding and some not.

Works and Contents

A more serious obstacle than unfamiliarity with working methods to discussing the problem of the uniqueness of creation is the apparently widespread confusion between *works,* on the one hand, and their *contents,* on the other. A play or a painting is a *work* of art, whereas a scientific theory or discovery is not a *work* of science but the *content* of a work, such as a book, paper, letter, lecture, or conversation. Thus, as formulated, Chargaff's proposition of differential uniqueness is not even false; it is nonsensical, because it compares a *work* of art itself *(Timon)* rather than its semantic content with the *content* of a work of science (the DNA double helix) rather than with the work itself, namely Watson and Crick's famous *Nature* paper.

Not only Chargaff but even Meyer, a theorist of the arts, seems unable to keep in mind the difference between works and their contents. For Meyer asserts that there is a profound and basic difference between scientific theories, which he says are *propositional,* and works of art, which he says are *presentational.* Meyer's antinomy between the propositional and the presentational is patently false, because *all* works, of science as well as of art, indeed all semantic structures, are "presentational" (in the sense of being a concrete pattern that can be occasion for human experiences that are found to be enjoyable, intriguing, and moving). By contrast, the quality of being "propositional" (in the sense of being a statement which affirms or denies something, so that it can be characterized as true or false) pertains not to works but to their contents. And here it is the case that not every "presentational" structure necessarily has a propositional content. For instance, Meyer rightly points out that a natural phenomenon, such as a sunset or Mt. Everest, is a presentational structure without propositional content. One of our principal agenda items will, therefore, have to be the question of whether the contents of works of art do or do not resemble the contents of works of science in being propositional. I will return to this central question.

Outer and Inner Worlds

Probably the most important of the reasons for the widespread acceptance of the proposition that artistic creations are unique and scientific creations are commonplace is the prevalence of an incoherent epistemological attitude toward the phenomena of the outer and inner world. The outer world, which science tries to fathom, is often viewed from the standpoint of naive realism, according to which phenomena and the relations between them have an objective existence independent of the human mind, and that this real world is as we see, hear, smell, and feel it. Hence the outer world and its scientific laws are simply there, and it is the job of the scientist to find them. At the same time, the inner world, which art tries to fathom, is often viewed from the standpoint of idealism, according to which phenomena and relations between them have no reality other than their invention by the human mind. Hence there *is* nothing to be found in the inner world, and artistic creations are simply cut from whole cloth. Here B or C or D could not possibly find tomorrow what A found today, because what A found today had never been there in the first place.

This incoherent epistemological attitude is held also by Meyer (1974), who asserts that only scientists discover truths; they do not create anything, except maybe intrinsically ephemeral theories. After all, "the structure of DNA was what it was before Watson and Crick formulated a theory of its structure" (p. 165). The reason for this is, according to Meyer, that "we assume evidently on good grounds, that while our theories explaining nature may change, the principles governing relationships in the natural world are constant with respect to both time and place" (p. 165). Artists, by contrast, he says, do not discover anything; they create their works, which had no prior existence.

In the 1960s and 1970s, Immanuel Kant's definitive resolution of this age-old epistemological conflict of naive realism versus idealism made its impact on the human sciences, under the general banner of structuralism. Structuralism emerged simultaneously, independently, and in different guises, in several diverse fields of study, for example in psychology, linguistics, anthropology, and biology. Both naive realism and idealism take it for granted that all the information gathered by our senses actually reaches our mind; naive realism envisions that thanks to this sensory information reality is *mirrored* in the mind, whereas idealism envisions that, thanks to this sensory information, reality is *invented* by the mind. Structuralism, on the other hand, provides the insight that knowledge about the world of phenomena enters the mind not as raw data but in an already highly abstracted form, namely as structures. In the preconscious process of converting the primary sensory data step by step into structures, information is necessarily lost, because the creation of structures, or the recognition of patterns, is nothing else than the selective destruction of information. Thus, since the mind does

111

not gain access to the full set of data about the world, it cannot mirror reality. But neither does it freely invent reality. Instead, for the mind, reality is a set of structural transforms abstracted from the phenomenal world.

Both Commonplace and Unique

This neo-Kantian, structuralist realism (it *is* a form of realism, in the sense that it does believe in an objective, albeit ultimately unfathomable, reality) leads to the recognition that every creative act in art and science is *both* commonplace *and* unique. On the one hand, every creative act is commonplace, in the sense that there is an innate correspondence in the transformational operations that different persons perform on the same primary data from inner and outer worlds. That is, we are all humans. On the other hand, every creative act is unique, in the sense that no two persons are quite the same and hence never perform exactly the same transformational operations on a given set of primary data. I therefore concluded my article by paraphrasing George Orwell's famous oxymoron, saying that even though all creative acts in both art and science are both commonplace and unique, some creations may nonetheless be more unique than others.

After these introductory remarks, I now come to the main point. Taking Meyer's essay as a typical contribution to the debate concerning the relationship between science and art, we can see that the source of the confusion is not so much the invocation of doubtful analogies of which Meyer accused me as the intractable nature of the underlying cognitive problems. To bring these problems into focus, let us first clear up the confusion inherent in the pronouncement made by Meyer that the term "discovery" pertains only to science, whereas the term "creation" pertains only to the arts.

Creation and Discovery

As I already noted, a scientific theory is an abstraction made from the "natural world," which presents our senses with a near infinitude of phenomena. Hence, in their work, scientists necessarily select only a small subset of these phenomena for their attention. Thus, contrary to the naive realist outlook on the discovery of the DNA double helix, the structure of the DNA molecule was *not* what it was before Watson and Crick formulated it, because there was, and still is, no such thing as the DNA molecule in the natural world. The DNA molecule is an abstraction created by century-long efforts of a succession of biochemists, all of whom selected for their attention certain ensembles of natural phenomena. In other words, the DNA double helix is as much a creation as it is a discovery, and the realm of existence of the double-helical DNA molecule is the mind of scientists and the literature of science, and not the natural world (except insofar as that world also includes minds and books). Hence, as applied to science, the distinction between discovery and creation is devoid of philosophical merit.

Complex Truths

However, Meyer's central objection to the explication of art and science as activities that endeavor to discover and communicate truths about the world lies in his claim that the concept of truth is simply not applicable to art. If Meyer's claim were valid, then the contents of works of art could not be propositional (inasmuch as they would not be statements that affirm or deny something that could be characterized as true or false), and hence artists could not be said to "discover" anything. Artists would merely create presentational structures without propositional content, just as God creates sunsets, no one of which has a content of which it can be said that it is true or false. All the same, Meyer (1974) admits that, unlike sunsets, "great works of art command our assent. Like validated theories, they seem self-evident and incontrovertible, meaningful and necessary, infallible and illuminating. There is, without doubt, an aura of 'truth' about them" (p. 184). But Meyer insists that in this connection "truth" is being used only in a metaphorical sense. Why? Because according to the naive realist standpoint from which Meyer approaches this deep problem, a literally true scientific proposition states what is actually, and objectively, the case, i.e., what is directly or indirectly observable, in the real world. And since there are no imaginable observations that could test the validity of the *content* of a work of art, it could be said to be "true" only in a metaphorical sense.

Viewing our cognitive relation to the world from the standpoint of structuralist realism, however, leads to a different literal concept of truth. Inasmuch as reality, to which truth relates, is something which each person abstracts from a world of things, the notion of truth has to be more relaxed. Namely, a scientific proposition is true (for *me*) insofar as it is in harmony with my internalized picture of the world (i.e., *my* reality) and commands *my* assent. This literal meaning of truth is obviously not an objective one but a subjective one. It leads to the concept of objective truth only as long as I am convinced that a proposition that is true for me would command also the assent of every other person qualified to make this judgment. Here the ideal of an absolutely objective truth is reached only if God also assents to the proposition. And so from the structuralist-realist viewpoint, the use of the term "truth" in connection with the content of a work of art is not metaphorical at all: It is the very same literal usage as that applied to the content of a work of science. It is exactly by their command of assent that we come to believe also in the truth of scientific propositions. In the 40 years that I have spent as a working scientist, I have personally validated (if indeed validation is at all possible), or even examined the published records of the validation by others, of only a small fraction of the scientific propositions which I believe to be true. The remainder simply command my assent, for the same reasons that Meyer cites as the basis of the aura of truth of great works of art.

A Continuum

Let us now reconsider the thematic continuum presented by art and science with regard to their principal foci of interest in inner and outer reality. Music, which appears to be the purest art form and has the least to say about outer reality, lies at one end of this continuum. Accordingly, music shows the least thematic overlap with science, which lies at the other end. The content of works of music is more purely emotional than that of any other art form, because musical symbolism very rarely refers to any models of outer reality, to which it could never do justice anyway; the meaning of musical structures thus relates almost exclusively to inner models. Musical symbolism is able to dispense with outer models because, according to Suzanne Langer (1948), "the forms of human feelings are much more congruent with musical forms than are the forms of spoken language; music can *reveal* the nature of feelings with a detail and truth that language cannot approach" (p. 191). Hence, music conveys the unspeakable; it is incommensurable with language, and even with representational symbols, such as the images of painting and the gestures of the dance.

Thus, the position of an art form on this continuum—i.e., how close it is to science and the extent to which it is addressed to outer reality—seems closely related to the degree to which its symbolism is embedded in language. The visual arts—painting and sculpture—are still relatively "pure" art forms, as is poetry, which, although it does resort to language as its medium, uses words in a quasi-musical form. But literature and drama, with their mainly linguistic symbolism and their close thematic ties to outer reality, while still addressing the inner reality of emotions, seem to lie halfway between music and science. Science is, of course, wholly dependent on language as its semantic modality.

All the same, the semantic transactions of art still pose a most difficult problem. What *is* the meaning of the propositions implicitly formulated in works of art? To what do the relationships exemplified by works of art actually refer? What are they about? Evidently the difficulty of answering these questions increases as we progress from science toward music in the thematic continuum. At the musical end of the continuum, where symbolism is incommensurable with language, these questions cannot be answered (verbally) at all. For instance, according to a legend quoted by Meyer, Beethoven, when asked what the *Moonlight Sonata* means—what it is *about*—went to the piano and played it for a second time. Meyer finds Beethoven's answer not only appropriate but compelling. But Meyer thinks that if a physicist were asked what the law of gravity is about and answered by letting some object fall to the ground, our inference would be that the physicist is trying to be funny.

Kinds of Meaning

I agree that Beethoven's response seems more reasonable than that of the uncooperative physicist, but not for the reason given by Meyer, namely that the *Moonlight Sonata* is not about the world and does not refer to anything, whereas the law of gravity is about the world and does refer to something. Rather, Beethoven's response is reasonable because he was asked a question for which there is no adequate verbal reply, whereas the physicist's response is unreasonable because he *could* have said something. This then is the paradox: logic demands that since the *Moonlight Sonata,* exemplifying a relationship, has some meaningful content—as opposed to a sunset, which has not—it must refer to something, must be about something. Yet we cannot say what that something is. In thus being generally speechless regarding the meaning of music, we resemble the split-brain patients studied by Roger Sperry; the patients can recognize familiar objects seen in the left half of their visual field but are unable to identify them verbally.

As we move away from music toward science in the thematic continuum, through the visual arts to literature and drama, verbal explanations of the meaning of art works, though still formidably difficult, become at least possible. Indeed, this is the very task to which hermeneutics is dedicated, the name given to the discipline originally concerned with the interpretation of sacred and profane texts but which has been extended more recently to also making explicit the implicit meanings that are hidden in a broad range of semantic structures (Gadamer, 1976). There would be massive unemployment among contemporary hermeneuticians—the Berkeley Center for Hermeneutical Studies would have to close—if Meyer's assertion were actually true that the contents of works of art do not refer to anything and are not about-the-world. Suppose, to stay with our original example, having just seen a performance of *Timon,* we asked a Shakespearean scholar, "What does the play mean—what is it *about?*"—and he simply took us back to the theatre to make us see *Timon* for a second time. Would we not consider his response as disingenuously witty and as nearly improper as that of the physicist? That is not to say that if the scholar did give us his verbal interpretation of *Timon,* it would fully capture the semantic essence of the play. Depending on his hermeneutic skills, he could go some considerable distance toward giving us an idea about the play's deep meaning, and not just about its plot. But what would be most likely missing from the scholar's verbal interpretation of *Timon* is precisely that part of the play's meaningful content which is not

115

embedded denotatively in the text and which arises from it connotatively, thanks to the contextual situation created by Shakespeare.

So we traveled a long way from Chargaff's reflections on the triteness of scientific autobiography to the bottomless depths of epistemology and cognitive philosophy. We saw en route that "art" and "science" are semantic activities that seek to discover and communicate truths about the reality in which we live our lives, with art addressing mainly the inner world of emotions and science mainly the outer world of objects. That explication allowed us to identify one common source of confusion in discussions of the relationship between the arts and sciences, namely the false antinomy of *works* of art as semantic structures and the semantic *contents* of works of science. Once that confusion had been cleared up it became plain that works are unique in both art and science.

We then noted one deep source of confusion in the discussion of the relationship between art and science, namely that the outer world, which science tries to fathom, is often viewed from the standpoint of naive realism, whereas the inner world, which art tries to fathom, is often viewed from the standpoint of idealism. This incoherent epistemological attitude leads to the false distinction that scientists merely discover what is already there—they do not create anything—whereas artists create something that had no prior existence—they do not discover anything. However, as soon as our relation to the world is viewed from the standpoint of structuralist realism, which envisages that the mind neither mirrors nor invents reality, but constructs it by a process of abstraction from the near-infinitude of phenomena, it becomes evident that in both art and science discovery and creation refer to the same process.

The most difficult aspect of the discussion turned out to be the nature of the semantic content of works of art. Does the concept of "truth," which clearly applies to the content of works of science, apply also to the content of works of art? Here we saw that, when viewed from the standpoint of structuralism, the concept of truth, as applied to scientific propositions, amounts to harmony with my internalized picture of the world, i.e., *my* reality and hence to command *my* assent. And since great works of art similarly harmonize with reality and command assent, this concept of truth would apply to their contents as well. Nevertheless, as concerns the communication of truth, there is an important difference between art and science: Works of science communicate their truths explicitly in language, whereas the truths of works of art are communicated implicitly in linguistic, tonal, or visual structures. But what *is* the meaning of the truths implicit in works of art? What is the *Moonlight Sonata* actually about? Here we finally encountered a deep semantic paradox: Although we can capture the meaning of the content of a work of art, we may not be able to *say* what that meaning is. So even if the marriage of the Two Cultures, art and science, *were* made in heaven, theirs would not be the first significant relationship in which the spouses turn out to have some difficulties in talking to each other. So maybe it would be a good idea after all to keep a hermeneutic humanist as an interpreter in the Arts and Sciences household.

References

Bronowski, Jacob. (1968). Honest Jim and the tinker toy model. *Nation, 206,* 381–382. (Reprinted in Watson, 1980, pp. 200–203)

Chargaff, Erwin. (1968). A quick climb up Mount Olympus. *Science, 159,* 1448–1449.

Comfort, Alex. (1968, May 16). Two cultures no more. *Manchester Guardian,* p. 10. (Reprinted in Watson, 1980, pp. 198–200)

Crick, Francis H. C. (1974). The double helix: A personal view. *Nature, 248,* 766–771. (Reprinted in Watson, 1980, pp. 137–145)

Gadamer, Hans-Georg. (1976). *Philosophical hermeneutics* (D. E. Linge, Trans.). Berkeley: University of California Press.

Langer, Suzanne K. (1948). *Philosophy in a new key.* New York: Penguin Books.

Medawar, Peter B. (1968, March 28). Lucky Jim. *New York Review of Books,* pp. 3–5. (Reprinted in Watson, 1980, pp. 218–224)

Meyer, Leonard B. (1967). *Music, the arts and ideas.* Chicago: University of Chicago Press.

Meyer, Leonard B. (1974). Concerning the sciences, the arts—AND the humanities. *Critical Inquiry, 1,* 163–217.

Stent, Gunther S. (1968). What they are saying about honest Jim. *Quarterly Review of Biology, 43,* 179–184. (Reprinted and expanded in Watson, 1980, pp. 161–175)

Stent, Gunther S. (1972, December). Prematurity and uniqueness in scientific discovery. *Scientific American, 227,* 84–93. (Reprinted in Gunther S. Stent, 1978, *Paradoxes of Progress,* pp. 95–113 [San Francisco: Freeman])

Watson, James D., and Crick, Francis H. C. (1953, April 25). A structure for deoxyribonucleic acid. *Nature, 171,* 737–738. (Reprinted in Watson, 1980, pp. 237–241)

Watson, James D. (1968). *The double helix.* New York: Atheneum.

Watson, James D. (1980). *The double helix.* In Gunther S. Stent (Ed.) *The double helix: A critical edition.* New York: W. W. Norton.

Part III

Programs: Certain Bases for Planning Curriculum and Instruction

The Early Environment of the Child: Experience in a Continuing Search for Meaning

Annemarie Roeper

My paper draws upon certain observations coming out of long study at the Roeper City and Country School for the Gifted, which I codirected for some 40 years (Roeper and Roeper, 1981). I will comment on the nature of our students' development of what Gardner Murphy (1961) calls their "innate capacity for effective reality seeking and testing" (p. 32); to me theirs is "experience in search of meaning" (Einstein's phrase). At appropriate points, I will relate this experience to what is learned and to the process of learning, including a consideration of the role of the teacher. My stress will be science in the broadest sense. Actually young children's attention is soon drawn to what we call science, for it provides a handle to help understand some of the world's structures and patterns.

What is the early environment at home and at school that encourages growth in all children? When a child is born, one may say it is an intervention in home and society. But it is also clear that when a child is born, a living learning environment creates itself. For it is from the child, the ever-learner, that all learning originates and proceeds. Children's growth is like the flow of a river, turning unexpected bends, running into hidden places and over sticks and stones, going fast and slow, bubbling up at times, and even going around in circles or backwards. Children develop their own methods of learning in a special hollowing out of the environment. But gifted children's learning environment grows rapidly from tiny beginnings to large spaces, visibly unique, or better, idiosyncratic. The impetus to learn, the process of it, the sheer force of necessity behind it are awe-inspiring to watch. It is like an ever-increasing whirlpool of bubbling water.

All children make their presence known but the gifted child often makes a special impact in a profound way. As Sandra W. Scarr and Kathleen McCartney (1983, April) see it, there seems to be an "avid reaching out" for new experience; what Murphy sees as striving to understand "reality . . . exactly as a rationalist would seem to demand" (p. 32). I would say "exactly as a scientist would seem to demand." In fact, I believe that the child is the original scientist. Just like the adult scientist, the child seeks the truth, the reality of life. Just like the adult scientist, the child is driven to find out. Just like the adult scientist, the child starts from where he/she is, according to his/her needs and the reality of the moment, and ventures into new explorations. Just like the adult scientist, the child needs a laboratory to do the work of discovery. The child's laboratory, however, is first home and then school. He or she is not yet sophisticated in the ways of the world and needs a guiding hand and a framework created by surrounding, nurturing adults. What is this inner learning environment like?

Early Environment: Early Experiences

In order to describe the environment, we must talk about the children—and the uniqueness of each child. Who is she? Who is he? What are his or her characteristics? All human beings share many characteristics and needs, and yet each is different. So children experience the world in similar and yet different ways. They are born with the same task, to master the world and make it their own. They are driven to learn physical and mental skills in order to build a foundation of trust in themselves and their environment by learning to understand both.

At no time in life is this motivation to learn and to understand as great as in early childhood. For what is learning, specifically, what is learning in science in its broadest sense, but the wish to create a structure from our experiences? This desire to learn about the physical and biological environment is as basic a need as that for food, love, and protection (Roeper, 1976). In order to feel safe, the children need to make sense of this strange chaos of sound, sights, odors, touches, tastes, vibrations, of hunger, thirst, cold, warmth, wet, and all the other sensations and experiences around them. From the beginning, the gifted show an even greater awareness of the complexities of the world than most, a greater sensitivity, and a greater desire to make sense. The gifted often use great skills to overcome the anxiety resulting from this awareness by trying to bring order into the apparent chaos around them.

In addition, the gifted experience genuine pleasure and excitement from knowledge, information, and understanding. All this is part of the motivation for learning and the reason for the rapid growth that leads young children to acquire new concepts daily. Their whole being concentrates on these intellectual, emotional, and physical tasks. Their intellectual ability helps them explore and understand sophisticated concepts, develop their self-images, and cope with life in a variety of ways. Their search for meaning goes in all directions. But in many children, particularly gifted ones, it expresses itself in a need to understand the laws of the physical world. For this reason, the learning of science becomes an inner necessity for many gifted children. This is why young children are fascinated by nature, by natural science, by all living things. But, in my opinion, there is a special place for physical science in the young, gifted child's search for meaning. My colleague Marian McCloud and I defined that quality as follows:

> Physical science provides a certain dependability that we do not find in nature or human relations. Water always turns to ice in the freezer. Ice always melts at temperatures above zero Celsius, but Johnny's behavior toward Ricky may be warm one day and cool the next without obvious reasons. (1965/1984, p. 3)

At Home

To live with a young gifted child whose self is developing well is a pleasure and a challenge. One sees a mind working overtime (Roeper, 1976, 1986). Soon the immediate surroundings appear to be exhausted—at least for the present. Frequently, necessary skills such as walking and talking are mastered quickly. Often, such children require little sleep, spending their waking hours learning or wanting to learn about life, about people. Gifted young want to know all about everything—the moon, the sky, the stars, the earth. Even as young as age four, they worry about politics, the environment, justice, hunger in the world, fairness, animals, all of nature, the universe, baseball players, etc. The tools for acquisition are available to the child from the beginning. How they are used differs with each child.

One of the most important tools is play. Play has a structure and a purpose of its own (Bergen, 1987). It contains risk taking, repetition, change, touching, tasting, and many other approaches that help the child to find out. As time goes on, play grows in complexity, mastery, excitement, and fun; the child gains concepts and general information through it. Never just play, it is growth; it is the way in which the young child fulfills a passionate and absolute need for mastery. Out of all this, slowly, certain experiences and concepts come into focus. Through playful interaction children develop many concepts of science. However, children should not be left on their own; they also need an open relationship with adults to avoid misconceptions and to get help in unraveling confusion.

How do they develop methods of exploration and discovery? At first children begin to have some kind of familiarity with daily rhythms and daily events. They may begin by learning to know their mother, her breast. Sucking sometimes stills hunger and sometimes doesn't: Sucking a thumb satisfies but does not feed. The child learns that a cry brings someone to pay attention. After a while, however, the child may find that some footsteps come to the crib and others pass it by. He or she begins to differentiate between a mother's and a sister's footsteps. Now crying stops only when the mother enters and nears the crib. Through this experience, the child has learned to understand that one event *usually* follows another—but only under certain circumstances.

This is a sophisticated concept (Roeper, 1976, 1986); now, the children grow by leaps and bounds in every way, each in a unique manner. As they grow in their understanding and begin to reach further and further, children may begin to wonder about day and night. This leads, at a later age, to an interest in the sun and the relationship between the sun and the earth, which is soon followed by an understanding of the movement of the earth around the sun. Learning about up and down and falling leads to an interest in gravity as children get older.

Experience related to mind has its own sequence. There is physical mastery, such as walking, climbing, and all types of movements; there are types of intellectual mastery, like talking, thinking, and learning—some of it academic—such as beginning reading, math, and science concepts. Children also begin to develop some ethical and social concepts. But most of all, as soon as gifted children can talk, which is sometimes very early, they begin to ask their many, many questions. As the inner world of the child expands, more and more of the outside one becomes an integrated part of self, and both worlds become familiar and manageable.

Everyone makes this integration in her or his unique way. Perhaps this is how we grow into people with different characteristics. No matter what, however, being young means being, in the first place, a learner. The amount of energy children expend on exploration and repetition has amazed generations of adults. To watch children encounter the world and attempt to penetrate its structure and meaning is to watch thinkers and probers at work. Often, one doesn't know how or where the search begins. It flowers in wonderful conceptualizations.

For example, one day a three-year-old boy came running up to me to say that he had just figured out how big his stomach was. (He showed me the tip of his finger.) When I asked him how he had found that out, he explained, "Well, my mother said my eyes are bigger than my stomach. If my eyes are bigger than my stomach, than my stomach must be smaller than my eyes, and, therefore, it would be about this size." In this instance, his logical conclusions led him to the wrong answer. But this is not always the case.

Once I asked a group of children what they would do if they wanted to find something out. These were the answers: First, "Make sure you know what you want to find out." "Ask the question." One child added, "Ask it clearly." Second, "You make a guess about what the answer might be." Another child: "You always have a thought of what the answer might be, even if you just make it up." Third, "There are many ways in which you can find the answer." "You can ask the question," one child pointed out. Another said, "Just asking the question isn't enough. Ask someone who might know the answer." "If you knew how to read," noted a third, "you maybe could find out from a book." The discussion continued: "Can you try out the answer?" "Go to the place where you might find the answer." "Talk it over with your friends. Sometimes they know answers that grownups don't."

Here is the basis of what has been called "the scientific method": statement of problem, hypothesis, examination of ways to solve the problem, proof.

In School

In early April, 1958, Paul F. Brandwein made one of many visits to the Roeper School and marvelled at the natural manner in which we taught concept seeking and concept forming. These aptitudes were part of every phase of the subject matter. The children learned to probe through convivial planning of what they wanted to encompass: the entire world.

Brandwein noted that many, if not most, of our gifted children were indeed conceptualizers, seemingly in all the areas of subject matter they considered. For example, he reports the following exchange about likeness and differences. Brandwein describes what happened as the teacher offered her class of six-year-old children

three each of six sets of material[s]: plastic forks, spoons, large clips, clothespins, paper boxes, and plastic bottles. [She asked] them to put like things in groups. Most children will have a set of six like categories of objects: forks, spoons, clips, clothespins, boxes, and bottles; concrete and somewhat conceptual operations, I suspect, in accordance with Piagetian prospects. But a few, say 2 or 3 out of 30, will find 3 groups: (a) forks and spoons, (b) clips and clothespins, and (c) boxes and bottles. When asked why, they respond somewhat as follows: one is to eat with, another is to hold things together, another is to put things in. Conceptual and abstract operations, perhaps. One or two out of a school of 200 or so young, eyes gleaming and merry, will have put all together: This is fairly rare. Asked why, one child told me, "All are things." One surprised me: "They are all made of matter." (1987, p. 35)

From his observations at our school and elsewhere, Brandwein has come to the conclusion—one with which I agree—that Piagetian constructs of sequential development should be modified for the intellectual and even artistic abilities of gifted children. For them, the sequences are either telescoped into shorter periods or are interdependent: That is, the concrete and formal operations may be meshed very early—much before the sixth and seventh year. (For further discussion of some of the limitations of Piagetian philosophy for those working with the gifted, please consult Steven J. Rakow's paper in this volume.)

To take the meaning of this lesson in conceptualization—via forks, spoons, clips, clothespins, boxes, and bottles—just a bit further: The model of the lesson on eating utensils applies equally to observation of objects and events in natural and physical science throughout the curriculum. Thus, a lesson on mammals, birds, amphibians, and fish, for example, will elicit that all these are in the common conceptualization "living things." But also the subconcepts of mammals, birds, and fish will be elicited. Similarly, metals will be distinguished from nonmetals—but both will be assigned to the concept *matter*. McCloud and I have formulated our perceptions of this ability to conceptualize in a book of discovery for children (*Physical Science for Young Children* [1965/1984]). Most of the activities and problems posed for children lead to the aptitudes and skills of categorizing and conceptualizing. Because, in our view, gifted children are so evidently inclined to "see" and "select" common attributes in objects and events as a basis for concept seeking and concept forming, our lessons in *Physical Science* incline to that end.

But, also, gifted children sometimes lag behind, because they soon see the complexities and the dangers surrounding them, and then they hold back. They may be the greatest explorers, but many of them also shrink from the unknown encountered in exploration. The gifted may learn skills differently from others. They may be early readers yet late swimmers or, at times, the other way around. Overall, their abilities help them explore and understand more sophisticated concepts but may also increase or decrease their self-images and lead them to cope with life in a variety of ways (Miller, 1979/ 1981). They are likely to find more alternative solutions than others. Emotions, attitudes, and understanding supplement each other, grow out of each other. Gifted children need to learn in order to understand so that they can feel safe, so that they can trust themselves and others. According to Erik H. Erikson (1950/1963), learning to trust is the first great task of all children. Understanding leads to fulfillment, if they can find the world trustworthy. The adult's job is to make it so, in order that children, and the learning environment in which they find themselves, are safe.

The Gifted Child—In the Group

All along the way of this process of growth, the child is not alone. Family, teachers, friends, society, chance, and many other factors play their part.

What are the ways in which we can support the natural growth and motivation of the child? What can we do to work for the child's growth rather than against it? How do we maintain the freedom for the child to grow from the inside out? All children need to feel understood. Children need to master the world as part of their growth of self, as part of developing a trust in self and family and society (Poole and Line, 1979). Children learn to trust themselves, to reach out and grow, if they feel loved, safe, and trusted by those around them. They can use their energies for exploration and accept their own personalities freely only if their surroundings, human and inanimate, permit. Just as the young need to touch and to play, they need others to respond to their overtures in kind. And all children need to be respected.

Within the Family

Parents of gifted children are likely to see the child as an extension of themselves. A gifted child often sees this parental need as an obligation and tries to fulfill it. The child serving as an extension of parental selves, the child trying to meet needs unfulfilled in parental childhoods, cannot feel safe and cannot fulfill his or her own needs. If, for instance, parents want their four-year-old daughter to show everyone who enters the house how well she can read, her reading ability becomes the parents' success and fulfills *their* need to be outstanding, not hers. However, if the child does not do well or is not interested in reading, it becomes *her* failure. She soon begins to feel that her parents will love her only if she can be the success they expect. In that case, parents need the child to make up for their own lack of success, and she becomes the child who believes, implicitly, that the world loves her only if she fulfills its needs. As she disappoints, she disappoints herself. In such a situation, young children will have neither the energies necessary for free exploration of their world nor, what is more important, those necessary for the constant restoration of self (Kohut, 1977; Freud, 1953–1974). Understandably, such a child is hindered by anger, anxiety, and inner restrictions. When failure becomes dangerous because the child fears that it will lead to loss of parental love, the child begins to look outside to figure out how to behave in order to please the parents and fulfill their expectations.

Therefore, a first indispensable condition for creating an appropriate environment for learning is that parents see their children as individuals with their own personhood, their own growth, their own successes, even their own failures. It goes without saying that parents need to love, protect, and guide their young until the children reach a period when they can protect and guide themselves. Moreover, children learn, then, to incorporate their love of parents not in that role alone but as teachers and protectors as well.

If parents understand and teachers give recognition for growth and success in the children's own terms, the children will feel accepted and understood. They will find themselves insiders in the world; they will see themselves as part of family, school, and society. Many children, especially the gifted, suffer from a feeling of being outsiders, of not being understood, of not being able to communicate their real selves and personal needs (Miller, 1979/1981; Roeper, 1976). Parents who can when appropriate set aside their own needs may achieve the empathy to fulfill one of their most difficult tasks, that of sensing the child's necessity of the moment. For example, the gifted child may require great freedom to explore in one instance and extensive guidance and protection in another. Sensitive parents respond carefully to such necessities.

Parents ought also to make the important realization that there is a clear-cut difference between allowing children to grow and treating them as though they were already grown up. It is obvious the young cannot yet take responsibility for their lives. Their freedom exists only within the security of adult care, and, if you will, leadership. Children need the protection of knowing that parents are in charge, knowing that there are clear-cut rules and regulations and that adults make decisions for their sake. But often children need to be free to make decisions about their own behavior. One of the difficulties of caring for gifted children is that parents may fail to sense when they, as parents, need to recede and turn the power of decision making over to the growing children.

Too often, parents of the gifted stand in awe of their children; then, parents can find it hard to maintain their responsibility as adults. Too often, the activities and concerns of the family revolve around these children, their personal interests, and their personal concerns. In that case, gifted children develop an unrealistic view of their own role in the family. They see themselves as different from other siblings (and other children), as not governed by the same rules. Sometimes gifted young consider themselves adults and believe that they have adult rights and responsibilities, a stance that often leads to anxiety and insecurity. Such children then need to maintain a position which they know is untenable and therefore frightening. Their energies, both psychic and physical, may be diverted from growing and learning.

In Society

Recall that children are not only surrounded by family but also by society where they are exposed to its expectations and traditions. The world expects something from them, just as they, in turn, expect something from the world. The flow toward them of social influences may change the direction of their growth. Others' behavior can be supportive and create a river of learning and growing, or their attitude can change the stream's course in the other direction, altering it, diverting it, at times slowing it down to a trickle. Or drying it up altogether.

128

Society's expectations come from the outside in: It has developed certain goals and expectations and certain methods of achieving them, and it expects children to adapt to them. Often society measures the responses of the young in terms of its personal, learned expectations. However, while *children may reach their goals of mastery through many, many different avenues,* society only allows a few. It has established definite sequences. Society has decided what, how, and when children will learn and what constitutes achievement in terms of "outside" success. But the child thinks and feels in terms of mastery, which means an "inside" success. Too often children become acceptable to society only if they succeed on society's terms.

Society also has decided on the method through which children are to learn. They are to be taught. They are passively to receive teaching in school. In reality, children learn actively by participation, experiencing with their whole being. Children do not all learn in the same way, at the same time, and at the same age. More often than not, society interferes with their basic motivation and their basic goals for learning. Thus, it may hinder the budding creativity and perhaps cripple the natural flow and development of a young child's individual pattern of growth.

The Gifted Child—Growing in the Schools

We need to recognize and respect the basic environment that already exists, namely the children themselves. We need to recognize that there is a legitimate reason for children's unique interests, for children's unique ways of learning. There is a real difference, then, in the learning society tends to expect in school and the ways of learning that are particular to children, especially to gifted children. See below for the contrast between the active or self-actualizing approach used by an increasing number of schools and the "traditional lecture" approach often found in schooling. In the first, the child is central as learner; in the second, the teacher is central as dispenser of information.

Children's motivation for learning may well be affected by these two approaches to schooling: The curriculum in the traditional elementary school is usually rigidly predetermined rather than developed through cooperation between teacher and children. In the traditional approach, textbooks are the basis of teaching, while in the child-centered approach, they provide useful resources or background. The curriculum of the elementary school is usually based on skill learning. The gifted are concept learners, often with specific interests in science or social studies, or humanities and arts, or mathematics, and their approach to learning is through what is often called "the inquiry method." These areas of study and the self-actualizing approach are infrequently central in the elementary school.

The curriculum for gifted children, indeed for all children, should be based on this active approach, which focuses primarily on concept learning with skills as necessary tools to achieve these goals. If concept learning through social studies, science, or the humanities were the focus, many gifted elementary schoolchildren would be able to share their amazing fund of knowledge with others and feel an accepted part of the community. That this opportunity usually does not exist is a loss both to the gifted and to the other children.

Self-Actualizing Approach	*Traditional Lecture Approach*
Education for life, self-actualization	Education for college, financial success
Emphasis upon the whole child	Emphasis on academics and sports
Learning central	Teaching central
Discovery, inquiry method for active learners	Facts, passive intake of information through listening to lectures or being informed or rote learning
Freedom within a framework—flexible curriculums	Rigid, linear curriculums
Concern with developmental phases	Traditional learning goals and methods, pressure
Conceptualization based on many causes, many effects	Skill development based on age and grade level
Self-definition by child	Definition by society, which also defines rate of learning and kind of growth desired.*

Children who grow up within the open, flexible atmosphere defined in the left column above usually want to learn, have developed a method of doing so, and have acquired a large, diverse fund of knowledge. Even though we look at this approach as preparation for life rather than the narrow goal of preparation for high school and college, many of these children excel in traditional academic environments.

We believe that children and teachers should interact. Often the curriculum is chosen by the teacher, but only after close observation of the development of interests in children does s/he create a framework within which the children can discover. (Examples of lessons and activities which inspire science interest in preschoolers can be found in Roeper and McCloud, 1965/1984.) Much of the planning is done cooperatively with the children. But in this sharing, the teacher does not relinquish responsibility as the one experienced in learning and teaching. As Brandwein writes, "It is the teacher who has the privilege, the responsibility, of helping youngsters in their decisions

---

*Chart based on Brandwein (1962), Poole and Line (1979), and Roeper (1986).

when and how to inquire so that they may choose with relevance" (1962, p. 19).

Expanding the Learning Environment

We need to find ways to expand the original learning environment to create a climate of trust, support, and opportunity for growth. How to create such a climate? Children need to feel acceptance, positive stimulation, and response from parents, teachers, and others in society. Children who are allowed to grow freely often develop some very specific, unusual interests and knowledge. They may acquire information, even wisdom, for which they need recognition and praise. (Too often, in reality, they receive a smile of disbelief.)

This quest for understanding is a particularly typical expression of the unique interests of the gifted young. For example, I have met some three-to-five year olds who want to know exactly how the toilet or the sewer system functions. Some collect roots, leaves, or stamps. One child wants to know how doorknobs function. Some children love to take apart small machines such as radios and clocks and put them back together again. There are probably countless interests and activities in which little children could participate but through which, as a rule, they are not allowed to develop and express themselves. The rocks found on the way to the museum may be more exciting to the small child than the beautiful displays in the museum. Were not rocks his or her discovery? Some of these inclinations may seem strange, but they are all valuable stepping stones toward mastery and growth.

These interests, better, fascinations, usually correspond to the children's developmental phases (Roeper, 1976). The three year old who has just been toilet trained is perhaps puzzled, perhaps frightened, by the disappearance of a part of "self." To understand how the toilet functions and where its contents go means familiarity and intellectual mastery; this in turn leads to emotional acceptance and control through understanding. Knowing how doors open and close, how the hinges work, what different types there are, what their purposes are all lead to a familiarity that helps to integrate the situation into one's capacity to cope. Knowing about doors relieves anxiety about being separated, about people leaving, about feeling shut up in a room. Finding out means taking, incorporating, integrating into the self.

Children want to know about origins. Interest in dinosaurs or roots goes back to children's need to find out where they come from, leading perhaps to an understanding of past and future. They want to know about where they themselves come from, how babies are made and born, why boys and girls are different. They want to know why adults, who as a rule are pleased when children want to learn, usually change the subject and seem to close their ears when questions about sex are asked. This is one area where adults, for their own reasons, often want to stop the child's flow of inquiry. This reaction in itself often becomes a hindrance to the natural curiosity of young children. They sense that some areas are forbidden territory, that certain questions must not be asked. But children's interest in sex is part of their desire to understand the world. In fact, it is a very important part. I believe that we should give children information on sex as soon as they seem to show any signs of interest; however, this interest may be unexpressed through direct questions—for the children sense many parents' reluctance even without actually asking and being rebuffed.

Some children may be preoccupied with death. Here again, they try to meet an emotional need to feel safe through intellection. What is death like? What happens to your feelings? One gifted four-year-old girl attempted not to move at all for a whole day trying to understand what death feels like (Roeper, 1976). In many cases, such concerns and questions are the starting point for ever-widening interests in many directions. I know gifted children, four years and older, whose interest in every part of the body has led them to name all its parts and their functions. This concern originated with the question, "What is the difference between life and death?" Out of this same question sometimes grows a preoccupation with all living things and their different physical characteristics.

Other gifted children may move in different directions, turning to religion and asking questions like, What is God? Is Santa Claus real? Is there life after death? (Erikson, 1950/1963). Part of the desire for mastery and protection develops into a need to understand the ethical framework of the interactions between people. Gifted children want to understand the reasons for the rules and regulations which govern our lives. Many of them have a deep sense of justice. They want to know what is right and wrong and why. Unfairness makes them feel unsafe, because it makes basic rules unrecognizable. Just as children need a structure of the world in general, they need a structure of the basic concepts of ethics. They are concerned with the moral questions that arise out of global interdependence, technological growth, and everyday human relations. They raise questions like these: Are we responsible for the problems of the environment? Is it up to us to change it?

Some Thoughts on the School Environment

What is the best school environment for the gifted child? Gifted children's eagerness to make the world their own soon grows beyond the confines of the home. They need to venture out, to relate to adults other than their parents, and to expand their opportunities to grow. Some children are ready for small group experiences at age one and a half. On the other hand, the gifted may have difficulty separating from home because of their greater awareness of the big step they are taking. The transition from home to school thus requires great support and empathy from parents. These children need to be given time to learn to learn, to feel safe in the new environment. They need a familiar person around while they begin to relate to others. They need to know that their parent trusts the new adult so that they may transfer some of their own feelings of safety to an environment away from home. If this move creates a loss of security, their sense of self may be impaired and they may withdraw energy from learning and growth, using the withdrawal for self-protection. We must remember that children at school are the same and have the same characteristics and needs as children at home. They need to be protected, loved, and respected.

Many children must develop a personal relationship with the teacher before they can become participating and secure members of a group. An essential factor in this personal relationship is recognition by the teacher as well as respect for the child's individual method of learning and his/her interests and goals.

Play and exploration remain the best learning tools for the young child. Children develop a sense of inner freedom and permission to reach out if they (and their goals and idiosyncratic ways of learning) are supported and cherished by the adults at the school. This security and freedom requires a flexible atmosphere with much opportunity for discovery, individualized and group learning, play, and stimulating, enthusiastic adults who are learners themselves.

Gifted children need adults who will serve as models, adults who will open doors with their own knowledge, attitudes, skills, resources, and desire to learn but who will not try to push children through those doors. These adults offer the first trusting relationship as surrogate parents outside the home. As children grow through various developmental phases, they often develop similar interests that can be discovered by a teacher and used as a basis for a framework of common learning experiences. For example, many children are interested in insects. Others might like to know what is real and what is pretend. To understand the difference between reality and fantasy allows children to cope better with each and creates opportunity for safe expression of normal hostility in fantasy without fear of any "real" consequences.

Knowing the difference between animate and inanimate objects is basic to the developing structure of the children's world. Learning about the laws of the physical world serves as a basic structure within which to develop further understandings. In an atmosphere of discovery, therefore, children often exhibit great enthusiasm toward learning about physical science or natural science. Nursery schools, more often than grade schools, offer this kind of atmosphere and activities. Preschools traditionally are keenly attuned to developmental phases and the needs of individuals or groups of children and provide appropriate experiences in cognitive learning as well as opportunities for social learning, interaction, and invitations to creative expression such as in music, dance, drama, and physical activities. Children in such groups maintain their passion for learning. Most nursery schools, however, do not stress the orderliness of physical science. This is regrettable, because it is at this age that children are so fascinated with science phenomena.

Through the open approach, the teacher becomes the facilitator. The emphasis is on learning, not teaching, and in this environment, a clear-cut, well-thought-out philosophy helps children use their real potential toward individual self-actualization in achieving a concept of life rather than narrow goals, such as traditional academic skills *only*. Skills are necessary as tools but not as an end in themselves. Teaching and learning to maintain the inner motivation and potential for learning are based on fundamentally different goals and methods from the traditional ones. According to Brandwein (1987, September), the elementary school curriculum should be an integrated one, particularly in the humanities, the social sciences, and the sciences. The curricular plans posited in his paper envision the humanities as concerned with concepts of ethics and aesthetics—truth, beauty, justice, love—expressed in music, painting, dance, literature, and the like. The social sciences examine the concepts and values of society (liberty, interdependence, values in government, economy, and in the development of nations). The sciences examine the concepts of the natural and physical environment (life, interdependence in ecosystems, resources of matter and energy).

Again, in my observation, gifted children are global thinkers; to repeat, they turn naturally to conceptualization. They are seekers after reality. Often, they need or want to understand the whole, the overall concept, before learning about details. For example, I once observed children being taught the names of different types of dogs. They learned them mechanically without a real concept of "dogness." The gifted need to begin with an overall concept of what is being "taught," whether stars, atoms, earthquakes, people, musical notations, equilateral triangles, or dogs as animals and mammals. Quickly, they probe likenesses and differences among them within a conceptual framework (Brandwein, 1962; Roeper, 1986).

As children grow older, the amount of information and the varied learning opportunities available may become overwhelming for them. What is the task of the teacher in that case? In addition to what I have described earlier, the teacher must make opportunities for learning available and build a bridge to the expectations of society without interfering with the child's self-direction. The conceptual and unit approach can provide a structure that makes information manageable without limiting the freedom for discovery for young children. Within the child-centered approach, such a unit representing real interests often grows out of discussions between children and the teacher. A unit that develops out of the questions asked by children and includes their approach to finding the answers allows individuals or groups of children to continue to create their own environment.

Some Special Traits of Gifted Children

As I have observed, a special characteristic of gifted children is their perfectionism. This trait sometimes interferes with their motivation for learning and exploration and requires special attention in any approach to their schooling. They become most impatient if they do not succeed at a self-imposed task, such as drawing a horse the way they see it. It is difficult for them to accept the fact that, often, their fine motor control does not keep up with their sophisticated ideas. This wanting to be perfect often has the effect of keeping gifted children from venturing into experiences that may expose them to failure. Helping them understand that failure is part of growth, that possibly their expectations of themselves are unrealistic, and/or assisting them to reach their goals are some of the ways in which adults can help children to cope with that particular difficulty. Perfectionism brings with it a fear of risk taking, a fear that again can block gifted children's freedom to seek new growth. They need an environment of support, of encouragement, in order to find the courage to try new experiences, especially when unsure about whether they can really achieve their goals.

Again, in my observation, many gifted children do not care about marks, honors, awards, etc., although parental or teachers' expectations may lead them in that direction against their real inner needs and personal desires. Such children may be faced with a dichotomy between their real interests and outside pressures, a dichotomy that creates internal conflicts. Although the gifted may enjoy competition in specific areas, they feel, in general, freer to grow without it. De-emphasis on competition allows them to pursue their own very original interests, which may not be shared by others, and, therefore, which may not be subject to competition. Many gifted children truly want to climb the mountain because it is there, not because the completed journey means success. With gifted children, even more than with others, the emphasis must be on their quest, their learning, what they take in, how they digest it, and how they use it. The curriculum should be flexible enough to permit this kind of active, individual learning, even as the teacher opens the door to useful acts of discovery in other areas.

Gifted children make friends with others of either sex whose behavior may seem unusual but who are also gifted. Such children may share specific or varied interests. Often, their friendships are discouraged, ridiculed, or laughed at, and further collaboration and common discoveries become impossible. For the gifted, the world is not divided into small subjects, or areas, or individuals. They learn concepts that apply to many subjects or individuals at the same time.

Gifted children are often daydreamers. Ideas and inspiration come to them at such moments. This trait is often misunderstood, even frowned upon, by the adults who observe what seems to be lack of interest. Nonetheless, gifted children need opportunity for making choices and for working with a variety of adults who may be experts in specific fields. For gifted children need the freedom to complete tasks, which they are passionately pursuing, rather than be disrupted to concentrate on something else. As such children grow older, they live more and more with the pressure of their multiple talents and interests, which cry out for creative expression in communication and discussion as well as more traditional avenues. Few such opportunities exist. If art, music, drama, dance, and other artistic expressions are considered frills in a school environment, have low priority, and are cut out as soon as money is tight, gifted children with their varieties of intellectual appetites and abilities may lose twice. These children will lack not only the opportunity to grow and to fulfill their potential but also lose interest and become bored or disenchanted with the particular type of "learning" expected in the school.

Why do children in general seem to lose their motivation for learning by the time they are about eight years old? What has happened in the years since the child was, in a sense, driven to learn at birth? What experiences created this change? Could it have been caused by the fact that the goals and methods of the child are at cross-purposes with the goals and methods of traditional teaching and learning? Instead of investigating this recurring phenomenon of loss of motivation, present practices in schooling seem to accept it as natural. Such practices seem to replace natural motivations with artificial ones such as grades, rewards, punishment, disappointment, and competition.

Gifted children seem to suffer greatly from the discrepancy between their expectations and those of society. They feel misunderstood, deprived of opportunities for growth; they may be stifled. They may react with depression, aggression, and misbehavior. On the other hand, in programs where approaches support gifted children's needs and methods, where concept seeking and concept forming through inquiry are central in instruction and in curriculum, where the study of the world in its great variety is the subject, children are motivated, knowledgeable, and learning. For such an environment, we must strive. For in children—and what we do to conserve their being—lies our future. Does it not?

Appendix

"You cannot bump your head on water, but you can on ice"

On a beautiful, crisp fall day, I decided to take a walk with the children. As usual, this suggestion was received with great enthusiasm. One of their favorite paths leads toward a shallow pond we call "The Lost Lake," which lies hidden in the bushes. This is where we went, as we had done on previous occasions. The moment we arrived, one of the boys remarked, "It looks different today. There is ice on it." Another child picked up a stone and threw it on the ice. Since there was only a thin layer of ice, the stone broke through. "How come there is ice on the pond?" asked one of the children. Another answered that it was because the winter was coming, and it was getting cold.

Then I asked, "Who knows what happens to water when it gets very cold?" Many of the children shouted, "It turns into ice!" Then a child asked, "How come the ice broke when Bobby threw a stone?" Here, another child said excitedly, "Last winter, I went on some ice, and it didn't even break, and my brother went skating on it." I asked if it were winter now. A child answered, "No, it is only fall. In winter, you have to wear a snowsuit all the time you are outdoors, and more and more water turns into ice, and it gets thicker and thicker, and finally you can walk and skate on it."

For a while, the majority of the children continued to concentrate on the ice, picked it up, held it, broke it, and commented on the fact that it was cold. The whole conversation did not take more than a few minutes. Every morning thereafter, the children noticed the puddles covered with ice.

After a few days, I said to a group of boys, who were stepping on a layer of ice which had formed on a puddle, "What do you think would happen if we took this piece of ice into the house?"

Some said that it would melt, and others said that it would turn into water. Others thought it would become snow.

I suggested that we wait and see. They eagerly filled a pail with some pieces of ice and took them indoors. They returned outdoors to their activities. After a while, I called them in.

"Shall we check and see what happened to the ice?" I called everybody to look. The children seemed to be as full of anticipation as when they are about to find out whether the "good guy" wins on television.

"It turned into water!" they shouted.

"Do you know why?" I asked.

Most of them knew that [the ice melted] because it was warm inside [the house]. Then I asked them what would happen if we took the same pail of water outdoors and left it there overnight. There were many guesses by the children.

The next morning hardly any of the children forgot to check on the pail to see what had happened. A few days passed by without much discussion about the subject. Then, one of the boys announced during a discussion period, "The thing really works, even at my house!" "What thing?" I asked.

"I took a cup of water outside and left it there all night. When I looked in the morning, it was ice. Then, I took it in, and when I came home from school, it was water again." Several of the children said, "When I get home, I will do that, too!"

A little girl said, "My mother forgot the milk outside, and it turned to ice and broke the bottle."

Then I asked them, "Do you know what an 'experiment' is? It is something people do to find out something, to study something. Tomorrow, I will do an experiment with you. We are going to heat some ice in a pot over an electric burner, and we will see what happens."

The next day, the experiment was carried out. [Before it], I told the children to guess what might happen. The answers showed the range of concepts present in young children.

"Nothing will happen."   "It will evaporate."
"It will turn into water."   "It will turn into steam."
"It will turn into snow."   "It will go away."

They watched with rapt attention. They saw the ice turn into water, saw how [the water] began to boil, and finally [saw the water] become steam. Some remembered having seen this at home when someone was cooking.

[A variation of] the experiment was repeated several days later; this time, a cold glass was held over the steam. The children saw the droplets of water that formed; in other words, [they saw] how the steam turned back into water.

In the meantime, winter had arrived. The children were taken on another walk to the Lost Lake. This time, it was frozen solid. They ran and played on it.

When one of the children fell, I started a discussion of the different consistencies of ice and water.

"You can slip on ice but not on water."

"You skate on ice and swim in water."

"Ice is hard; water is not."

"You cannot bump your head on water, but you can on ice."

All these concepts were developed mainly by the children.

More "experiments" were made [to investigate] the same general subject. "I am going to put a toy car into this pail of water," I said. "Can you take it out of the water?"

All the children tried and found that they could.

"Now tomorrow morning when you come back, you will not be able to pull it out." I left the pail outdoors overnight where it froze and was covered with ice, and the children found they could not remove the car.

Another day, the children had a special treat. They took their juice and crackers outdoors. When they were finished eating and drinking, I burned the paper cups on the ice. The children could see how the fire melted the ice, and how we then used the water to put out the fire. "If you want more water, do you have to melt more ice to get it?" asked one of the children. Another experiment was carried out the next day to answer this question.

One day, a child reported that he saw a truck pouring salt on the ice streets. We poured salt on a block of ice to find out what would happen. Most of the children asked their parents to repeat this experiment at home. One little girl arrived at school holding a huge piece of ice, which had a hole in the middle made by salt. She had taken the ice all the way from home on the school bus; she was proud.

The climax of the project came when we took hot maple syrup outdoors and poured it on the snow. It hardened and made delicious candy. The children were delighted, and their comments showed that they understood the process.*

---

*This excerpt from Roeper and McCloud (1965/1984) describes a typical first probe, made at the Roeper School by approximately three-to-five year olds, into early levels of understanding in science.

References

Bergen, Doris. (1987). *Play as a medium for learning and development.* Portsmouth, NH: Heinemann Educational Books.

Brandwein, Paul F. (1962). Elements in a strategy for teaching science in the elementary school. In *The teaching of science* (with Joseph J. Schwab) (pp. 105–144). (The Burton Lecture at Harvard University.) Cambridge, MA: Harvard University Press.

Brandwein, Paul F. (1987, September). On avenues to kindling wide interests in the elementary school: Knowledges and values. *Roeper Review, 10*(1), 32–41.

Erikson, Erik H. (1963). *Childhood and society.* New York: W. W. Norton. (Original work published 1950)

Freud, Sigmund. (1953–1974). J. Strachey (Ed. and Trans.), *The standard edition of the complete psychological works* (24 vols.). London: Hogarth Press. (Original work published 1923)

Kohut, Heinz. (1977). *The restoration of self.* Chicago: University of Chicago Press.

Miller, Alice. (1981). *The drama of the gifted child.* New York: Basic Books. (Originally published 1979 as *Das Drama des begabten Kindes [Prisoners of childhood]*)

Murphy, Gardner. (1961). *Freeing intelligence through teaching.* New York: Harper and Bros.

Poole, Linda, and Line, Laura. (1979, March). An open classroom approach. *Science and Children, 16*(6), 34, 39.

Roeper, Annemarie. (1976, December). Normal phases of emotional development and the gifted child. *Roeper Review, 12*(4), 2–4.

Roeper, Annemarie. (1986). *What I have learned from gifted children.* Paper presented at a conference of the California Association for the Gifted, Oakland, CA.

Roeper, Annemarie, and McCloud, Marian. (1965). *Physical science for young children: A guide for the teacher.* Bloomfield Hills, MI: Author. (Abridged edition published 1984)

Roeper, Annemarie, and Roeper, George. (1981). *Philosophy of Roeper School.* Unpublished manuscript.

Roeper, Annemarie. (1988). *Education for life: Fostering self-actualization and interdependence.* Monroe, NY: Trillium Press.

Scarr, Sandra W., and McCartney, Kathleen. (1983, April). How people make their own environments: A theory of genotype-environment effects. *Child Development, 54*(2), 424–435.

The Gifted in Middle School Science

Steven J. Rakow

Like the middle child in a family, the middle school is struggling to find its identity. The baby of the family (the elementary school) and the firstborn (the high school) have their places. The middle school is considered by some an extension of the primary years and by others a training ground for secondary education. Rarely have middle schools achieved that unique blend of philosophy and program for which they were originally conceived. This is not to say that such a philosophy does not exist. Essentially, the emphasis is on programs planned to meet "physical and cognitive, affective and social needs of 10 to 14 year olds" (Hurd, Robinson, McConnell, and Ross, 1981, p. 4). These authors emphasize the importance of balancing the psychomotor and affective domains as well as tending to the cognitive domain in middle school curriculum and instruction.

In general, however, the schooling and education of the gifted in the middle school have been neglected. Accusations of elitism have hounded efforts to develop programs for these students. Mistaken notions that gifted children will succeed no matter what is (or is not) done for them have served as the rationale for cutting the funding for development of innovative curricular and instructional models adapted to their needs. Marsha M. Correll (1978) presents the following justification for the existence of programs for the gifted:

> A universal system of education is ultimately tested at its margins. It functions fairly well in educating most students in the middle or normal range of abilities, but it has a tendency to be less effective with exceptional groups with either high

or low abilities. Beginning in the late sixties, education policymakers in the United States committed themselves to the improvement of instruction for millions of previously neglected students through legislation aimed at the disadvantaged and handicapped. Less emphasis has been placed on improving instruction for those who possess exceptional abilities or talent. These are the "extraordinary few" whose lifework will improve their disciplines, their societies, and perhaps humankind. (p. 7)

It is rare to find an effectively differentiated science program for gifted middle school students. Paul deHart Hurd and others write, "Rarely is it possible to have special classes for very slow or talented students in science" (Hurd et al, p. 12). However, the characteristics of an exemplary science program for the middle school have been identified by the NSTA Task Force on Excellence in Middle/Junior High School Science Programs (Reynolds, Pitotti, Rakow, Thompson, and Wohl, 1984). This Task Force defined an exemplary middle/junior high school science program under five domains: goals, curriculum, instruction, teachers, and evaluation. Later (1985), I surveyed a group of middle/junior high school science teachers to collect their perceptions of an exemplary middle school science program. Making use of the same five categories, these teachers described a science program that corresponded well to the description of the middle school program given by Hurd and his colleagues. Some examples of the teachers' responses to the first three categories follow:

The GOALS of an exemplary middle/junior high school program will
• meet both the social and emotional needs of middle/junior high school students
• generate interest in science through activities that are meaningful to the students
• be developed and taught at a cognitive level appropriate to the students being taught
• broaden science education for everyone
• develop students' problem-solving skills
The CURRICULUM of an exemplary middle/junior high school science program will include
• an approach in inquiry that offers opportunity for problem solving, data recording, hypothesizing, predicting and concluding, rooted in hands-on inquiry
• students' use of various tools and instruments of science
• the introduction, explanation, and demonstration of new terms and vocabulary, which are relevant to the subject
• the use and application of science information by students within a cultural/social environment
• a multifaceted focus with the involvement of local community resources and current societal concerns
• the meeting of the needs of students of all abilities
• unified and sequential curriculums for grades K–12, which include an interdisciplinary focus
The INSTRUCTION in an exemplary middle/junior high school science program will

- use "motivators" to generate interest (e.g., current events that relate to the students' experience, discrepant events, etc.)
- have an organized and structured format
- be student centered, not teacher centered
- focus on inquiry including investigation and experimentation, various levels of questioning, the use of wait-time, and a consideration of the scientific method
- move from the concrete to the abstract (Rakow, 1985, pp. 632–633)

This report, among others, paints a picture of the middle school as a place for active inquiry that takes into account the unique characteristics of middle school students and develops a program to engage their interest and abilities. Such are characteristics of an exemplary middle school program for the gifted as well.

The Middle School Student

While the phrase "middle school" is often used to refer to an administrative structure, it also refers to the particular age group of students that middle school serves—10 to 14 year olds. What are the unique characteristics of these learners?

Physically, middle school students are developing at a rate faster than at any time in their lives except infancy. This rapid growth may bring with it clumsiness and poor coordination. Often the young compound their anxiety; growth rates among classmates might differ by several years, leading some children to wonder if they will ever catch up with their rapidly maturing classmates. Early adolescents tend, therefore, at times, to be very self-conscious and may become "behavior problems."

Socially, middle school children seek to establish their own identity and their own self-concept. As they seek to "fit in," their peers become an increasingly important influence on their lives. Along with this newly established identity is the need to make their own decisions. Often, middle school students are reluctant to act in accordance with the demands of their parents or those of other adults. While this stage of development may bring a certain anxiety to parents and teachers, it is an indication that the early adolescent is developing a greater capacity to conceive of alternative approaches to solving personal problems.

Middle school classrooms contain a fascinating array of cognitive abilities. Bärbel Inhelder and Jean Piaget (1958) provide one model that attempts to describe the intellectual development of the young.* They suggest that individuals progress through four states of intellectual development: sensori-

---

*Piaget's years of studies are the basis for the combined work with Inhelder cited here.

motor (ages birth to 2 years), preoperational (ages 2 to 7), concrete operational (ages 7 to 11), and formal operational (ages 11 to 14). In brief, Inhelder and Piaget propose that, at different stages of cognitive development, people have a repertoire of patterns of reasoning—identifiable and reproducible thought processes directed at a narrow task—which individuals may apply to solving problems. The "concrete operational" student will be able to carry out logical thought processes using concrete objects, while the "formal operational" student will be able to engage in the manipulation of purely mental constructs. Several other differences also distinguish these two categories, such as the ability of "formal operational" students to use proportional and probabilistic logic patterns, to consider the influence of many variables simultaneously, and to examine many possible combinations of a group of objects. Inhelder and Piaget's conclusions about patterns of reasoning could have implications for the kind of cognitive demands that we place upon students in science classrooms.

Subsequent research, however, has questioned Piaget's formulations, particularly as applied to specific age groups. It has been suggested that an individual's ability to approach a problem may be as much a factor of experience as of his or her generalized cognitive development. Thus J. Paul Guilford (as early as 1956), Robert H. Ennis (1976, December), and Marcia C. Linn (1982, December) have criticized Piaget's empirical approach to his findings. They have also questioned the use of particular terms to identify stages of development and the emphasis on the cognitive aspects of development, without an appraisal of the affective ones. Moreover, studies by Daniel P. Keating (1973) and Roger A. Webb (1973) demonstrate that gifted middle school students often operate at a pace of cognitive development ahead of that projected for the "patterns." Gifted students advance rapidly in their cognitive development and, therefore, often appear quickly to go beyond "operational skills." This ability can make them early able to use the higher-order thinking skills required in scientific inquiry.

In any case, successful middle/junior high schoolteachers need to be aware of and sensitive to the physical, emotional, affective, and cognitive characteristics of their middle school students. Because of the heterogeneity of abilities and physical characteristics, and because of the rapid rate at which these early adolescents are developing, the identification of special ability is particularly difficult.

Finding Gifted Middle School Science Students

Simple reliance upon global measures of intelligence or ability is misleading, because individuals may possess a potential for giftedness in areas not readily assessed by these instruments. Personal recommendations by teachers or school staff members are also limited by a basis that tends to equate "bright and polite" with "gifted and talented," often missing the contributions of the unconventional but sometimes brilliant maverick. One of the more widely accepted definitions of giftedness is that proposed by Joseph S. Renzulli (1979). He suggests that gifted students are those who are at the intersection of three traits: *above-average general ability, task commitment,* and *creativity.* While this definition helps to refine our notion of giftedness as a general characteristic, it still fails to address the specific potential of students gifted in science.

Certain studies have begun to isolate characteristics which appear to be exhibited by gifted students in their approaches to the particular aspects posed in scientific inquiry. Some of these characteristics are based on the personal characteristics of successful scientists. One should, however, heed the warning of Gerard F. Consuegra (1982a) in using such lists:

> One must keep in mind that any list of characteristics of gifted students relates in a general sense to the designated group. Certainly there will be individual students who do not match all the characteristics. The implication for educators is to use a list of characteristics to gain a holistic sense of the students as a group. (p. 2)

In answering the question—"Does your child have scientific ability?"—the (now defunct) Office of Education for the Gifted and Talented of the U.S. Department of Health, Education, and Welfare (1978) listed six identifying characteristics:

1. Good motor coordination
2. Devotion to investigations and personal projects
3. Persistence despite failure
4. Interested in cause-and-effect relationships
5. A tendency to read science-related materials
6. Enjoyment of scientific discussions (as cited in Consuegra, 1980, p. 12)

For the middle school, Consuegra (1982a) identified three characteristics for those who show potential for giftedness in science. These characteristics are a questing nature, personal drive, and an investigative nature. A *questing nature* asks about the world and accepts answers only when they seem to have scientific validity. *Personal drive* leads to the willingness to spend long hours conducting routine and repetitive experiments on topics of interest. An *investigative nature* is willing to conduct investigations despite failure. The characteristics identified recently by Consuegra in middle school students are close both to personality traits and work habits typical of successful scientists and to those traits Paul F. Brandwein (1955/1981) found in gifted students of high school age. Brandwein identified four constellations of traits related to the investigative skills in science that his students (a sample of 354) exhibited. These characteristics were a high level of verbal and mathematical ability, adequate sensory and neuromuscular control (constituting, largely, a *genetic* factor); a factor termed *persistence* (a willingness to spend time in investigative work beyond ordinary schedules, to withstand discomfort and fatigue in pursuit of objectives, and to face failure); and a factor termed *questing* ("a notable dissatisfaction with present explanations of the way the world works" [p. 10]). All these were interdependent with an *activating* factor—a key teacher and an environment combining to give access to equal opportunity for individual work in investigation.

This match in the traits posited by these two investigators for the middle school and the high school respectively may indicate that programs for the science talented at these two levels may be readily articulated in curriculum and instruction. But we are at the beginning of the preparation of instructional programs that build on the experience of the gifted young in the middle school: a critical period in their identification, instruction, and guidance.

Selection Procedures: *External Methods*

While lists of personality and behavioral characteristics may be of very limited use in defining the potentially gifted in science as a group, many programs for middle school students presumed to be talented in science depend upon an initial means of "identification." Generally, a three-step procedure of *screening, identification,* and *selection* is suggested as an appropriate procedure. (As we shall see, however, another method may set this procedure aside.)

Screening involves the use of standardized achievement tests (such as the Iowa Test of Basic Skills, the Stanford-Binet test, or the Wechsler Intelligence Scale for Children—Revised), or teacher, parent, or self-nominations. Although group tests have the potential to be misleading, they do serve as a screening mechanism *when used with other procedures.*

A second level of identification tries to pinpoint the talents of middle school science students. Some schools use instruments such as the Test of Science Related Attitudes (TOSRA) or the Test of Integrated Process Skills (TIPS) to identify specific science-related skills. These tests seem to be less dependent upon a student's particular exposure to science content than is evident in standardized science tests (such as Educational Testing Service's Sequential Tests of Educational Progress [STEP]). TOSRA and TIPS attempt rather to measure the student's ability to use science-related thinking patterns. The recommendations of teachers who have been prior observers of the students' specific performance in science are useful here.

The third level of identification is *selection.* At this point, the school staff should work with individual students and their parents to develop an educational plan to meet specific interests and needs. This process may take several forms, from individualized research to in-class enrichment. Its direction depends in part upon the students' needs and wishes and in part upon the school system's resources.

The Council for Exceptional Children (1978) has suggested three questions that parents and teachers might use as devices to identify potentially gifted middle school science students for purposes of the selection procedure. They are

1. Is the child bored by in-grade science and math material?
2. Does the child score at or above the 95th percentile on a standarized science and mathematics test?
3. Does the child seem eager to learn more and at a faster pace? (p. 30)

Because middle school children are such a rapidly changing, heterogeneous group, the use of standardized tests with arbitrarily chosen numerical criteria is an unsatisfactory method of identifying the gifted among them. Unfortunately, as the previous cautionary note from the Council for Exceptional Children points out, reliance on standardized tests is a procedure used by too many school systems. Instead, giftedness in science should be assessed by the students' ability to "do science." Perhaps the most reliable measure of the students' ability to do science is the ability to use the investigative processes of science to solve problems arising from the students' own questions. Still, there is clearly a need for a specific and reliable method to assess the ability of individual students to conduct scientific investigations.

Performance

One such method does not depend on testing per se. It relies on observations of demonstrated work in science by the young in the elementary and middle school years. It attempts to develop methods of self-assessment, of self-selection for *all* students in a given school. In effect, those students interested in science identify themselves through their competence in science *and* their abilities in the investigative modes of inquiry; that is, they select themselves.

The Vista Verde Year-Round Elementary School in Irvine, California, probed such a device at early middle-school grade levels in the early 1980s.* Essentially, all fourth- and fifth-grade students in heterogeneous groups were involved in elementary science study based on conceptual approaches and rooted in laboratory activity. Certain aspects of the laboratory were open to individual work on problems selected by students who wanted the experience. In the 6th and 7th grades (students of 11, 12, and 13 years of age) were given an opportunity to select science programs grouped under the rubric

---

*Principal Barbara Barnes headed the program.

VOPIS (Vocationally Oriented Program in Science). The program was freely chosen, *and it was open to all those with an interest in science as a possible career no matter their "ability."* The program, centered richly in science and technology, naturally required a decent level of abilities in mathematics (as a tool of science), in English (as a tool of science in communications and writing of reports), and an ability to carry through individual work in investigation (problem solving). Students who found the work too demanding were allowed free exit to the "regular" course work. They could return at the beginning of a new semester if they were so inclined. (On this, see *The Revolving Door Identification Model* [Renzulli, Reise, and Smith, 1981].)

Early assessment indicates that there were at least two classes in the eighth grade who were of exceeding competence in science. That is, they were competent in *acquiring* scientific knowledge and skilled in *inquiry* by the scientific mode (including the laboratory and computer). In brief: There was no need to identify the young at the early age of 9 or 10; after all, children do develop at different rates. The entire elementary school could then be utilized to permit children to "sort themselves out." The key: a rich program that appeals to the young so that they may *select themselves* without threat or penalty of "failure."

There is, however, clearly a need for specific and reliable methods to assess the ability of individual students to conduct scientific investigations. We turn next to various programs that offer such opportunities.

Environments That Promote Expressions of Giftedness

Science programs designed especially for gifted middle school students occur in a variety of locations, ranging from the classroom of a teacher who successfully challenges his or her one or two gifted children, to summer camp programs that involve groups of gifted children interacting with each other and the environment. Jay McTighe (1979) has identified five criteria that serve the needs of the science talented at the middle school level. He suggests,

1. Emphasis on methods and processes of scientific investigation
2. Exposure to topics and concepts that interest and challenge the highly able
3. Actively engaging in "discovery-oriented" student-designed experimental studies
4. Sharing acquired knowledge with others
5. Observing and working with practicing scientists (p. 26)

One of the oldest public school programs designed exclusively for the gifted is Pine View in Sarasota, Florida (Woolever, 1979). The program serves students in grades 4 to 12. A typical 4th grade science class meets 8 times weekly (40–50 minutes) with double periods for lab on alternate days. Field trips and hands-on activities form the heart of the program.

Summer programs for the gifted are also popular. The Summer Discovery Institutes in Warren County, Ohio, provided the opportunity for 90 gifted students from grades 5 to 8 to study water pollution (Hern and Hern, 1979). Each student identified a particular problem and spent the week investigating it. They were motivated in part by the knowledge that on the fifth day they would present their work to their classmates.

Another summer program, which was sponsored by the Maryland Department of Education, was held at St. Mary's Center (McTighe, 1979). This two-week program for students in grades six through eight allowed the students to choose to investigate two of the following areas: artistic and creative expression, problem solving, historical and cultural exploration, and environmental studies. In the latter session, students developed an original research project.

Programs for the gifted are also found as part of the regular school program, but sometimes at unconventional sites. For example, the Clear Creek Independent School District (near Houston) has a special program for gifted students in grades four to six. Students identified as gifted by means of a combination of "objective" tests of general "ability," along with recommendations, have the opportunity to select several courses, one for the fall and one for the spring. These classes meet at the University of Houston—Clear Lake, to which students are bused from their individual schools. Course offerings in science include science skills, physiology, earth science, marine biology, field ecology, and environmental history. The last three courses are taught at the Armand Bayou Nature Center adjacent to the campus.

Another program that makes use of an unusual location is the Science Magnet School Zoo Component in Buffalo, New York (described in Rakow, 1984). This program for students gifted in science, cited by NSTA as exemplary through the Search for Excellence in Science Education, uses the local zoo as the site for study and research. Students work with the zoo staff both in the general day-to-day functioning of the zoo and on specific research projects.

The Montgomery County (Maryland) Public Schools recently offered an extensive three-year science project for gifted middle school students (described by Dragoo, McCullough, and Emmanuel, 1980). Funded under Elementary and Secondary Education Act Title IV-C, the project built resource banks in the county schools to enrich the science studies of individual students in grades three to eight. Students identified by screening and interviews as potentially gifted would meet with an adult counselor, who would use the Project Resource File to locate materials for the students and make arrangements for use. A list of possible resources gives some indication of the breadth of experiences available to these students, as well as ideas for those who would develop their own programs: government agencies; corporations, factories, and industries; state or local academies of science; libraries; environmental groups; organizations with health concerns (cancer, heart, lungs); water research stations; weather stations and meteorologists; speakers' bureaus; 4-H nature centers; scout leaders; museums; military organizations; airports; teachers (retired or active, at all levels); science fair judges; state departments of education (p. 10).

This review of the educational environments supporting giftedness gives some indication of the range of programs useful in stimulating the interest and intellect of scientifically gifted students. Whether in a classroom setting or an informal learning center, whether during the school year, on weekends, or during summer vacations, middle school students should have the chance for ample opportunities to stretch themselves to the limits of their abilities in the area of their interests. Surely, students' freedom to quest should extend to science in a world dependent upon it and upon technology.

In Reflection: Certain Personal Experiences

It is 9:15 a.m. as I sit in my laboratory awaiting my new group of gifted middle school science students. I wait with some apprehension because I am never sure what to expect from them. If I have learned anything after three years of teaching gifted students, it is that they are characterized more by their differences than by their similarities. Each is unique. This makes teaching them exciting. It also makes identifying and neatly classifying them nearly impossible.

Whom did I expect to come through the door? As I think back upon the gifted middle school students I have taught, many faces come to mind.

I remember the shy girl from Taiwan. She could barely tell me her name on the first day of school. Getting her to speak in class was difficult. But she participated enthusiastically in the laboratory activities and seemed to have better-than-average manual dexterity. Soon her success with lab manipulations gave her the courage to become a leader in the group, and, by the end of the semester, she was actively joining in the class discussions. Her growth over 12 weeks was amazing.

I also think of a boy with those special problems that test a school in its basic thrust: to meet the needs of all young. On TIPS, he earned the highest score of any student that I have taught. Overweight, with poor social skills, he lived for science; he used his knowledge to compensate for his perceived inability to relate to the other children in a social setting. And, within an environment designed for the gifted, he was accepted. He brought technical skills and knowledge that the group needed to succeed. I believe that the program contributed to his growth.

And I remember others: a boy who would leave my class to practice for his violin concerto with the local symphony; a girl, younger than most of the others, who while intelligent was socially immature; their classmates, each different and each special. How could I possibly begin to challenge and excite the imaginations of each of these individuals? Was there a way?

Over the years, administrators, parents, and faculty developed a program in my city that seems to work. Middle school children from the 14 elementary schools in our local school district are identified as gifted on the basis of a number of standardized tests and personal recommendations; these students are then eligible to participate in a pull-out program. Each Wednesday, for four hours, they leave their home schools and are bused to the University of Houston—Clear Lake, where they participate in the program for the gifted. There, each semester, they have the option of participating in one of many courses such as geology, physiology, problem solving in math, computers, writing, archaeology, cultures and communications, and, at a local nature center, marine biology, ecology, and environmental history. The children select on their own their particular area of interest for in-depth study. Thus, students who have a keen interest in science are able to select from a number of science courses to develop that interest.

One such course, designed to teach the students the process skills that scientists use while carrying out their investigations and experiments, is called "Science Skills." An intensive, hands-on laboratory course, it offers activities chosen to illustrate a variety of concepts and science disciplines. The common thread is development of process skills more than content. Each week a process skill is the central focus of the day's activities.

A typical day will have the student systematically identifying substances using techniques based on a variation of the Elementary Science Study unit "Mystery Powders." Another day might find groups of students building towers with 25 drinking straws and 10 straight pins to see who can build the highest tower. After this contest, each student receives 50 straws, 25 pins, and the challenge to build the strongest bridge in the class. Visit on another day and students might be designing experiments to compare brands of paper toweling to determine which is the best buy. This day closes with each group's presentation of a "commercial" to promote the best brand.

What makes a good lesson for gifted middle school children? I believe that it has these characteristics. The lesson
- focuses on the processes of science
- encourages the students' problem-solving skills
- is active, hands-on oriented
- deals with concepts and materials that are relevant to the students' lives and are of interest to them

- challenges the students' abilities in origination and innovation (sometimes called creativity)
- involves the students in sharing with each other

While the lesson is an important ingredient in developing a stimulating program for middle school gifted students, at least two other factors are also essential—the environment and the teacher.

The environment where giftedness may be expressed is not only physical but also social. I am fortunate to have a beautiful teaching laboratory with state-of-the-art lab tables, equipment, and supplies, but these aren't mandatory. To capture the students' imaginations, the room should be inviting and stimulating. There should be attractive posters on the walls. Live animals in aquaria or terraria, materials for exploration, and books for reference all contribute to a physical environment that spurs the children's interest and curiosity.

The social environment is also important. Many gifted children have difficulty with social skills. They have always been able to do much by themselves and sometimes find it difficult to muster the patience to work with another person. Many of them have been taught that asking anyone for help is a sign of weakness and that they should be able to figure things out alone (perhaps this message comes from their regular classes, where a gifted student is often given a project to complete alone). The bottom line is that many of the students with whom I work simply do not know how to cooperate with others.

For this reason, developing a cooperative climate is a primary concern from the first day on. Before the students come to class, I have divided them into five groups of four trying to maximize heterogeneity. Boys and girls are mixed together as are students from different schools. Each student gets a name tag with a group color and a shape indicating his or her specific job.

The development of cooperative skills involves helping the students to understand the importance of working together. We spend quite a bit of time the first day discussing why sharing with others is useful. One technique that appears to be especially effective in helping the students to share responsibilities in the group is the assigning of specific roles. I divide duties into four primary areas: principal investigator, recorder, materials manager, and checker. These roles are rotated weekly so that each student does each job several times during the semester. Each role has specific responsibilities that must be carried out for the group to function effectively. For example, the principal investigator is the group leader. This student manages the group's work. The recorder completes the data sheet and turns it in to the teacher. The materials manager gets the laboratory materials and resource materials and directs the cleanup. The checker monitors the involvement and understanding of each group member so that no one is left out. Research by Roger T. and David W. Johnson (cited in this volume, 1975, and elsewhere) has demonstrated the effectiveness of using a cooperative learning strategy to increase the achievement, motivation, and socialization of students of all ages. Along with these noble goals, I like the cooperative approach because the students work more efficiently and effectively in their groups when they have had this exposure to group social skills.

The final ingredient is the teacher. Obviously, teachers need to have a good background in science but as important as their knowledge is their attitude toward science. Teachers are important role models for young students and, as such, need to exhibit an inquiring attitude. Teachers should be willing to say, "I don't know. How can we find out?" if students are also to do so.*

Conclusions

The middle school years are a time of rapid physical, social, cognitive, and emotional change. Teachers of this age group face a heterogeneous population that not only has vastly differing characteristics but also is rapidly changing. These two conditions make it difficult to identify qualities of giftedness; they may be hidden by the complex changes that these children are undergoing. The need to nurture differences is critical, however. It is during the middle school years that some of the most important lifelong attitudes and perceptions are formed. We cannot afford to neglect the full development of even one student simply because of the difficulty of identifying special needs.

Middle school science programs in general can go a long way to promote the expression of talent in science, and programs that embrace ideas promoted by the Search for Excellence in Middle/Junior High Science will engage students' minds and attentions, gifted or not. These hands-on, problem-centered programs will help all students to recognize the central role that science plays in the lives of people and nations. But first and last, students will see science as a mode of self-expression and conduct, and a way of contributing to the advancement of the culture of which they are a part.

References

Brandwein, Paul F. (1955). *The gifted student as future scientist: The high school student and his commitment to science.* New York: Harcourt Brace. (1981 reprint, with a new preface [Los Angeles: National/State Leadership Training Institute on the Gifted and Talented])

---

*Science programs in middle and junior high schools can make effective use of the support offered by science centers. On this, see Sid Sitkoff's paper.

Consuegra, Gerard F. (1980). *Education for the gifted in science and mathematics*. Raleigh: North Carolina Department of Public Instruction. (ERIC Document Reproduction Service No. ED 199 938)

Consuegra, Gerard F. (1982a, November). *Developing gifted programs [sic] in science*. Paper presented at the regional NSTA meeting, Baltimore, MD.

Consuegra, Gerard F. (1982b). Identifying the gifted in science and mathematics. *School Science and Mathematics, 82*(3), 183–188.

Correll, Marsha. (1978). *Teaching the gifted and talented*. Bloomington, IN: Phi Delta Kappa Educational Foundation. (Fastback Series No. 119)

Council for Exceptional Children. (1978). Fact sheets on the gifted and talented. Reston, VA: Author. (ERIC Document Reproduction Service No. ED 179 035)

Dragoo, Mary Carol, McCullough, Nancy, and Emmanuel, Elizabeth. (1980). *Gifted science project: Final report*. Rockville, MD: Montgomery County Public Schools. (ERIC Document Reproduction Service No. ED 226 564)

Ennis, Robert H. (1976, December). An alternative to Piaget's conceptualization of logical competence. *Child Development, 47*(4), 903–919.

Guilford, J. Paul. (1956, July). The structure of intellect. *Psychological Bulletin, 53,* 267–293.

Hern, Dennis E., and Hern, Ann. (1979). Summer discovery program for academically gifted students in grades 5–8. *Clearing House, 52*(9), 441–444.

Inhelder, Bärbel, and Piaget, Jean. (1958). *De la logique de l'enfant à la logique de l'adolescent*. Paris: Presses Universitaires de France.

Johnson, Roger T., and Johnson, David W. (1975). *Learning together and alone*. Englewood Cliffs, NJ: Prentice-Hall.

Keating, Daniel P. (1973). Precocious cognitive development at the level of formal operations. Unpublished doctoral dissertation, Johns Hopkins University, Baltimore, MD.

Hurd, Paul DeHart, Robinson, James T., McConnell, Mary O., and Ross, Morris M., Jr. (1981). *The status of middle school and junior high school science: Summary report to the National Science Foundation* (Vol. 1). Louisville, CO: Center for Educational Research and Evaluation. (Biological Science Curriculum Study)

Linn, Marcia C. (1982, December). Theoretical and practical significance of formal reasoning. *Journal of Research in Science Teaching, 19*(9), 727–742.

McTighe, Jay. (1979, March). The summer program at St. Mary's Center. *Science and Children, 16*(6), 24–26.

Rakow, Steven J. (1984, February). Excellence in m/jh science: Science Magnet School zoo component. *Science Scope, 7*(3), 18.

Rakow, Steven J. (1985). Excellence in middle/junior high school science: The teacher's perspective. *School Science and Mathematics, 85*(8), 631–635.

Renzulli, Joseph S., Reis, Sally M., and Smith, Linda H. (1981). *The revolving door identification model*. Mansfield Center, CT: Creative Learning Press.

Renzulli, Joseph S. (1979, March). What makes giftedness: A reexamination of the definition. *Science and Children, 16*(6), 14–15.

Reynolds, Karen E., Pitotti, Nelson W., Rakow, Steven J., Thompson, Thomas, and Wohl, Sandy. (1984). *Excellence in middle/junior high school science programs*. Washington, DC: NSTA.

Webb, Roger A. (1973). *Concrete and formal operations in very bright (IQ greater than 160) six- to-eleven-year-olds*. Baltimore, MD: Johns Hopkins University Press.

Woolever, John. (1979, March). Pine View's departmentalized program. *Science and Children, 16*(6), 30–33.

Equality, Equity, Entity: Opening Science's Gifts for Children

Bill G. Aldridge
Deborah C. Fort

Equality, equity, and entity—American goals as old as the Declaration of Independence—but hardly self-evident and by no means always realized: witness the Civil War and the end of slavery followed by segregation, the women's movement triumphant for suffrage but now in the service of fair treatment more broadly defined, and Martin Luther King, Jr.'s—among many others'—need to *dream* of personal, philosophical, professional fulfillment. While all teachers frequently face these concepts and their opposites, science teachers meet them in a special way and with special urgency. Equality—and differences, equity—and injustice, and entity—and compromises are all present in our classrooms. We aim to give an *equal* shake to *different* children; we try to help the young develop *individually* and as responsible members of family, group, and society, while they also learn to compromise; we strive for justice, but our students must learn that unfairness also exists.

In any event, adherence to the principles of equality, equity, and entity takes on paramount, even superordinate, importance in elementary schools, where love for all kinds of learning can be born. Or, if the school is inadequate to nurture children in the spirit that greatness in teaching requires, they may be lost. For, even with the necessary resources that few schools have in this present political climate, remediation, even extensive remediation, may not be enough.

After briefly tracing some of the sources of disadvantage in this land of plenty, we will try to set a course where equality, equity, and entity can flourish. We shall describe communities that nourish classrooms with broad curriculums in safe and friendly environments that encourage the discovery of the wide-ranging skills that may become "science." In so doing, we are fortunate to be able to inform our discussions by drawing upon the work of the scientists and teachers who have contributed to this volume.

But first, some definitions.

Equality

We are using equality* to mean *access to equal opportunity* in schooling—not a simple concept at all. It might at first seem that merely to give every student the same chance at the same materials and the same instruction would offer access to equal opportunity. But such a "fair" distribution does not bring true equality because every child does not bring the same genetic and environmental makeup to school. Different children come from different *educational* milieux that comprise their socioeconomic and familial preparation as well as their schooling. Thus, we must distinguish between the entire educational configuration of children's learning, personally, at home, in society, and their *schooling*—the instructed learning they do through the agencies of schools. Because educational and genetic backgrounds differ, we cannot achieve equality through simple division of resources into equal parts—one per child. One of our goals in the service of equality of opportunity we have borrowed from Hippocrates to at least do no harm. We at least must not *increase* the gap between opportunities.

To narrow this gulf, children from disadvantaged educational backgrounds need extra help in schooling. So do children with gifts. As R. Haskins and James J. Gallagher (1981) put it and Gallagher elaborated (1986), we need to offer "unequal treatment of unequals in order to make them more equal" (p. 234). Speaking of the need for equal opportunity for slow learners, G. Orville Johnson offers a summary useful for all students:

> *Every child has the right to an equal opportunity for an education.* This does not mean that all children will receive the same or identical educational experiences. This means that the educational experiences provided each child will be those that will promote learning for him[/her] in the best way and to the highest degree possible. (1962, p. 33)

---

*Absolute equality is, of course, not only impossible but also undesirable even if attainable. While we strive for equal opportunity, we affirm our differences. What a confusing and boring world if we all looked alike, thought alike, acted alike, and aspired alike, if we not only all could be president or compose a sonata or play goalie or bake a cherry pie but also all wanted to. Access to equal opportunity does, at least, give a certain visibility to different competencies, skills, and talents.

Equity

Equity—*fairness*—is also a complex matter. We will be using the concept here as it applies to both the macrocosm of the young's environment at large and the microcosm of their world in school. A number of thinkers have pointed out that asking schooling alone to foster equity is fruitless, because the school is only one part of education, ideally an important part, to be sure, but impotent on its own to ensure fairness, to make the unequal more equal in opportunity.

Statistics show that what the fairest, the kindest, the most competent teaching can do for most children of disadvantage is not enough. Lawrence A. Cremin documents the noble efforts of American schooling to serve *all* the children by the movement called progressivism (1966). Richard H. de Lone (1979) in *Small Futures* (alas, a perfect title) discusses the *dream* of Horace Mann, growing from Locke's philosophy and embraced positively or negatively by reformers of American "education" since our nation was born, that through schooling children of poverty can be given an equal chance at American opportunities. Schooling is often asked to reform society.

In this task, the schools must fail. Progressivism asked superhuman feats of its teachers; the Horatio Alger *myth* was not even a reality for Alger's own heroes, to say nothing of real office boys who want to become corporate presidents (Kawelti, 1968). Part of the rags-to-riches aspect of the American dream has come to include the poor hero's use of his diligent application at school to rise to wealth and power. But this feature is also a myth, and often a destructive one in terms of what schooling *can* do to fight unjust inequality. De Lone's first paragraphs define the reality all too clearly:

> Jimmy is a second grader. He pays attention in school, and he enjoys it. School records show that he is reading slightly above grade level and has a slightly better than average IQ. Bobby is a second grader in a school across town. He also pays attention in class and enjoys school, and his test scores are quite similar to Jimmy's. Bobby is a safe bet to enter college (more than four times as likely as Jimmy) and a good bet to complete it—at least twelve times as likely as Jimmy. Bobby will probably have at least four years more schooling than Jimmy. He is twenty-seven times as likely as Jimmy to land a job which by his late forties will pay him an income in the top tenth of all incomes. Jimmy has about one chance in eight of earning a median income.
>
> These odds are the arithmetic of inequality in America. They can be calculated with the help of a few more facts about Bobby and Jimmy. Bobby is the son of a successful lawyer whose annual salary of $35,000 puts him well within the top 10 percent of the United States income distribution in 1976. Jimmy's father, who did not complete high school, works from time to time as a messenger or a custodial assistant. His earnings, some $4,800, put him in the bottom 10 percent. Bobby lives with his father, mother, three brothers, and two sisters. . . .
>
> In the United States, as elsewhere, it is a penalty to be born poor. It is a compounding penalty to be born to parents with little education. It is a further penalty to be born to parents who are frequently unemployed and whose employ-

ment opportunities are limited to relatively uninteresting, dead-end jobs. All these penalties are increased still more for children in racial minorities. They are further increased for girls. Some of the penalties are immediate—poor housing, inadequate medical care; some accumulate slowly, influencing the development of adult skills, aspirations, and opportunities. Together, they produce the odds that make Bobby's probable future a vista rich with possibilities and Jimmy's probable future a small door into a small room. (pp. 3–4)*

It is *society's* job to reform society. This is not to say, however, that schools cannot be and should not be part of the overall reform. In Japan, education, long an effective way of rising socially, has become an even more prominent source of upward mobility with postwar demands for economic and industrial expertise. (See Pinchas Tamir's paper in this volume.) But our national purpose is not unified to the extent Japan's is; thus, schooling cannot alone guarantee success in this country. Kenneth Keniston (1979) and others note that remaking society to ensure a more equitable *educational* background for all children is the far-reaching means to offering a fairer stake in the future. De Lone focuses on economic redistribution as the means to this end; other philosophers suggest other roads.

Recognizing the importance of larger social and economic structures, we still point out that one of the ways in which teachers can act to help their students enjoy equal access to opportunity in schooling in general and science in particular is to offer as broad an educational environment as possible. Wide instructional and curricular approaches should help to offer students with many different strengths (and weaknesses) the most equitable chance at discovering gifts in science. And finding a gift, finding a liking, *may* open the door—even to a child of disadvantage—to the later development in high school or college of a talent in science. Paul F. Brandwein writes:

> To give opportunity not only for all children so that they may (a) reduce disadvantage in development . . . to (b) reduce disadvantage in knowledges, values, and skills and to (c) open avenues to selection of interests in a rich world of work, an elementary school curriculum appealing to the diversity of talent found in almost any culture becomes a requirement of the culture that wishes to advance the life and living of its young. (1987, p. 36)

Teachers can help to foster the equity they cannot alone guarantee, then, first by offering different kinds of attention to differently prepared students

---

*See note on pp. 209–210 for the sources of de Lone's "odds" for Jimmy and Bobby. (Note that the amounts are in 1979 dollars; in 1988, Bobby's father would make about $56,595; Jimmy's, $7,761.) Excerpt from *Small Futures* by Richard H. de Lone, copyright © 1979 by Carnegie Corporation of New York, reprinted by permission of Harcourt Brace Jovanovich, Inc.

(tending, for example, both to those whose background calls for remediation and to those whose gifts open them to wider experience) and then by providing the widest possible curriculum.

Science calls for an enormous variety of skills, as we shall see not all of them part of traditional curriculums—that is, reading, writing, and arithmetic; thus, a broad curricular approach can both serve student populations with varied strengths and weaknesses and begin to awaken science interest. Mixing students with *educational* advantages with those whose backgrounds are different may, as we shall see, also serve this purpose. In this, we endorse the findings of many researchers, as well as our own experiences in teaching both selected and mixed groups of students. The heterogeneous classroom also affirms the benefits James S. Coleman—originally strongly, then with somewhat less vigor—believed accrued from mixing children of different academic and social backgrounds (1966, 1972). More recent studies "robustly document" (Cole and Griffin, 1987, p. 28) the detrimental effect upon "low-track" students by "ability" groupings; disagreement remains about the effect of tracking upon the "high-ability" students (Persell, 1972, cited in Cole and Griffin, sees a slight improvement in their progress; others find a diminution of expression).

Finally, the National Science Foundation has recently announced its conviction that offering science to the widest variety of students possible will open some fields previously closed to women and minorities (1987, May). A report to the National Science Foundation (Knapp, Stearns, St. John, and Zucker) suggested that the Foundation concentrate on an educational model that would bring into science careers a larger population, rather than skimming the cream.

The present sequencing of courses and tracking of students reinforces the accumulation of advantage for some students and effectively precludes others from later selecting science as a career. A child who has not enrolled in algebra in the eighth grade and has not subsequently taken the other more advanced math courses in sequence is unprepared for the series of advanced science courses required in high school. In addition, if he or she wishes to major in a physical science in college, a further handicap can occur: Since these advanced high school courses often duplicate introductory college courses, the children who took them enter college with advanced knowledge of the subject, which is perceived as greater ability; thus, their prior advantages are rewarded, leading them to continue in these fields. Others, who did not pursue such courses early but later show an interest, are at such a disadvantage that, without special assistance or extraordinary motivation and persistence, they may not successfully compete with their more advantaged peers. Since disadvantage in physical sciences often is associated with women and minorities, it is not surprising that we find them in such disproportionately low numbers among physical scientists.

How then do we help the gifted child who has not yet achieved significantly in science or math? Motivation is well-established as a major variable
in learning. Self-esteem and self-image are strongly affected by success.
Many children incorrectly believe that they lack the ability to learn science
or math when they compare their own achievement with that of their more
experienced and advantaged peers. This may make the underprepared avoid
taking such courses, neglecting science and math experiences because of
their belief that some special ability is required, some talent that they fear
they do not have. Yet most of the achievement in these courses is tested
through the device of measuring acquisition of information and facts.

Few young in high school understand science in terms of "What is the
evidence?", "How do I know?", "Why do I believe?", or "How do I find
out?" These questions of thought and action are prerequisite to scientific
inquiry and scientific literacy. The 1987 recommendations to the National
Science Foundation mentioned above incline to the conceptualization of
science-as-inquiry undergirding and preceding the acquisition of relevant
data and information, leading in turn to the attainment of stable concepts.
These concepts, then, form the underpinning for further understanding of
the knowledge, attitudes, and skills that are the cumulative body of science.

This study, as well as the conclusions of most other major reports, proposes
that we must focus on science education for the general public. By offering
our resources and attention to *all* the young, those who will be outstanding
scientists will show themselves, all in their own time.

Entity

A concern for entity, *for full development of unique individual promise,* will
guarantee that the search for fairness and openness does not lead us to
schooling that levels rather than encourages the development of unique gifts
and talents. With Carl Rogers (1969), we believe that schools should offer
students a harbor where they can find and try their skills without danger to
themselves or others. Of course, children must be *physically* safe as they
grow and learn. But Rogers also develops the notion of *psychological safety.*
To move toward *entity,* students must be able to pursue their individual
interests without being stigmatized for being different. While all students
can legitimately be asked to learn certain basic parts of the curriculum—to
read, to write, to figure, for instance—the school must also nurture individual differences as desirable parts of a whole. James Boyer, Rudolph E.
Waters, and Frederick M. Harris (1978) put this philosophy well:
"Individuals ... by their very nature represent *difference, diversity,* and
complexity. Within the framework of human difference one must prepare
for the ultimate quest of the good life" (p. 175).

160

We make this respect for differences a cornerstone of our educational philosophy in general, but we insist upon it in science in particular. As we shall see, many historians and sociologists, among them Derek de Solla Price (1961/1975; 1986), Harriet Zuckerman (1978), and Roger T. Johnson and David W. Johnson (in this volume and elsewhere), have commented upon the collaborative nature of science. But teams do not mean clones—science flourishes, often, when highly individual talents and insights combine to make new knowledge. Teachers in general and science teachers in particular can neutralize inequity and encourage entity, paradoxically, by embracing *differences* as they help children discover and exercise *individual* abilities. *Neither equality nor equity need demand the sacrifice of entity.*

Increasing and Decreasing by Multiplication

A number of scholars have examined the almost exponential effect—both for advancement and for failure—that selection, usually predicated upon some kind of ability grouping, can have upon learning and achieving. Harriet Zuckerman (1978) calls this phenomenon the "accumulation of advantage"; Ann Bastian, Norm Fruchter, Marilyn Gittell, Colin Greer, and Kenneth Haskins (1986) term it "meritocracy"; Robert Merton and others have defined it as the "Matthew Effect" (also, actually, the Mark and Luke Effect as well). Merton, who apparently coined the term, used the "Matthew Effect" to describe the sciences' system of rewards (1968, 1975). More recently, Herbert J. Wahlberg and Shiow-Ling Tsai (1983) have mustered statistical evidence to show its applicability to students in general, and Keith E. Stanovich has discussed its consequences in literacy (1986). Zuckerman's analysis of the "accumulation of advantage" vis-à-vis the "scientific elite" is relevant here:

> Advantage in science, as in other occupational spheres, accumulates when certain individuals or groups repeatedly receive resources and rewards that enrich the recipients at an accelerating rate and conversely impoverish (relatively) the nonrecipients. Whatever the criteria for allocating resources and rewards, whether ascribed or meritocratic, the process contributes to elite formation and ultimately produces sharply graded systems of stratification. (pp. 59–60)

Similarly, students from homes with parents who care about and are knowledgeable about their children's education come to school with an environmental "head start" over their less fortunate peers from backgrounds where education is not a high priority. The children of advantage are also likely to have parents who work to see that their offspring receive the best possible schooling; those of disadvantage may not. Wealth is not *essential* to the creation of advantage (the value of education among certain minority groups seems to have little to do with their economic status [see Jim Cummins' analysis cited below]). However, the advantaged students are *likely* to come from more prosperous homes than those of the disadvantaged.

These individual inequities among educational backgrounds seem inevitable. Different parents care differently for their young with different resources. The children from privileged homes often start with advantage accumulated by "addition." (On this, see de Lone, passim, but particularly pp. 113–139.) Science elites, Zuckerman writes, begin "with certain ascribed advantages [and] continue to benefit, to receive resources and rewards on grounds that are 'functionally irrelevant' [Merton's term] that is, irrespective of their occupational role performance" (p. 60). How similar her thesis seems, on another ground, to de Lone's.

There are a number of ways a child can accumulate *dis*advantage with regard to a future career in science:

• One can be female. Jane Butler Kahle (1983) points out that, although women make up 50 percent of the work force, they make up only 6 percent of the scientists and engineers.

• One can come from a home (often but not always a poor one economically) that does not emphasize traditional academic education.

• One can be a member of a "dominated" minority group. (See Cummins' discussion—briefly summarized below—of immigrant, autonomous, and dominated minority groups.) However, the National Center for Education Statistics reports that of 362,369 bachelors' degrees awarded in the physical and social sciences and psychology in 1980–1981, 338,271, or over 93 percent, went to white non-Hispanic students. (For comparison: The 1980 census reports that white non-Hispanics make up 83 percent of the total U.S. population and 73 percent of the secondary and elementary school population.)

• And, finally, one can attend an inadequate school.

The last circumstance can put at hazard any gifts that have survived these original disadvantages. But an effective school *can* work to begin amelioration. Reducing disadvantage takes long, careful work.

Changing the Formula

Perhaps the greatest mischief in accumulating disadvantage for some while guaranteeing success for others has been done by sometimes entirely well-meaning people who embraced a now rather widely discredited view of the fixity and genetic nature of "intelligence," particularly as expressed in numerical IQ. (Other tests can also be unfair for elementary schoolchildren with limited educational backgrounds, but perhaps misuse of IQs has done the most damage.) Stephen Jay Gould (1981) entitles a chapter of *The*

Mismeasure of Man "Hereditary IQ: An American Invention." Richard C. Lewontin (1976) is equally blunt, attacking as does Gould the experimental techniques that produced hypotheses on *fixed* IQ and noting, in addition, "it is totally incorrect to equate terms such as 'inherited' or 'genetic' with a term such as 'unchangeable' although this equation is often made" (p. 8). Of the " 'facts' from which the heritability of IQ has been calculated, ... " Lewontin comments, "the gathering and presentation of these facts have been so scandalously bad as to constitute a veritable Watergate of human behavioral genetics" (p. 10).

If one sees "intelligence" as a multifactorial trait resulting from the interaction of heredity and environment, one can turn to schooling and education for certain remedies to deprivation. Like most scholars, those who affirm the importance (not the omnipotence but the importance) of excellent schooling agree that environment interacts with heredity in the expression of individual gifts. This is not to deny the importance of genes to intelligence, merely to admit our relative ignorance of genetic effects on the multifactorial trait we call intelligence. But, with our admission of our relative powerlessness to change genes, we also posit our relative power to change environment—in this case that of the school—in such a way as to help potential intelligence emerge.

Giving children of disadvantage the boost that could move them toward the originally unequal position occupied by their luckier classmates can take a number of forms. Ideally, disadvantage should be attacked in the whole configuration of a child's education—parental training, attendance to physical health, building of self-esteem from the earliest possible age. Stanley Coopersmith (1967) notes four areas of psychological strength that should be fostered both at home and at school. Children who value themselves believe in their own significance (as mirrored in the acceptance and affection of others), their own power (as it appears in their relations with their peers), their own virtue (as expressed in self-developed ethical standards), and their own competence (as reflected in successful achievement).

Unfortunately, far from all children come to schooling secure in this kind of education. One remedy, which has seen widespread success internationally (Austin, 1976), has been the institution of "Head Start" programs which help children starting in the preschool years to transcend poor environments. Although all children seem to profit from good preschool programs, Austin finds, they are "crucial" for disadvantaged children. A later study of poor children (between 9 and 19 years of age) who had attended preschool programs confirm Austin's findings. Irving Lazar and Richard B. Darlington (1982) found that preschool programs "had long-lasting effects in four areas: school competence, developed abilities, children's attitudes and values, and impact on the family." Another longitudinal study tracing the effects of preschools on disadvantaged young at 19 years who were enrolled at 3 and 4 is equally hopeful. Lawrence J. Schweinhart, David P. Weikart, and Mary B. Larner write (1986),

> High-quality early childhood education for disadvantaged children is a highly effective way of improving their life chances. High/Scope's Perry Preschool study is a good example. Disadvantaged children who had attended a good preschool program, compared to a randomly assigned control group, were significantly more likely to graduate from high school, enroll in postsecondary education, and find themselves employed. They were significantly less likely to be assigned to special education classes, commit crimes, have children themselves during their teenage years, or receive welfare assistance (Berrueta-Clement, Schweinhart, Barnett, Epstein, and Weikart, 1984).*
>
> So it was found that early childhood education can work. High-quality programs for disadvantaged children, the ones most in need and most difficult to reach, can have positive effects. It is clear, however, . . . that even highly effective programs will not *cure* the problems of all the children, their families, or of society—far broader efforts at social reform involving more than education [sic, *schooling*, as we have been using the term] alone are necessary for that end. (p. 16)

Widening the environment of the disadvantaged should begin to open for them some of the environmental riches of the more fortunate. One way of doing so, as we have seen, seems to be the Head Start programs. Another is through programs that attempt to encourage groups—like women, for example—to try their hands at science and math by establishing special programs. The American Association for the Advancement of Science (1984) has studied a number of such programs, among them the successful EQUALS program at the Lawrence Hall of Science (Berkeley, California). Another approach opens—by special scholarships or other dispensations—excellent schools to students whose finances or locations would ordinarily disqualify them from entrance. One thinks of magnet schools or out-of-boundaries placements in the public schools and financial aid to send disadvantaged students to private schools and colleges. Open environments should also begin to introduce all students to science, which encompasses not only our wide world but also worlds beyond and within it.

Broad Horizons and Wide Curriculums

NSTA has lamented on numerous occasions the number of underprepared science teachers at the secondary and postsecondary levels. The profession simply lacks science teachers—as such—at the elementary school level. Few science teachers, let alone specialists, are attracted to the first six grades.

---

*(See also Berrueta-Clement, Schweinhart, Barnett, Epstein, and Weikart's [1984/1985] *Changed Lives: The Effects of the Perry Preschool Program on Youths Through Age 19* for a full summary of the studies.)

There are approximately 1.2 million teachers at the elementary school level in the United States. The average elementary schoolteacher has studied one course in science. It is not surprising, therefore, that such teachers feel inadequate to teach science in the elementary school setting where experiences developed for children include inquiry through "hands-on" processes in phenomena embracing both biological and physical sciences. It is in the latter—chemistry, physics, geology—that elementary school science teachers have inadequate preparation; thus, they are inclined to give these over to treatment by reading, often reading alone.

This is not at all to deny the importance of reading; scientists often spend as much time in the library as in the laboratory. So students who read as well as experiment in other ways do so in the mode of scientists; they do not "reinvent the wheel" but use the discoveries of others in the time-honored tradition of collaboration in scientific labors. But, without knowledge of the general sciences, teachers will tend to focus on the textbook. At the elementary school level, this is especially true. There is, however, a difference between reading science and doing science in relation to reading. And there is a difference between reading readiness in subject areas outside of science and reading readiness in science. The latter happens when students have direct real experience with a phenomenon before the words or symbols are introduced through reading. As Lorraine J. Daston writes elsewhere in this volume, "World views begin with in-the-fingers knowledge."

"Hands-on" science—the actual experience with materials prior to the concept-formation that also requires *reading* of prior scientific work—is thus the *science readiness* that is prelude to understanding concepts. Teaching scientific vocabulary requires skills similar to teaching reading in other areas. And many elementary schoolteachers are highly skilled in helping their students learn to read. In science, however, reading and direct experience in duplicating and/or inventing ways of inquiry into phenomena *together* supply the "brains-on" and "hands-on" modes of scientific thought and action. Young children benefit particularly from this happy combination.

There just aren't very many general science teachers in the first six grades, let alone specialists. But, paradoxically, we do not see this lacuna as being as crippling to the future scientists of the nation as the shortages at later levels of schooling when the young's gifts for science begin to become talents.

How can this be so? Because to incline as a young child to science is to sample the widest possible range of subjects, although not—at the elementary school level—to specialize in any aspect of science (much less, say, in biology or in chemistry). To become a scientist, one needs the broadest possible schooling and education: one must have the opportunity to observe, write, speak, understand, reason, guess, connect, separate, manipulate, judge, do math, make ethical decisions, . . . more. Only as the child develops, some think in middle or high school (see Kopelman, Galasso, and Schmuckler; and Eilber and Warshaw in this volume; and Brandwein [1955/1981]); others in university (see Humphreys [1985]), does a *gift* for science mature into a *talent,* perhaps attached, probably attached, to a specific field, such as biology or geology or physics (see Wallach, 1985). Then, because an essential way in which science talent is expressed is in originative work, the absence of a trained teacher who could serve as mentor can become a crucial lack.

The loss to society as well as to the individual student of any child gifted *in* science *to* science is twofold and increasingly serious. We must not let this loss occur because of socioeconomic disadvantage; we must not let it happen to the very young because of limited environments in schooling or in education. For, in what Daniel Bell (1973) and others have called our post-industrial society, scientists are of paramount importance. This society

1. . . . strengthens the role of science and cognitive values as a basic institutional necessity of the society;
2. By making decisions more technical, it brings the scientist or economist more directly into the political process;
3. By deepening existing tendencies toward the bureaucratization of intellectual work, it creates a set of strains for the traditional definitions of intellectual pursuits and values;
4. By creating and extending a technical intelligentsia, it raises crucial questions about the relation of the technical to the literary intellectual. (p. 45)

Because scientists have become even more essential than in the past, we cannot lose individuals who may desire a lifework in science; we cannot inadvertently "grind up the seed corn." But we will do just that if we allow children without wide experience to have their gifts overall—but particularly for science—lie largely undiscovered. As Brandwein writes (1987, September),

It is abundantly clear that we cannot "select" children for their developed interests, attitudes, and aptitudes in the arts, the sciences of society, and the natural sciences, let alone other areas of societal obligation, unless we afford them opportunity to develop their interests in an environment rich in opportunity. This must occur early in elementary school. It would be fine if it could begin and continue in some recognizable way in the home. . . . Full minds do not develop on restricted experience and in an indifferent environment that does not augment a genome-driven organism that depends on its full development in interaction with the environment. (p. 40)

To build such an environment in the elementary school, Brandwein suggests three curriculums based on the widest literacy (reading and writing) and numeracy (mathematics). Then, he adds constructs in three: the humanities, the social sciences, and the sciences. (These three curriculums are summarized on tables reprinted in Brandwein, 1987. Note: He is *not* prescriptive, nor is he arguing for standardized curriculums: The conceptual and cognitive schemes he devises are purposefully broad to encourage particular schools to make choices appropriate to their specific needs.)

Many Paths

Thus, we need not only broad curriculums affecting *what* is taught but also wide paths of instruction affecting *how* that subject matter is taught. And the eclectic nature of science could help us combat original disadvantage by affirming differences that could lead to entity. We can and do, often without knowing, welcome many talents—Howard Gardner (1983) has defined seven ways to be gifted: linguistic, logical-mathematical, musical, spatial, bodily-kinesthetic, interpersonal, intrapersonal. Rather than searching via tests for a homogeneous group of "gifted" students to whom to minister, we should embrace the rich and varied contributions of many of our nation's young to our science classrooms. In this way, we can be not only fairer but we can also teach science more effectively.

Indeed, a homogenous group—however selected—unless artistically and effectively taught within a broad curricular design may deprive children of early and promising interaction with the wide population of individuals with whom they will later spend their working lives. And it seems ever more clear that explicit and implicit nurturing of our nation's precious variability in peoples, in cultural diversity, in propensity and promise, may well guarantee our political, social, and economic freedom. And a homogeneous group of high achievers (again, however defined) can, *if badly taught*, fail in their courage to question, achieving, therefore, only incomplete understanding.

Many children from backgrounds of disadvantage perform well in activity-based science programs. Also, through participation in hands-on activities that ask them to *make observations*, the young who tend to science are themselves *made observable*—visible—to their teachers and peers. When the activities involve observing and asking important questions, all students, advantaged and disadvantaged alike, are within the realm of science. Learning to question makes their search for necessary information valuable rather than useless. And both the advantaged and the disadvantaged can learn together.

A dangerous corollary assumption to the practice of rewarding achievers with the best resources—that because some students are "gifted," they are best served by isolation in groups with each other away from their chronological peers—needs some specific attention in its application to the treatment of gifted young in science. Because outstanding accomplishment in science requires a galaxy of skills and aptitudes, often shared among a *team* of researchers and technicians, any school or other educational environment concerned with science must emphasize that individual research scientists, however individually engaged in tracking an elusive concept or datum, require assistance from individuals with a variety of skills. And today scientists also often require a modern technology wondrously developed outside their laboratories by many kinds of technicians, engineers, and technologists. For example, an academic chemist engaged in research often depends on substances supplied by industrial chemists working worldwide. Students can learn this conception of scientific work in any school laboratory and can demonstrate their knowledge in individual problem-solving investigations in which they engage, as individuals and as collaborators, as scientists in embryo.

Even more than in the case of other aptitudes, "giftedness" in science is not a single quality that can be increased by identification, isolation, and attention.* So, the specialized science high schools clearly turn out some young who become outstanding scientists. Their success is not at issue. Rather, the question we raise is this: What would happen if the same resources were devoted to improving science education for a larger audience?

It is true, of course, that to gain an advanced degree in science an individual must demonstrate the ability to conduct an investigation; this process requires a kind of isolation. Naturally, a mentor should be there for consultation, but the aspiring scientist *must* also show an ability to use prior researches to undergird his or her labors in a thorough and honorable way. All scientists personally have *individual research ability in the arts of investigation,* in whatever context or specialty. Even when researchers are part of a team in a laboratory, each member of the group works individually on a subset of the problem, pooling the findings. Of course, one person may in the course of the investigation be required, again individually, to confirm another's data.

Thus, scientists use books, test tubes, microscopes, space vehicles, rats, computers, telescopes, the research of contemporaries and predecessors—all or none of the above and much, much more to discover or uncover or create new knowledge. In the end, each major contribution to modern science achievement depends upon the work of many scientists, working individually or in groups.

We believe that these abilities should not necessarily be stacked, and in fact are not always stacked, in hierarchies. It would be silly to argue that Jonas Salk and his laboratory assistant deserve or should have the same

---

*Some talents may develop more quickly through isolation and/or pull-out groups. A pianist who plays Carnegie Hall at 11, a child who does calculus at 5— prodigies like these may profit from intensity and selection—assuming that careful assessment of emotional cost is monitored during the "precocious" years.

"credit" for the vaccine. But in the kind of collaborations Zuckerman describes, the weight of the contributions is not as clear.

She notes that

> Quite contrary to the twin stereotypes that scientists, especially the better ones, are loners and that important scientific contributions are the products of individual imagination . . . the majority of investigations honored by Nobel awards have involved collaboration. Altogether, as many as 185 or almost two thirds of the 286 laureates named between 1901 and 1972 were cited for research they did with others. (p. 176)

We may easily postulate that most scientists work with assistance in laboratories—or assist others. So science *is* collaborative. Scientists do form a special society.

Jacob Bronowski, in *Science and Human Values,* argued this point eloquently: Not only do scientists indeed form a society, but "the society of scientists must be a democracy" (p. 62). He continues,

> Science confronts the work of one man* with that of another, and grafts each on each; and it cannot survive without justice and honor and respect between man and man. Only by these means can science pursue its steadfast object, to explore truth. If these values did not exist, then the society of scientists would have to invent them to make the practice of science possible. In societies where these values did not exist, science has had to create them. (p. 63)

Nonetheless, Zuckerman's study has certain implications for schooling and education, if only because the models of investigation lead to applications in curriculum and instruction of benefit to the young as they study and learn. However, we need *not* stamp a single matrix upon the way we learn or solve unknowns. We find that individuals with various personalities and capacities are successful within various settings—on teams, in collaboration with others, in distant lands, even alone, in the work they have chosen, whether in field or laboratory. (Witness the lone naturalist in a special environment.)

Hands On, Off, Together—and Minds On

According to a study by Ted Bredderman (1982), all children doing activity-based science became more active learners and improved in their understanding of science process skills. However, he reported that the students who are helped most by such a program are the disadvantaged:

---

*Man, that is, *human.* Were Bronowski writing today, he would surely have written of *men and women.* © 1956 by Harper and Row. Reprinted by permission.

It seems that with an activity-based science program, we have a form of teaching that particularly facilitated learning for disadvantaged students—a consistent finding that has been ignored for many years. Lower-ability, inner-city, lower-socioeconomic, and rural students benefit the most from activity-based science when compared with average-ability and advantaged students. This is true for all outcome areas for which evidence is available: science process, science content, attitude, creativity, and language development. (p. 75)

Elsewhere, Bredderman (1985) reports that other studies have tended to show that activity-based science programs helped students master science content as well as process.

Mary Budd Rowe (1975) asserts that the group helped most by activity-based science programs were the children of the poor, many of them minority students. She bases her statement on two sets of data. Some years ago, disadvantaged Harlem children did well during her trial runs at Columbia Teachers College of the hands-on devices of the Science Curriculum Instruction Study (SCIS) and the Elementary Science Study (ESS), developed in the late 1950s and early 1960s with National Science Foundation assistance.* Her 1978 National Institutes of Mental Health study showed similar achievement. In activity-based classrooms, programs offer a significant opportunity for otherwise underachieving children to develop valuable skills that have long-term academic payoffs. Rowe found that a hands-on investigatory approach, pursued with much conversation among students and with the teacher, leads to satisfying gains in attitude as well as increased competence in certain other areas. She observes, further, that "without science experiences, disadvantaged children tend to be easily frightened and frustrated by simple problems. Their problem-coping skills simply do not develop satisfactorily ... With [science experiences, children] usually learn strategies for attacking problems" (1975, p. 25).

Ruth T. Wellman (1978), in a study that doesn't specifically single out disadvantaged students though it mentions their strong progress, finds that

---

*Some of these curricula, among many others including Science—A Process Approach (SAPA), are available through *Science Helper K–8,* a compact disk product containing public-domain science and mathematics lessons for elementary and middle school students. Directed by Mary Budd Rowe and supported by a grant from the Carnegie Corporation of New York, the program offers almost a thousand activities, including early versions of the lessons developed in the 1950s and 1960s with National Science Foundation help. Many of these carefully conceived programs have lain largely unused for years. *Science Helper K–8,* a compact read-only memory laser disk (CD-ROM), can be used by teachers, supervisors, researchers, and publishers with appropriate computer support. When attached to the appropriate reader and printer, it can provide hard copy as well as screen versions. For further information, contact Dave Anderson, PC-SIG/ASC CD-ROM Publishing Group, 1030E East Duane Avenue, Sunnyvale, CA 94086.

"research . . . builds a strong argument that the study of science helps young children to develop language and reading competencies" (p. 7). Science demands mastery of vast amounts of subject matter; it demands process skills of many kinds, as well as mastery of broad ranges of content.

So studying and doing science in mixed groups creates the kind of complex environment for both the haves and the have-nots in which "real" science will later exist. Elsewhere in this volume, Johnson and Johnson have explained that the system of cooperative learning they have adapted from America's early schools—the one-room schoolhouses of the pioneers—works well in learning a variety of subject matters. In small groups of boys and girls, mixed in "abilities" as well as socioeconomic and ethnic backgrounds, students can cooperate toward one goal, one breakthrough on a model similar to those of scientists in laboratories. Such a system encourages the recognition of individual differences as students work toward entity in their separate contributions to the group's project. It also encourages cooperation rather than competition, working together rather than alone. As each child makes his or her contribution, he or she explains it so that the others understand. But the next contribution further develops the whole and provokes further understanding.

For example, say the goal of the teacher (in this case, one "Mr. F.") is to help third graders understand something about the nature of gases in preparation for a study of weather. Brenda Lansdown, Paul E. Blackwood, and Brandwein describe the following interaction. A special kind of class discussion called a "colloquium"*

revealed to Mr. F. that his children considered air as a kind of light fluid, a homogeneous substance which could flow and push very much like a stream of water. Therefore, he decided that an emphasis on the particulate nature of matter would be appropriate. . .

He gave each group of four children a berry basket containing a banana, some cinnamon, perfume, and peanut butter. Then he opened a bottle of ammonia and left it in one corner of the room. The ammonia attracted most of the attention and started the colloquium. The children were able to establish the order in which groups of children noticed the odor. It was a straight line from the corner of the room. Silence! Then one child said, "Something must be coming from the ammonia bottle to us." "Bits of ammonia," said another. "What about the berry baskets?" was Mr. F's redirective question. "Bits must be coming from them,

---

*By colloquium, the authors do *not* mean a discussion where a teacher asks questions and the children try to produce the "right" answers. "The outcome of a colloquium cannot be preordained," they explain. "Each child is not expected to discover the same set of facts nor to see the same hidden likenesses. . . . Individual reactions to . . . common experiences are never completely similar, resulting in the richness of speaking thoughts together, engendering thoughts by the verbal expression of various observations" (p. 119).

too." There were a number of elaborations of this theme. "If the bits move, how do they get through the air?" The children said, "*You* can get through air, so why can't tiny bits?" "Air gets out of your way when you move; it's like water when you swim, only easier." (Still the analogy with a liquid!)

"The bits you are talking about are *molecules*," said Mr. F. "Will you try to use this word and describe a model of what you think really happens?"

[The children then volunteered and all endorsed these statements:]

Molecules came off the ammonia and moved through the room.

Molecules came off the banana, cinnamon, peanut butter, and perfume. We could smell these objects across the room. Molecules must have moved through the air.

Just as Mr. F. was writing down the last sentence, one quiet child came up with, "But what I want to know is why the molecules move in the first place." (1971, pp. 198–199)

This kind of cooperation works well in science learning. So does the approach Annemarie Roeper takes in her school for gifted children (described in this volume) which focuses in contrast on the unique need of the individual child to move, at his/her pace, toward self-actualization in a safe, reassuring environment with the support of gentle, sensitive teachers. Indeed, there are children with different temperaments and attitudes; as we know, adults, too, vary in these traits. Some are or were loners and require or required the opportunity to work alone. Think of Albert Mendel, of Isaac Newton, of Emily Dickinson, of Virginia Woolf—history provides unlimited examples from many fields.

Some Alternative Approaches

Some alternative approaches to instruction, to testing, and to progress through levels of schooling might help us to better serve *all* students, with their individual needs.* Brandwein's design for self-selection and evaluation

---

*The recently organized (1985) Washington-based National Science Resources Center is compiling a collection and computer database of science teaching materials for the elementary grades. The Center has also prepared a comprehensive inventory of elementary school science teaching resources, which includes annotated listings of many of the materials in its database, together with bibliographic data. *Science for Children: Resources for Teachers* (1988) also includes sources of information and listings of periodicals, museums, and professional organizations. Both the directory and the inventory will be updated periodically.

Sponsored by the National Academy of Sciences and the Smithsonian Institution, the National Science Resources Center is building a nationwide network of teachers and scientists interested in improving the teaching of elementary school science. The Center develops and disseminates resource materials for science teachers and offers a program of outreach and leadership development.

For further information, write to the National Science Resources Center, Smithsonian Institution, Arts and Industries Building, Room 1201, Washington, DC 20580.

in terms of students' originative work is persuasive, and not only for the gifted. (See "Apprenticeship to Well-Ordered Empiricism" in this volume.) With Bastian and her colleagues, we "reject that the alternative to quality for the favored few is egalitarian mediocrity or worse; instead, we identify the ways in which quality can be made available to all" (p. 31). One way of doing so is to provide what James Reed Campbell (1983) calls "horizontal enrichment" opportunities to all students, who can then choose to develop their gifts in many directions.

While it is clear that not *everyone* can become a great and talented scientist, schools can create heterogeneous atmospheres that foster the varied gifts that combine to make science happen. Individual rather than standardized instruction and evaluation techniques could also help the "bottom" 20 to 25 percent of the students, who get most of the failing marks and who often drop out of school. Some of those 25 percent are the slow learners whose case G. Orville Johnson so eloquently outlined (1962). Among his suggestions are differentiated curriculums and instruction for this group. Slow, we think, is not necessarily bad; lock step, we believe, is also not usually good.

Cummins (1986, February), discussing "minority students," another "group" that is often part of the bottom quarter of the class (until its members drop out), makes some recommendations for individual treatment of, and respect for, minority cultures. He offers a three-pronged framework for "empowering" minority students that encompasses both their schooling and education. Beginning with a careful definition of "minority" which notes the vast differences among such groups (are they low status ["caste"] or high [some "immigrant" or "autonomous"] for example?), Cummins notes that

> minority students are disabled or disempowered by schools in very much the same ways that their communities are disempowered by interactions with societal institutions. Since equality of opportunity is believed to be a given, it is assumed that individuals are responsible for their own failure and are, therefore, made to feel that they have failed because of their own inferiority, despite the best efforts of the dominant-group institutions and individuals to help them (Skutnabb-Kangas, 1984). (p. 24)

Cummins would "empower" rather than disable such students by changing three sets of power relationships: "classroom interactions between teachers and students, ... relationships between schools and minority communities, ... and the intergroup power relations within the society as a whole" (p. 19).

Like William Glasser's (1969) "schools without failure," which, by honoring different learning speeds and different approaches to subjects, could help each child succeed in his/her own way and time, Cummins would harness the forces of both schooling and education to help actualize the strengths of minority students rather than label them and helplessly watch them fall.

The Stakes

In a world so complex that, as quickly as we learn, we risk falling under the avalanche of information produced every day, all children must be helped, through the widest possible understanding, to offer *individual* contributions. Not an equal contribution, not the same contribution, but a unique contribution of both each child's ability not only to take in, to understand, a snowflake of the avalanche but also to make of it a part of the structure of the changing world. As each snowflake is unique, so must be each student. We cannot afford to lose a single child—much less one in four.

Many of us—both children and adults—are tempted to shrink before technology that can seem mysterious and threatening rather than enriching and helpful. James John Jelinek (1978) believes that "basic among [the former] assumptions and values is the belief that advances in the technologizing of culture more and more deprive the individual of choice ... [and notes that] Marcuse, Ellul, Whyte, Kafka, Toynbee, and Orwell expostulate this value with force and clarity" (p. 213). Those are eloquent voices against technology. But Jelinek goes on to cite its powerful potential for virtue as well:

> Basic ... is the belief that transience, novelty, and diversity become increasingly greater for individuals in a society as the culture of that society becomes increasingly more technological. The writings of Malinowski, Boas, Lederer, Ogbrun, Medawar, Chase, and Toffler provide vigorous and powerful elaborations of this assumption. The consequence of advanced technology ... is not a deprivation of individual choice but rather a plenitude, a complexity, a surfeit of individual *over* choice. (pp. 214–215)

Children are not the only ones who must be able to embrace the positive offerings of technology and use it to help make the unique contributions on which our very survival may depend. A world so knit by science and technology that a gunshot in Asia can be heard in Brooklyn needs a literate, not a frightened populace. Adults too, especially but by no means exclusively teachers, must continue to open themselves to the scientific and technological information that comes to us daily. Through continuing education of all kinds, through retraining, through development of our fullest entity—not through sameness—our belief is that equity will be achieved: a truly equal chance that affirms individual rights to move from different beginnings freely to different destinations.

References

American Association for the Advancement of Science. (1984). *Equity and excellence: Compatible goals: An assessment of programs that facilitate increased access and achievement of females and minorities in K–12 mathematics and science education* (AAAS Publication 84-14). Washington, DC: Author.

Austin, Gilbert R. (1976). *Early childhood education: An international perspective.* New York: Academic Press.

Bastian, Ann, Fruchter, Norm, Gittell, Marilyn, Greer, Colin, and Haskins, Kenneth. (1986). *Choosing equality: The case for democratic schooling.* Philadelphia: Temple University Press.

Bell, Daniel. (1973). *The coming post-industrial society: A venture in social forecasting.* New York: Basic Books.

Berrueta-Clement, J., Schweinhart, Lawrence J., Barnett, W. S., Epstein, Ann S., and Weikart, David P. (1984/1985). *Changed lives: The effects of the Perry Preschool Program on youths through age 19.* Ypsilanti, MI: High-Scope Press.

Boyer, James, Waters, Rudolph E., and Harris, Frederick M. (1978). Justice, society, and the individual. In James John Jelinek (Ed.), *Improving the human condition: A curricular response to critical realities* (pp. 163–182). (1978 Association for Supervision and Curriculum Development Yearbook.) Washington, DC: Association for Supervision and Curriculum Development.

Brandwein, Paul F. (1955). *The gifted student as future scientist: The high school student and his commitment to science.* New York: Harcourt Brace. (1981 reprint with a new preface [Los Angeles: National/State Leadership Training Institute on the Gifted and Talented])

Brandwein, Paul F. (1981). *Memorandum: On renewing schooling and education.* New York: Harcourt Brace Jovanovich.

Brandwein, Paul F. (1987, September). On avenues to kindling wide interests in the elementary school: Knowledges and values. *Roeper Review, 10* (1), 32–41.

Bredderman, Ted. (1982). The effects of activity-based science. In Mary Budd Rowe (Ed.), *Education for the eighties: Science* (pp. 63–75). Washington, DC: National Education Association.

Bredderman, Ted. (1985). Laboratory programs for elementary school science: A meta-analysis of effects on learning. *Science Education, 69* (4), 577–591.

Bronowski, Jacob. (1956). *Science and human values.* New York: Harper and Row.

Campbell, James Reed. (1983, April 7). Horizontal enrichment for precocious high school science students. Paper presented at the 56th annual meeting of the National Association for Research in Science Teaching, Dallas, TX.

Cole, Michael, and Griffin, Peg. (1987). *Contextual factors in education: Improving science and mathematics education for minorities and women.* Madison: University of Wisconsin, Wisconsin Center for Education Research. (Prepared for Committee on Research in Mathematics, Science, and Technology Education, Commission on Behavioral and Social Sciences Education, National Research Council)

Coleman, James S. (1972, March). Coleman on the Coleman Report. *Educational Researcher, 1,* 13.

Coleman, James S., et al. (1966). *Equality of educational opportunity.* Washington, DC: U.S. Department of Health, Education, and Welfare.

Coopersmith, Stanley. (1967). *The antecedents of self-esteem.* San Francisco: W. H. Freeman.

Cremin, Lawrence A. (1966). *The genius of American society.* New York: Vintage Books, Random House.

Cummins, Jim. (1986, February). Empowering minority students: A framework for intervention. *Harvard Educational Review, 56*(1), 18–36.

De Lone, Richard H. (1979). *Small futures: Children, inequality, and the limits of liberal reform.* New York: Harcourt Brace Jovanovich. (Carnegie Council on Children)

De Solla Price, Derek. (1961/1975). *Science since Babylon* (enlarged ed.). New Haven: Yale University Press.

De Solla Price, Derek. (1986). *Big science, little science and beyond.* (rev. ed.). New York: Columbia University Press.

Gallagher, James J. (1986, May). Equity vs. excellence: An educational drama. *Roeper Review, 8*(4), 233–234.

Gardner, Howard. (1983). *Frames of mind: The theory of multiple intelligences.* New York: Basic Books.

Glasser, William. (1969). *Schools without failure.* New York: Harper and Row.

Gould, Stephen Jay. (1981). *The mismeasure of man.* New York: W. W. Norton. (And in other works)

Haskins, R., and Gallagher, James J. (Eds.). (1981). *Models for analysis of social policy: An introduction.* Norwood, NJ: Ablex.

Humphreys, Lloyd G. (1985). *A conceptualization of intellectual giftedness.* In Frances Degen Horowitz and Marion O'Brien (Eds.), *The gifted and talented: Developmental perspectives* (pp. 331–336). Washington, DC: American Psychological Association.

Jelinek, James John. (1978). The learning of value. In James John Jelinek (Ed.), *Improving the human condition: A curricular response to critical realities* (pp. 183–225). (1978 Association for Supervision and Curriculum Development Yearbook.) Washington, DC: Association for Supervision and Curriculum Development.

Johnson, G. Orville. (1962). *The slow learner—A second-class citizen?* (The J. Richard Street Lecture for 1962, Syracuse University.) New York: Harcourt, Brace, and World.

Kahle, Jane Butler. (1983). Do we make science available for women? In Faith K. Brown and David P. Butts (Eds.), *Science teaching: A profession speaks* (pp. 33–36). (NSTA Yearbook.) Washington, DC: NSTA.

Kawelti, John G. (1968). *Apostles of the self-made man.* Chicago: University of Chicago Press.

Keniston, Kenneth. (1979). Foreword to de Lone, Richard H., *Small futures: Children, inequality, and the limits of liberal reform* (pp. ix–xiv). New York: Harcourt Brace Jovanovich. (Carnegie Council on Children)

Knapp, Michael S., Stearns, Marian S., St. John, Mark, and Zucker, Andrew. (1987, May). *Opportunities for strategic investment in K–12 science education: Options for the National Science Foundation.* Menlo Park, CA: SRI [sic] International.

Lansdown, Brenda, Blackwood, Paul E., and Brandwein, Paul F. (1971). *Teaching elementary science: Through investigation and colloquium.* New York: Harcourt Brace Jovanovich.

Lazar, Irving, and Darlington, Richard B. (1982). Effects of early education. *Monographs of the Society for Research in Child Development, 47.* (2–3 Serial No. 195)

Lewontin, Richard C. (1976, March/April). The fallacy of biological determinism. *The Sciences,* 6–10.

Merton, Robert K. (1968). The Matthew Effect in science. *Science, 159,* 56–63.

Merton, Robert K. (1975, September). *The Matthew Effect in science II: Problems in cumulative advantage and distributive justice.* William S. Paley Lecture at

Cornell Medical School, New York Hospital, New York. (Further revised in series of subsequent public lectures)

Merton, Robert K. (1988, December). The Matthew Effect in science II: Cumulative disadvantage and the symbolism of intellectual property. *ISIS, 79* (299), 606–623.

Rogers, Carl. (1969). *Freedom to learn.* Columbus, OH: Charles E. Merrill.

Rowe, Mary Budd. (1975, March). What research says to the science teacher: Help is denied to those in need. *Science and Children, 12*(6), 23–25.

Rowe, Mary Budd. (1978, September). Externality and children's problem-solving strategies. (National Institutes of Mental Health Grant R01MH25229)

Schweinhart, Lawrence J., Weikart, David P., and Larner, Mary B. (1986). Consequences of three preschool curriculum models through age 15. *Early Childhood Research Quarterly, 1,* 15–45.

Stanovich, Keith E. (1986, Fall). Matthew Effects in reading: Some consequences of individual differences in the acquisition of literacy. *Reading Research Quarterly, 21*(4), 360–407.

Wahlberg, Herbert J., and Tsai, Shiow-Ling. (1983, Fall). Matthew Effects in education. *American Educational Research Journal, 20*(3), 359–373.

Wallach, Michael A. (1985). Creativity testing and giftedness. In Frances Degen Horowitz and Marion O'Brien (Eds.), *The gifted and talented: Developmental perspectives* (pp. 99–123). Washington, DC: American Psychological Association.

Wellman, Ruth T. (1978). Science: A basic for language and reading development. In Mary Budd Rowe (Ed.), *What research says to the science teacher* (Vol. 1, pp. 1–13). Washington, DC: NSTA.

Zuckerman, Harriet. (1978). *Scientific elite: Nobel laureates in the United States.* New York: Free Press.

Science Centers—An Essential Support System for Teaching Science

Sid Sitkoff

During the past few generations, many science programs have evolved. The great majority of these programs emphasize an investigative, "problem-doing/solving" approach where students are active in the learning process. They are free to be curious and creative; they are free to question, to explore, to inquire, and to seek experience in search of meaning. Basic to this approach is the need to provide teachers with those essentials necessary to have the science classroom function as an environment in which investigation is a clear purpose. The laboratory is once again coming to the fore.

The Science Center Facility and Its Functions

A major purpose of the science center is to provide direct support for the teacher and the classroom. A science center makes available science materials and training related to the instructional program and the student activities being planned. Other operations of the center include instruction for visiting students and special individual projects. After-school activities may also be a part of the program; of course, this feature is based upon facilities and support staff available. The center functions in this regard as a resource center by providing

• Science supplies and equipment for science instruction—these materials may be purchased from commercial vendors and/or produced by the science

179

center staff. For example, a multitude of items, such as heat conduction apparatus, gravitational test apparatus, battery holders, pulleys and pendulums, mineral sample kits, and planetarium models can be readily produced at low cost.

• Specialized instruction for visiting students that motivates, reinforces, and extends learning in science—visiting teachers are trained to assist and guide participating students.

• Staff development for teachers and school principals—the training of the school's staff needs to be considered as a continuing process that bends to the improvement of the instructional program in science.

The training provided by the center staff should be carried out in cooperation with key teachers at the school level.* This training may take the form of workshops at the science center and school, inservice classes for academic credit, and miniconferences. The science center staff should not bear the entire responsibility for training. Leaders among the teachers and the principal at the school site need to be offered instruction in the problems of implementing the instructional program, as well as in laboratory procedures, experimentation, and safety measures. Once trained, these faculty members should actively participate in teaching other staff.

Getting Started:
The Initial Planning Advisory Committee

One of the most effective approaches to mobilize support for the science program is to bring together leaders in the community and teaching staff. In this regard, the formation of an advisory committee is essential to establishing a science center facility. Such a committee usually consists of teachers, principals, administrators from staff and school district business offices, as well as community and local industry members. The committee does initial planning covering the center's

• purpose
• location
• budget
• staffing
• functions and services
• communication to and from the field
• evaluation of its operations

The plan should develop specific objectives and tasks that are achievable within an allotted period. Because adequate funding will probably be un-

---

*Science centers often offer key teachers—for example, heads of departments or chairs of curricular and instructional committees—the opportunity to attend week-long seminars, summer training courses, or workshops.

available the first year of operation for all desired services, it is worthwhile to consider a long-range or master plan. The master plan should encompass short- and long-range goals commensurate with the school's needs and related budget available for the operation. It is self-evident that the support of the district's board of education, administrative staff, and teachers is essential to the project's success. However, it may also be desirable to invite interested parents to the planning sessions.

The Ongoing Advisory Council

As the science center becomes established, there are definite advantages to expanding the committee's participation with the parent-teacher association and the school-advisory council in particular and the community in general. Involving the community will encourage cooperation from representatives of local industry and university scientists, as well as parents. This expanded science-teaching/community-advisory council should then advise the science center director in matters pertaining to the work of the science center as well as its school program. The term *advising* means inquiring, informing, suggesting, recommending, and evaluating. Thus, the council serves as a resource to the center and to its director, who remains responsible for decisions necessary to the administration and supervision of the center.

A council should be a vehicle for increasing communication with as many groups in the community as possible. If successful, this discussion requires a frank and open exchange of information. The council's advice should serve the best interests of students and teachers. Although advice, in whatever form, should represent the views of the majority of the council and should be based on objective information, in no way does this imply that minority reports or views would be unwelcome.

If the science center is to best serve its function, its director would seriously consider advice from any additional sources such as staff, students, community organizations, area administrators, and, of course, the board of education. In making decisions, the director should consider such factors as available facilities, personnel, funds, and the objectives of the science curriculum. In summary, the director is bound to consider the effect of his or her decisions upon the students, staff, and the community.

Staffing the Center

Personnel assigned to a fully operating center may include a science resource teacher or science specialist who directs the center's operations. The center will also require a clerk and support staff or science aides. The latter are responsible for constructing and maintaining equipment and preparing specialized materials as needed, such as putting together chemical solutions, growing cultures, caring for live animals, and organizing materials. Depending upon the size of the particular district and the nature of services rendered from kindergarten through grade 12, more than one science aide may be necessary. Volunteer help from the local community should be encouraged; college students employed on a part-time basis are especially welcome and important. They may become future science teachers.

The science center should provide essential supplies and equipment items used by schools for the implementation of a full science curriculum. These materials may be identified and provided individually, in quantities of particular items, or in classroom kits.* In the case of the latter, the center's staff—as with the commercial kits described briefly below—should also restock the kits returned after use to ready them for the next classroom delivery.

Commercially Prepared Kits and Their Refurbishment

Many schools purchase commercial science kits to accompany their programs—either supplementary kits designed to go along with textbooks or kits meant to stand alone. An important function of the center may be to refurbish the science kits after classroom use. The process of restocking the kits at the center should include taking an inventory of kit contents, replacing from the science center's central inventory materials used or missing, and remodeling the kit as necessary. This function can save teachers and school staff considerable time, effort, and money. In addition, the kits can be placed back into service within a far shorter period than would be necessary were each teacher or school aide to check the returned kits, requisition missing materials, and reorganize.

Kinds of Materials in the Center

A science center can make available science materials such as those listed below, which are useful in elementary and secondary science programs. As a

---

*A number of school districts buttress their science curriculums with such kits of science materials, which are assembled locally and then delivered to individual classrooms, either routinely according to a prearranged schedule or upon request. Some districts with particularly successful science kit delivery programs are Anchorage, Alaska; Mesa, Arizona; Jefferson County, Colorado; Schaumburg, Illinois; Carroll County, Maryland; Monroe and Orleans Counties, New York; Springfield, Ohio; Monmouth/Independence and Multnomah, Oregon; Midland, Texas; Fairfax, Virginia; and Highline, Washington.

matter of course, legal prescriptions guaranteeing the safety of both students and specimens must be followed rigorously by staff responsible for collecting and maintaining plants and small invertebrates.

Elementary Science Supplies [a]

Type of Supply	Examples
Aquarium Plants	Duckweed, elodea, foxtail
Land Plants	Those plants native to the region or specialized plants used for terrariums and laboratory work. Examples are ivy, mosses, clearweed, boxwood, philodendron, small podocarpus, and cones of gymnosperms
Aquarium Animals	Guppies, snails, tadpoles, crayfish, frogs, and marine life indigenous to and legally available from the region
Other Animals	Earthworms, mealworms, lizards, crickets, praying mantises, butterfly larvae, protists, and small animals, which are indigenous to the region and for which teacher and students may care in health and safety
Chemicals	Benedict's reagent, bromthymol blue, dilute iodine solution, vinegar, baking soda, cornstarch, plaster of Paris, yeast
Seeds/Bulbs	Carrots, beans (various kinds), peas, onions, radishes, sunflowers
Various Soil and Fertilizer Materials	Soil, sand, commercial fertilizers, potting containers, flats for propagating seeds
Physical and Earth Sciences Materials [b]	Magnifiers, electrical leads, batteries, bulbs, bulb sockets, prisms, thermometers, weather instruments, rocks and minerals, pulleys, transformers for low-voltage use, plastic test tubes, test tube racks and holders, microscopes, illuminators, extension cords, microprojectors, telescopes, planetarium models, globes, land form models, portable planetariums for auditorium/classroom use
Science Center-Prepared Kits of Materials	Many of the above items may be assembled in larger quantities within a single container to serve an entire class. This type of package (a kit) can relate to a specific unit of study such as the use of microscopes for the study of microorganisms, or instruments for the study of meteorology. In addition, kits (accompanied by printed instructions) may be assembled with materials relating to a specific topic, such as the growth of seeds, the metamorphosis of butterflies, magnets, or electrical circuits. [c]

[a] Many of these materials are useful at secondary as well as elementary levels.
[b] Some of these materials also have application to the study of biology.
[c] See note to page 182 for further information on such curriculums.

Secondary Science Supplies [a]

Type of Supply	Examples
Animals, Preserved	Ascaris, crayfish, frogs, grasshoppers, fetal pigs, squid, starfish
Animals, Tissues [b]	Calf or sheep brains, beef eyes, lamb shanks, lamb "plucks"—lungs, heart, trachea, liver, and gall bladder
Animals, Live	Chameleons, land hermit crabs, crayfish, crickets, fertilized chicken eggs, earthworms, frogs, lizards, mice and/or rats, garden snakes, brine shrimp, snails, newts, guppies (as well, of course, as food for the animals)
Plants	Bryophyllum pinnatum, cobra lily, coleus, Mimosa pudica, moss, tomato, Tradescantia, Venus flytrap
Seeds, Bulbs, Spores	Barley, bean, corn, lettuce, geranium, melon, peanut, radish, sunflower, tobacco seed, onion bulbs, fern spores
Cultures	Amoeba proteus, Blepharisma, Paramecium, Chlamydomonas, marine plankton, Daphnia, Didinium, Hydra, Penicillium notatum, Planaria, pond water, Chlorella, Spirogyra, Euglena, Volvox
Stains and Indicators	Acetocarmine, Benedict's reagent, Biuret reagent, Bromthymol blue, Congo red, Eosin, Methylene blue, Methyl orange, Methyl red, Lugol's solution, Phenolphthalein, Phenol red, universal indicator, Cobalt chloride paper, litmus paper
Kits	Study of brine shrimp, Drosophila culture, chemical control of plant growth, microbial techniques, bacteriology, minerals, study of X-irradiated seeds, yeast population study, model rockets

[a] The examples given are biological, perishable materials supplies. Similar lists may be developed for chemistry, physics, and geology. Laboratories used for these sciences are, like the biology laboratory, stocked with nonperishable supplies and equipment for normal classes.

[b] May be preserved frozen.

Housing for the Science Center Facility

Housing for a science center facility can range from a classroom-sized facility to an individual building with several rooms. Some school systems choose a complex of buildings offering indoor and outdoor science facilities. The available space determines, of course, many of the center's capabilities and functions. If space permits, students at the elementary level may visit the center for lessons that extend concepts taught in the classroom. The experiences and concepts developed at the school, in turn, should relate to and build upon those activities experienced at the center. At the secondary level, students can also use the center for designing and developing individual projects.

Teachers with appropriate training may teach their own classes visiting the center, basing the lessons on materials assembled by the center staff. And the instructional area of a center may be divided into science investigation areas, or stations, allowing considerable individualization of instruction based upon the students' interests and aptitudes. After-school activities are a tried and excellent way of stimulating individual instruction. Volunteer community resource people, among them scientists, are often very willing to assist students in their endeavors.

Staff Development

The science center may also serve as a key site for the training of teachers. This training should relate to materials available from the center but, with proper planning and design, should correlate with the basic curriculum that teachers are obliged to pursue in their own classrooms. Science center staff may be designated to assist teachers at their individual schools; the custom is to design workshops and inservice classes that address special problems. Centers should plan extensive training for beginning teachers to make sure they receive enough help. A long-term training program is particularly useful when a school needs the leadership of science center staff, working cooperatively with the administrator and teachers, to establish a total science curriculum. Of course, during such curriculum design, the significant relationships that exist between science and other subjects, such as mathematics, reading, oral and written English, and the social sciences, should be emphasized.

Budget

The funding required to support science center staff, equipment, and materials varies. The amount necessary depends in large part upon the number of staff employed and the availability of part-time salaried college students and volunteers from the community. While the cost of salaries represents a major part of the budget, the science center should also have money for the initial purchase of both consumable and nonconsumable science materials.

The initial investment for the center need not be overwhelming. It is quite possible to begin with a budget adequate for the introduction of the program and develop that budget more fully over several years. Then, the staff to enlarge services can keep pace in number and skills with a growing curricular and instructional program. Once the center is operating, a proportion of the budget can be transferred to the center from schools receiving the services. For example, when schools order certain consumable science materials, funding for them could be transferred from the school account into the science center budget. The amount of transfer should be based on the actual amount the science center initially paid for the materials, plus a small percentage to cover future costs when the same material is repurchased. The cost to a given school would still be far less than individual teachers would spend at supply houses, because the center orders in quantity; further, the center can and does build and collect some of its own materials. A small additional override might also be added to the price of the materials to cover delivery in large school districts. In this regard, a minimal turnaround time for the ordering and delivery of materials to the classroom is essential. The center should be able to operate, in part, on a self-supporting basis with such transfers in funding. This economic process also places the center in the position of supplying those materials that schools really need. Thereby, accountability is automatically built into the center's operation.

Evaluation

Schools, teachers, principals, students, and the community using the science center should be able to evaluate the kinds of services provided and the quality of those services. A simple science center questionnaire, along with personal contacts and advisory committee perceptions, can provide this information. The evaluation should allow teachers to identify additional science materials their classes require and to request particular types of training workshops. In addition, the center should maintain some kind of record identifying materials checked out, school visitations to the center, and science center staff visitations to schools. A full description of the effect of seminars and workshops upon the teaching program is also a good idea. And ongoing and annual reports are valuable ways of analyzing science center functions.

In summary, a science center is a valuable asset for any school district. It can be cost-effective. It can provide for science teachers' immediate needs. It can be an indispensable support for improving curricular and instructional programs, thus enlarging the teacher's capacity to meet the needs of the great variety of students. In these ways, the science center can be part of the fulfillment of students' special abilities as they embark upon their special searches for excellence.

Avenues to Opportunity for Self-Identified, High-Ability Secondary School Science Students

Sigmund Abeles

For some, curiosity and the delight of putting the world together deepen into a life's passion.

—Horace Freeland Judson
The Search for Solutions (1980)

In the 1971 study released by the United States Office of Education on programs for the nation's gifted and talented youth, then Commissioner S. P. Marland, Jr., reported that, of the approximately two million gifted youngsters in the United States, fewer than 4 percent were receiving appropriate services. Just as Paul F. Brandwein's study of gifted science students (1955/ 1981) served some three decades ago as a prelude to a period of intense interest in the education of gifted and talented youth, so did the Marland Report initiate yet another period of concern for this important group of individuals.

Ultimately, the education of any student lies in the hands of the teacher. According to the National Science Board, "The teacher is the key to education—the vital factor in motivating and maintaining student interest in mathematics, science, and technology" (1983, p. 27). The importance of this principle has come to the fore once again after a period of some decline.

The intense efforts today to assure that students have the best and the brightest teachers available is recognition of what should always have been a self-evident truth. Yet, one is mindful of the materials developed during the 1960s that were advertised as being "teacher-proof."

The teacher is the key, but many organizations and agencies have also contributed time, resources, and action. For example, federal and state governments have been helpful from at least two points of view—one, in providing financial resources either to encourage or, in some cases, to mandate programs for gifted students; and two, in helping to support research on the gifted and to disseminate findings. Beyond this, the federal and state governments have provided information about model programs and have assisted in the creation of new programs at local, regional, and state levels.

Definitions of Giftedness

As a case in point—the federal and state governments played a significant role in promulgating an appropriate definition for gifted and talented. This definition has broadened considerably over the years. Half a century ago, giftedness was defined almost solely in terms of a monolithic concept of "intelligence." In many cases, performance on IQ tests was the sole parameter used for determining acceptance of youngsters into programs for the gifted. As research pointed to other components of giftedness, such as productive and creative thinking, task commitment, and leadership ability, the definition of "gifted and talented" began to change. In the minds of some, the federal definition of giftedness has helped to shape the state conceptions currently in effect. Jack Cassidy and Nancy Johnson (1986) quote that definition as follows:

> . . . . the term "gifted and talented children" means children and, whenever applicable, youth, who are identified at the preschool, elementary, or secondary level as possessing demonstrated or potential abilities that give evidence of high performance capability in areas such as intellectual, creative, specific academic, or leadership ability, or in the performing and visual arts, and who by reason thereof require services or activities not ordinarily provided by the school. (p. 15)

Cassidy and Johnson believe that the impact of the evolving definition of giftedness as having multiple attributes has led to the fact that presently only four states still use a cutoff score on a test or tests as the sole parameter in the identification of gifted youngsters.

The changing definition of giftedness has been guided by much research and discussion in the field of human intelligence. The works of J. Paul Guilford (1967), Joseph S. Renzulli (1978), Howard Gardner (1983), Robert J. Sternberg (1984), and Calvin W. Taylor and Robert L. Ellison (in this volume and elsewhere) all propose a multiplicity of talents and intelligences, a position that leads to the notion of a broader definition for the identification of the gifted and talented.

Special Opportunities for the Gifted

As the definition changed, so did the opportunities for gifted youngsters. The number of studies considering methods of schooling for the gifted and talented has also proliferated. The Richardson Foundation of Fort Worth, Texas, conducted an extensive survey of such methods in 1985. This study, which collected information from 1,172 school districts across the country, identified some 16 practices for helping gifted students. Since, of the 16,000 school districts solicited for information, only about 7 percent chose to respond, the data reported cannot be considered representative of the nation as a whole. Nonetheless, the large number of returns provides considerable information about the structures and methods in use. The practices reported were as follows:

1) enrichment in the regular classroom
2) part-time special classes
3) full-time special classes
4) independent study
5) itinerant teachers
6) mentorships
7) resource rooms
8) special schools
9) early entrance
10) continuous progress
11) nongraded schools
12) moderate acceleration
13) radical acceleration
14) The College Board's Advanced Placement Program
15) fast-paced courses
16) concurrent or dual enrollment

Of these practices, it was reported that those most often used were the part-time special classes where students are pulled out for special instruction (72 percent of the reporting districts), enrichment in class (63 percent of the reporting districts), independent study (52 percent), and resource rooms (44 percent). At the other end of the spectrum, the following practices were used the least: fast-paced courses (7 percent of the reporting districts), special schools (4 percent), and nongraded schools (3 percent) (Richardson Study Q's and A's, 1985, January-February).

Identifying the Gifted

This same study found that 97 percent of the districts reporting used a formal identification procedure to select students taking part in the programs or practices for the gifted and talented. The most often used procedure for identification was teachers' recommendations (91 percent), followed by achievement tests (90 percent), and then intelligence tests (82 percent). (This last finding supports a 1981 study [Alvino, McDonnel, and Richert], which found that the most commonly used methods of identification for gifted students were *based* upon intelligence tests, which are highly correlated with achievement tests. Teachers' recommendations often take both tests into account.) Student grades were not used as much (50 percent) as these three methods, nor were parent recommendations or self-nomination (6 percent).

While self-nomination into a program by gifted or talented youngsters does not occur nearly as extensively as some of the other procedures mentioned above, entrance by self-identification is, nevertheless, useful. However, since the comprehensive work on self-identification conducted by Brandwein (1955/1981), research on the methods of self-nomination has been relatively sparse, as Theodore J. Gourley points out (1984). Brandwein showed that, when parameters such as IQ, achievement in mathematics and science, and achievement on standardized mathematics and reading tests were held constant, students who were "predisposed" (being persistent and questing) to science in their high school years were in fact more likely to pursue science at higher levels of education and as a career than those who did not display this predisposition as strongly (pp. 30–32).

Still, given the broad range of programs that are available, some questions still beg answers. For example, what percent of the students in a given program are there because of self-nomination? What percent meet with success in the program? Gourley also asks what percent meet other more traditional criteria for identification. Further, self-identification does not necessarily happen in isolation. Parents, teachers, friends, school counselors, and printed materials help a student decide what course to follow, project to undertake, or path upon which to embark (Delisle and Galbraith, 1987).

While many sifting practices involve self-selection, one, the Westinghouse Science Talent Search Identification Model, uses this process more intensely than do many others. Joyce Van Tassel-Baska has written,

> The talent search focuses much more sharply than most identification protocols on self-election or the volunteerism principle. The commitment to talent search and to follow-up procedures must be made by students and parents in order for the identification to occur. It is this volunteerism principle that has been an important aspect in making the model so adaptable to a variety of school and geographic settings. (1984, p. 175)

Clearly, self-nomination can be an important aspect of a program for gifted youngsters and, therefore, should appear as a goal within such a program. For gifted students, teachers should be concerned about developing "ability for self-appraisal, identification of special abilities and interests, finding oneself by tryout; in short, goal setting and self-concept building" (Morgan, Tennant, and Gold, 1980, p. 37).

At the secondary school level, self-nomination plays an important role expressed in various programmatic ways. Some important ones are:

- advanced course electives
- honors programs, off-campus
- independent study, mentoring, and internships
- special sites and schools
- programs using new technologies

Some of the many examples of these types of practices and programs in science will be described below. There can be considerable overlap between and among these categories.

Advanced Course Electives

Many students take courses in science at the high school level that go beyond the "big four"—earth science, biology, chemistry, and physics. There are a number of advanced courses in areas such as anatomy and physiology, geology, ecology, meteorology, astronomy, marine biology, electricity and magnetism, organic chemistry, and biochemistry. These elective courses, often taken in the 12th grade, are usually selected at the discretion of the student—an example of self-identification.

Another type of advanced course is one that carries with it the possibility of college credit. Many arrangements between local high schools and neighboring colleges and universities allow students still in high school to earn college credit. Some universities award credit in specific subjects to a number of high schools within their state or region, if the high school has met the university's requirements for course content and teacher background.

The most notable program in this area, however, is the Advanced Placement Program, an activity of the College Board. This program also offers students still in high school the opportunity to take college-level courses. Upon completion of the course or courses, students, for a fee, may take examinations and get credit at the participating colleges where they enroll. This program has grown dramatically over the years. In 1955–1956, 1,229 students from 104 high schools took 2,199 advanced placement examinations. During that same year, 130 colleges took part in the program. In 1985–1986, 231,378 students in 7,201 high schools took 319,224 examinations. In that year, 2,125 colleges participated. In the sciences alone, in 1985–1986,

25,931 students wrote examinations in biology, 15,191 in chemistry, 5,124 in physics, 3,730 in physics (mechanics), and 2,280 in physics (electricity and magnetism) (statistics from the Advanced Placement Program of the College Board, 1985–1986).

Honors Programs, Off-Campus

Among the increasing number of activities in off-campus honors programs are

• The Junior Science and Humanities Symposium. Participation in the Symposium is competitive. The program allows the presentation of research papers by students who have developed an interest in a particular topic and who have done considerable study and research about it. Students read papers before those in attendance—mainly other students, but also teachers, college professors, and representatives from business and industry. If winners of the some 40 Symposia held around the nation become national winners, they may then go abroad to attend other science activities such as the Youth International Science Fortnight held in London.

• The Department of Energy Honors Program. This program is highly competitive. It offers students the opportunity to spend several weeks during the summer working with prominent scientists at outstanding science laboratories. Currently, only six students from each state are selected*—by methods that vary from state to state—to attend. In 1988, six activities were offered as a part of this program:

1. The High School Supercomputing Honors Program—Lawrence Livermore National Laboratory

2. The High School Honors Research Program—Brookhaven National Laboratory

3. The High School Life Sciences Honors Program—Lawrence Berkeley Laboratory

4. The High School Honors Program in Particle Physics—Fermi National Accelerator Laboratory

5. The High School Honors Program in Superconductivity—Argonne National Laboratory

6. The High School Environmental Science Program—Oak Ridge National Laboratory

• The National Youth Science Camp. This summer program, in operation for more than 20 years, accepts two outstanding students from each state for its camp in West Virginia. At the camp, among other experiences, students

---

*This number has been rising annually.

hear presentations by scientists, engineers, and representatives of the state and federal governments.

- The Westinghouse Science Talent Search. Participation in this well-known program is quite competitive. About 1,400 students engage in research and prepare a report of their findings for judging. At the finals in Washington, D. C., 10 winners are selected from 40 finalists, who were themselves selected from the 300 students who received honorable mentions.
- The International Science and Engineering Fair. This program serves students who participate in state or regional science fairs affiliated with the International Science and Engineering Fair. Students conduct research, prepare demonstrations, and write papers on topics of their own choice. Among the some 500 awards given at this event is an all-expenses-paid trip for two to the Nobel Prize ceremonies in Sweden.

Independent Study, Mentors, and Internships

Individual work opportunities offer the possibility for greater one-on-one relationships between the student and the adviser, mentor, or teacher than those available in most other school situations. Many schools offer independent research projects within their programs. Others, such as the Bronx High School of Science, make research an intrinsic part of their curriculum.

A number of businesses, industries, and government agencies offer students the opportunity to experience an activity of interest to them through part-time employment during the academic year or over the summer months. The Executive High School Internships of America places youngsters from a number of states with various businesses, agencies, or interested professionals. These experiences allow the students to learn about various fields and give them guidance in the determination of their career aspirations.

Opportunities to work with mentors are available to many students. Often a faculty member from a nearby college, a scientist or engineer from a local industry, or a physician will assist a student involved in a research project of mutual concern.

Special Sites or Schools

Interest is growing in using special sites or opening special schools for gifted and talented youngsters. Entrance into these schools is often competitive and can involve student self-identification. A few of the many programs designed to foster science and mathematics education under way across the United States are described below.

- The Iowa Summer Science Training Program—The University of Iowa. For 28 years this program has provided students with the opportunity to engage in many different areas of scientific research. Students are involved in independent research work with mentors and write reports.

- Operation S.M.A.R.T. (Science, Mathematics, and Relevant Technology). This project, operated by the Girls Clubs of America, holds its programs in girls' clubs, in museums, and at various other sites across the country. The purpose of Operation S.M.A.R.T. is to provide experiences that will encourage girls to develop their interests in science and mathematics. MuseumLink, a part of this program, is a collaborative effort with the Boston Museum of Science, the Association of Science-Technology, and the American Association for the Advancement of Science to develop a model that connects community-based organizations with the expertise and resources of science centers. The program has been piloted in Springfield, Pittsfield, Greenfield, Holyoke, and Lynn, Massachusetts, and in Schenectady and Syracuse, New York, and is now available nationwide.

- National High School Institute—Northwestern University, Evanston, Illinois. For more than 56 years, this program has provided summer opportunities for high-ability high school students. It offers many subject fields, including engineering science.

- Project Oceanology has its site on the coast of Long Island Sound at Avery Point in Groton, Connecticut. It provides marine programs for students in several states in the Northeast. Many students engage in research projects related to environmental issues in their own or nearby communities. Some of these students have become so expert in these fields that they have been called upon by towns to assist in the study of environmental concerns.

- Special high schools—local, regional, and state-run. Some areas with large populations of students who are particularly able academically offer opportunities for such students to attend schools with programs designed to meet their needs. New York City, for example, provides students with the option of attending the Bronx High School of Science or Stuyvesant High School. In some states, year-round or part-time resident institutions (like those in operation in Georgia, Illinois, Louisiana, North Carolina,* and Pennsylvania, and those starting up in California, Mississippi, New York, South Carolina, and Texas) serve a purpose similar to that pursued by specialized local high schools. However, these residential public schools—devoted by no means entirely to science and mathematics but also offering fine programs in many fields—draw their applicants from across their particular state.

- The Talcott Mountain Academy for Science and Mathematics in Avon, Connecticut, is a school for students in grades five through eight. In addition to the regular academic program, students have the opportunity to work with

---

*For more information on residential high schools, particularly the one in North Carolina, see the paper by Charles R. Eilber and Stephen J. Warshaw in this volume.

specialists in various fields of science who are members of the staff of the Academy. These individuals, and the outstanding facilities at the site, make this a rich experience for self-nominated youngsters.

Programs Using New Technologies

The invention of the microchip and the applications that have resulted from its use make this innovation an important aspect of schooling and education for all students, including those who incline to science. This basic technology, combined with others such as those in the communication field, is making opportunities available that have not existed in the past. Students from sparsely populated areas, which lack the diversity of programs and practices available in large school districts, can vastly broaden their experience through the offerings available via computers. Further, many science experiences that these technologies permit are already increasing the number and types of topics of interest available to self-identified science students.

Distance Learning

Schooling over long distances is not particularly new—radio and television have provided course work for many years. Many will remember televised early morning college courses. What is new is interactive or two-way communication, often through television but also by computers over telephone lines. Small, rural schools sometimes have difficulties in offering students the educational opportunities available at larger urban institutions (Hagon, 1986). Distance learning projects can help bridge such gaps by providing instruction to students in schools that do not have courses or teachers for particular subjects. This is accomplished by providing—transmitting over television—the work of a teacher from one school to students at a receiving school or schools and allowing the students to talk with the teacher through various electronic devices.

There are variations on this theme. One distance-learning project, "Project Circuit," initiated in Wisconsin over cable television, was developed largely to overcome geographic barriers interfering with the television signals carrying educational programming. The Talcott Mountain Science Center in Avon, Connecticut, has a program titled "Shoulders of Giants." Through it, the presentations of eminent scientists are beamed via satellite all over the country. Schools that have access to a "dish" can receive the program. Students watching can also talk via telephone to the scientists during their presentations and thus get answers to the questions that may occur during the "class."

This model is in use elsewhere. Texas has an extensive system through which "students in the network watch lessons on the special monitors and use cordless phones to ask questions of their teachers and one another" (Brown, 1985, p. 31). Many other states are exploring and installing interactive TV systems. In some instances teleconferencing techniques are used. By using both TV cameras and receivers in each classroom, this technology allows a simulation of face-to-face contact.

Wider Computer Use

Computers have a number of uses for youngsters motivated to learn in many topic areas that would not otherwise be readily available to them. Tutorials and simulations for a host of science topics have become available over the past few years. Many are designed to assist in problem-solving and critical-thinking approaches. Others use the computer to allow youngsters to access large data bases via telephone. Many data sources are already available, providing significant amounts of information tied to particular research needs.

A more recent development, and indeed a most promising one, is the microcomputer-based science laboratory. Janice Mokros (1987) believes that this approach to science investigation can have an effect on teaching methodology as well as the types of activities undertaken. Essentially, the computer does the sometimes tedious task of collecting data during the laboratory experience. Sensors or probes such as thermistors, photocells, and microphones measure many of the phenomena met in the laboratory. The information from these probes, when presented in tabular or graphic form by the computer, can assist the student to understand and analyze the phenomenon under study.

Temperature, voltage, speed, time, force, sound, and light intensity are but a few of the variables for which programs for use in laboratory activities are already available. Linked to the computer, these probes form an instrument which detects, measures, and then displays or records the measurement. Time formerly spent in collecting, reducing, and recording data can now go into analyzing the results of the investigation, understanding the underlying concepts, and changing the parameters to address "what if?" situations (Abeles, 1985). Still in its infancy, the microcomputer-based science laboratory holds considerable promise for all students, as well as for those who, through self-identification, embark upon activities that lie beyond the typical classroom program.

Schooling's Responsibilities

Whether through state, federal, or local encouragement, or through the favorable circumstances occasioned by mentorships, special sites, or the new technologies, or by the many activities initiated or supported by businesses and industries concerned about the viability of their present and future work forces, the opportunities for students who have identified themselves as competent and motivated are there in greater number than in the recent past. However, the findings of the 1971 Marland Report mentioned at the outset of this paper should not go unheeded.

Only a small percentage of gifted and talented youth, including those able self-identified students who show an interest in science, are receiving opportunities appropriate to their abilities. Therefore, it continues to be incumbent upon those in the profession dedicated to the transmission, survival, and improvement of our culture to strive to fulfill the aspirations of all these able youngsters.

References

Abeles, Sigmund. (1985, Fall/Winter). A new era for the science lab? *Connecticut Journal of Science Teaching, 38*–40.

Advanced Placement Program. (1986). *National summary information (1985–1986).* Princeton: College Board Press.

Alvino, J., McDonnel, R. C., and Richert, S. (1981). National survey of identification practices in gifted and talented education. *Exceptional Children, 48*(2), 124–132.

Brandwein, Paul F. (1955). *The gifted student as future scientist: The high school student and his commitment to science.* New York: Harcourt Brace. (1981 reprint, with a new preface [Los Angeles: National/State Leadership Training Institute on the Gifted and Talented])

Brown, Francis C., III. (1985, November 15). Televised classes help rural high schools offer fuller, more demanding curricula. *The Wall Street Journal,* p. 31.

Cassidy, Jack, and Johnson, Nancy. (1986, November/December). Federal and state definitions for giftedness: Then and now. *G/C/T* [Gifted/Creative/Talented], 15–21.

Delisle, James, and Galbraith, Judy. (1987). *The gifted kids survival guide, II.* Minneapolis: Free Spirit Publishing.

Gardner, Howard. (1983). *Frames of mind: The theory of multiple intelligences.* New York: Basic Books.

Girls Clubs of America. (1986). *Operation S.M.A.R.T. (Science, Mathematics, and Relevant Technology).* New York: Author. (Annual Report)

Gourley, Theodore J. (1984). Do we identify or reject the gifted student? *Gifted Child Quarterly, 28*(4), 188–189.

Guilford, J. Paul. (1967). *The nature of intelligence.* New York: McGraw-Hill.

Hagon, Roger. (1986, January). Two-way cable TV. *TechTrends,* 18–21.

Judson, Horace Freeland. (1980). *The search for solutions.* New York: Holt, Rinehart, and Winston.

Marland, S. P., Jr. (1971). *Education of the gifted and talented: Report to the Congress of the United States by the U.S. Commissioner of Education* (Vol. 1); *Background papers* (Vol. 2). Washington, DC: U.S. Government Printing Office.

Mokros, Janice. (1987, Winter). Emerging MBL research. *Hands On!, 10*(1), 5, 21.

Morgan, Harry, Tennant, Carolyn G., and Gold, Milton J. (1980). *Elementary and secondary level programs for the gifted and talented*. New York: Teachers College Press, Columbia University.

National Science Board Commission on Precollege Education in Mathematics, Science, and Technology. (1983). *Educating Americans for the 21st century*. Washington, DC: National Science Foundation.

Renzulli, Joseph S. (1978). What makes giftedness? Re-examining a definition. *Phi Delta Kappan, 60*(3), 180–184.

Richardson Study Q's and A's. (1985, January/February). *G/C/T*, [Gifted/Creative/Talented], 1–8.

Sternberg, Robert J. (1984). *Beyond IQ: A triarchic theory of human intelligence*. New York: Cambridge University Press.

Van Tassel-Baska, Joyce. (1984). The Talent Search as an identification model. *Gifted Child Quarterly, 28*(4), 172–176.

The author would like to express his gratitude to William G. Vassar, formerly consultant for programs for gifted and talented, Connecticut State Department of Education, for his guidance in the development of this paper.

North Carolina School of Science and Mathematics: The Special Environment Within a Statewide Science High School

Charles R. Eilber
Stephen J. Warshaw

Soon after assuming office in 1977, Governor James B. Hunt, Jr., of North Carolina established the North Carolina School of Science and Mathematics (NCSSM), a public residential high school for academically talented 11th and 12th graders. Governor Hunt acted for two main reasons. One was to provide the resources necessary to improve science and mathematics education for academically talented high school students. The second was to provide sources of new teaching methods, curriculum materials, and teacher training.*

The school has given the nation a model for excellence in science and mathematics education. As it begins its eighth year of operation, two other states are operating similar schools, and a number of other states are planning them. The Louisiana School of Science, Mathematics, and the Arts opened in 1983; the Illinois Mathematics and Science Academy, in 1986;

---

*Portions of this paper first appeared in Charles R. Eilber (1987, June), *Phi Delta Kappan, 68*(10), 773–777.

and the Governor's School for Science and Mathematics (South Carolina), the Mississippi School for Math and Science, and the University of North Texas Academy of Math and Science, all in 1988. Other states, among them Maryland, New York, and Oklahoma, are in various stages of discussing the possibility of such a school or are actively planning one.

The NCSSM was established on the grounds of a 27-acre, 15-building former hospital in Durham, North Carolina. The school opened with its first class of 150 juniors in 1980. Since then, enrollment has grown steadily, reaching 475 juniors and seniors in 1987–1988 from 87 of North Carolina's 100 counties, from 123 of its 140 school districts, from 209 high schools altogether.

The students do not pay tuition, room, or board, so parents' economic circumstances are not a factor in the admissions process. Students come from a diversity of socioeconomic backgrounds. Sixteen percent are black, Native American, or Hispanic; 8 percent are Asian; 53 percent are male and 47 percent are female. Admission to the school results from a multifaceted selection process. We develop a portfolio for each prospective student, including the verbal and mathematics scores on the Scholastic Aptitude Tests taken during the 10th grade; a nonverbal test of critical thinking; the student's grade, in mathematics and science from the 9th grade and the first semester of the 10th grade; a checklist, prepared by a teacher or counselor, which rates the student on characteristics attributed to academically talented students; an essay written by the student under controlled circumstances; letters of recommendation from teachers; and a statement from the student explaining his or her reasons for seeking admission to the school. We expect indications of strong interest in science and mathematics to show up in many ways on the applications of the students admitted.

From the usual 800 applicants, we select 400 for an interview on campus. The results of that interview are incorporated into the other data, and the Admissions Selection Committee then chooses the final enrollees for the coming year. The school follows some general demographic guidelines that recognize the diversity of educational opportunities throughout the state, but it seeks to enroll only qualified applicants.

Curriculum

Although the name of the school might suggest otherwise, the instructional program includes more than science and mathematics. While we emphasize those subjects, the School of Science and Mathematics is a comprehensive college-preparatory high school with a great diversity of course offerings. The school offers 20 courses in mathematics and computer science, 16 in biology, 8 in chemistry, and 8 in physics. Students must select at least one course from each of the three main science areas. English, with an emphasis on writing, is required in both years. So is a foreign language—students can choose French, Spanish, German, Russian, or Latin. We also require American history; and other classes in social sciences, though elective, are well attended. The school provides a variety of electives, including activities in music, the visual arts, and a number of interdisciplinary courses.

Most courses are taught at an advanced level and explore content with a conceptual thrust that is usually found in college courses for freshmen and sophomores. For those among the student body who desire Advanced Placement (AP), we offer some courses that include material covered in AP syllabuses. Some students do take the AP tests. But the exemption of NCSSM students from college course requirements is not a primary emphasis of our school.

In addition, there are students with strong interests in fields that cross traditional disciplinary lines. Biophysics, bioethics, and human sexuality are interdisciplinary, team-taught courses for such students. The various departments determine the level of the courses in response to the goals of the school and the needs of this special population of students.

Virtually all NCSSM students plan to attend college; beyond that, however, they have a wide range of goals. Many of them will use their knowledge of science as writers, voters, or government officials, rather than as bench scientists, engineers, or science teachers.

In order to ensure that all students are knowledgeable in biology, chemistry, and physics, the school offers first- and second-year survey courses in each. It requires students to take a one-year course or demonstrate competence to graduate by passing an exemption exam in each discipline. On the other hand, students with more specialized interests and those who just can't get enough science have a variety of upper-level courses from which to choose. In biology, these courses include genetics, cell biology, embryology, ecology, taxonomy, anatomy, and behavioral science. In chemistry, students can take classes in chemical instrumentation, organic chemistry, environmental chemistry, and polymer chemistry. In physics, we offer astrophysics, modern physics, and electronics.

Research

For students who want to find out what it is like to do scientific research, the school provides research courses in biology and chemistry, as well as research mentorships at nearby universities and laboratories. NCSSM's research courses are good examples of the kind of educational opportunities that may only be practical in "magnet" or other special schools. Students who have already developed an interest in research through science fairs and the efforts of their previous teachers can pursue that interest here in a program designed to create, albeit on a smaller scale, the sort of experience they might encounter in graduate school or in a scientific laboratory. Other less experienced students, at a stage in their education before they must choose a college or a major, have the opportunity to try research for the first time. Overall, about 15 percent of our students take either the biology or the chemistry research course.

Through the generosity of the area universities and corporate donors, the school's research resources include a modern recombinant DNA laboratory, magnetic resonance imaging, gas chromatography, and extensive computing facilities. Clusters of networked personal computers are available around the campus, as are terminals for a new superminicomputer. These and other facilities are available to students in nonresearch courses as well, for a variety of uses including word processing, drilling, data analysis, and simulation.

In addition to the research courses on campus, NCSSM students have access to year-long mentorships with researchers and other professionals in the Research Triangle area (Durham-Raleigh-Chapel Hill). Approximately one-third of our students, usually as seniors, work with mentors. Our students have worked on such diverse projects as the development of ion-exchange material to be used at home by dialysis patients, analysis of photometric and spectroscopic data concerning the galaxy's oldest stars, characterization of a computer "windowing" system, and computer modeling of processes in a chemical plant. Some students have received summer positions in their mentors' laboratories, while other mentors have helped students to their first scientific publication or taken their charges to national meetings. The high level of participation testifies that this program is very popular with the students. In essence, the program provides opportunity for individual or group research.

Special Projects Week provides another opportunity for students to pursue original investigations. Each spring, we suspend classes for a week. During this time, students study a subject or participate in some educationally valuable activity that is not required in any of their regularly scheduled courses. A few are attracted to artistic endeavors, while others pursue scientific investigations. In recent years, the latter have been a diverse array—the mechanics of bubble formation, the construction of a sampling boat for a limnology class, and the development of a chemical magic show, for example. The proximity of many research facilities to the campus makes field trips an important component of the NCSSM science program, and Special Projects Week offers the chance for more extensive trips than those regularly scheduled. Sometimes these visits amount to minimentorships in such fields as electron microscopy and biomedical imaging.

Choices

Within the constraints of budget and faculty workload, the science department strives to meet the diverse academic needs of a unique student body.

We sustain diversity of NCSSM's offerings in science and other subjects through continuing innovation by the faculty and staff. Because the school's teaching and other resources are finite, however, we must balance our push to create innovative new programs with an evaluation procedure that determines effectiveness and relevance to the overall goals of the school. Through this evaluation process, we revise or eliminate less successful and less valuable programs from the curriculum.

Students enter NCSSM with a broad range of preparation in science. Most arrive having had a life science course, but those courses vary widely depending upon the resources available to their home schools. Those students for whom testing shows a deficient background meet the biology requirement by taking a self-paced survey course. Those who arrive better prepared in biology can choose among the topical courses (e.g., behavior, embryology, research, or a course more oriented to the AP examination). The biology curriculum is a popular one, because it offers students a wide range of options and lets teachers teach some specialized courses that allow in-depth coverage.

On the other hand, most students enroll without a prior high school course in either chemistry or physics. Only 10–15 percent of the entering class has studied chemistry; only a handful of students, physics. In each discipline, we offer two introductory survey courses, one of which calls for mathematical rigor; this focus is necessary because students arrive with an even wider variety of preparations in mathematics than they do in science. We also provide second-year survey courses in chemistry and physics, advising students intending to take AP exams to enroll in these courses. Of course, as with biology, students with sufficient background in chemistry and physics can take advantage of more specialized upper-level electives, such as organic chemistry or astrophysics.

Nonacademic Learning

At any residential school some of the most important learning takes place outside of the classroom. Away from home, probably for the first time, students need to adapt to a community that is at once more structured and restrictive than that in most homes and yet encourages more independent decision making. Students learn how to manage time, what new friendships to initiate and develop, how to accommodate the needs of a roommate and the expectations of a dormitory community, and how to assume responsibility for some of the ordinary tasks of daily life that their parents may have previously done.

Many aspects of the nonacademic program contribute to the valuable learning that takes place in the NCSSM community. Students organize in groups of about 35, each with a resident advisor. These groups work together in orientation at the beginning of the school year and in many social activities, as well as assume the responsibility for cleaning all of the areas where they live. The resident adviser is both counselor and disciplinarian, offering advice and sympathy as well as enforcing quiet hours, curfews, and various other rules.

In addition to cleaning his or her living quarters, each student does three hours per week of unpaid work or service campuswide, and, during the summer, juniors contribute 60 hours of service in their home communities. NCSSM has an active student government, an intramural and an interscholastic athletic program, many clubs and organizations, a school newspaper and yearbook, and numerous recreational and cultural activities.

Commitments: Past and Future

Almost every graduate goes on to a college or university, two-thirds of them attending one of North Carolina's outstanding postsecondary institutions. The one-third going out of state are widely distributed among some of the most selective colleges and universities, including the military academies. Many begin with advanced course standing and credit awarded by various means of evaluation. Two classes have moved through four years of undergraduate study and are now in graduate school or have taken jobs. Early indications are that 80 percent of these graduates work in fields related to science or mathematics.

The commitment of the school to sharing its resources with other schools throughout North Carolina is serious. We hold workshops every summer for secondary schoolteachers in science, mathematics, computer science, and the use of libraries and instructional media. More than 4,500 teachers involved in these workshops, either on the campus or in various locations around the state, have returned to their home classrooms with up-to-date software, fresh laboratory experiments, and new ideas. And, of course, through their visits to the school, the teachers gain a new understanding of the role of science in society to share with their students.

In 1984, NCSSM came under the governance of the University of North Carolina as an "Affiliate School of the University of North Carolina." One immediate result of this action was that summer programs for high school students were established on six different campuses of the university system. These programs, taught by teams of university faculty members and local high school teachers, are administered and coordinated at NCSSM. Each summer, more than 700 North Carolina 11th- and 12th-grade students take part in 5 weeks of a research- and laboratory-oriented experience. Their parents pay only for transportation.

To steal a phrase from one of our summer teacher-training workshops, "Science is a Verb" at NCSSM. Students and teachers teach and learn science through a wide range of activities and through formal and informal interaction among faculty, students, and mentors. Because NCSSM is residential, sciencing goes on many hours of the day and many days of the week.

Although only a very few public schools can provide the advantages of a residential campus for their students, we encourage administrators and teachers of science everywhere to look for resources of time, expertise, and even donations of equipment and funds beyond the usual constraints of the school day, the school building, and the school budget. We know this is not an easy task, but the satisfaction that comes from seeing the expanded learning that results makes the effort entirely worthwhile.

A Program for Stimulating Creativity in a Citywide High School: The Bronx High School of Science

Milton Kopelman
Vincent G. Galasso
Madeline Schmuckler

In the 1987 Westinghouse Science Talent Search, out of 300 semifinalists nationwide, 27 were seniors from the Bronx High School of Science. This was not a random occurrence. Bronx Science students have consistently made a strong showing in this prestigious competition since its inception 46 years ago. In fact, the Bronx High School of Science leads the nation in the total number of finalists and semifinalists produced—with 106 finalists over the years, almost twice as many as its nearest competitor. (For further discussion of the Search as a means of fostering science talent, please see "Science: In an Ecology of Achievement" and "Apprenticeship to Well-Ordered Empiricism" elsewhere in this volume.)

Like composers of symphonies or authors of poems, these young people have demonstrated their creativity by their work. The definition of creative gifts has, like a living thing, changed with time. Louis M. Terman and his associates (Cox, 1926) defined giftedness as "the top 1 percent in general intellectual ability, as measured by the Stanford-Binet Intelligence Scale or a comparable instrument." Virtually since that moment, the use of so-called IQ tests has caused considerable controversy. Many of Terman's contemporaries challenged the ability of any single test to assess something as complex and subjective as "intelligence" and objected that such tests rewarded arbitrary correct answers, leaving no room for unexpected responses, however creative and promising. Indeed, Terman himself shortly recognized the limitations of IQ (Gould, 1981). A quarter of a century after Terman's original work, Paul A. Witty (1958) took a different approach, stating that "there are children whose outstanding potentialities in art, in writing, or in social leadership can be recognized largely by their performance" (p. 62).

A Brief History of Bronx Science

It is largely this latter spirit that has served as a philosophical point of departure for the Bronx High School of Science. Celebrating its 50th anniversary in 1988, the school was conceived by Morris Meister, its first principal. Meister hoped that intelligent, creative youngsters, stimulated by one another and by a carefully planned educational program, would come to love learning and would take imaginative leadership roles when they became adults. (For further information, consult Taffel, 1987, and Kopelman, 1988).

Changes in society have, over the last half-century, been reflected in changes within the school. Women were admitted in 1946 to what was initially an all-male institution; a move to a new, custom-tailored building occurred in 1959; and the nature of the student body has altered to reflect the changing background of New Yorkers. Basic goals and philosophies have remained the same, however: Bronx Science strives to produce capable, creative leaders of the intellectual community. How can we achieve these aims? If we accept as a corollary that scientists do in fact demonstrate talent through research, then it follows that gifted youngsters who are potentially creative in the sciences can have that potential developed significantly by working in a planned educational program.

Bronx Science Objectives

The main objectives of this program as it has grown at Bronx Science are the identification of gifted, motivated students in science and the development of these students to the point where they can carry through original, creative, independent pieces of research. Toward this end, a three-year sequence of courses has been developed. The approaches and strategies employed emphasize providing "hands-on" experience with a variety of scientific equipment and techniques, exposing students to the inquiry approach used in science, and stimulating them to handle problems in a rational, scientific way.

The students benefit from this program, not only by developing their ability to solve complex scientific problems but also by taking a positive attitude toward handling many other kinds of difficulties. Furthermore, it has been our experience that college, and even graduate and medical school acceptances increase when students can provide concrete evidence of having done independent work. Achieving such goals must involve in synergistic interaction the students, the staff, and the educational program itself.

Ways of Selecting

The Bronx High School of Science is a specialized high school for gifted youngsters. Students living within the five boroughs of New York City are eligible to take an entrance examination, passage of which is the sole criterion for admission. The exam consists of verbal and mathematics portions. The verbal part includes objective questions on vocabulary, sentence completion, and reading comprehension. The mathematics segment tests mathematical concepts, computation, and problem solving.

There is much evidence, however, to support the assertion that pretesting is not required to implement the program. Any comprehensive high school can use it by allowing one segment to act as a screen for the subsequent part. To investigate this theory, in September, 1976, the biology department applied for, and was granted, model program status by the Office of the Gifted and Talented (then a part of the U.S. Office of Education).* The purpose of this grant was to allow the biology department, over a three-year period, to disseminate nationwide its "Model Program for Developing Creativity in Science" so that schools could replicate and/or adapt its program. The department prepared sample materials describing the general approach and orientation, as well as some of the more specific aspects of day-to-day activities and model lessons. An 11-page brochure was sent out to the over 1,000 school districts that expressed interest. Some educators availed themselves of on-site visitation opportunities to observe the program in action.

Since the time that the biology program was designated as a model, its philosophy and technique has spread to other departments in the school, including physical science, mathematics, and social studies. This happened because the philosophy underpinning a carefully planned program, whose goal is to get students to produce a piece of independent research, has universal application. Original projects appeal to parents who want their children to be able to work through problems, interact with people, take the initiative, and see a task through to completion. In addition, the English department has added a course in the writing of a research paper. This

---

*This office no longer exists in the reorganized U.S. Department of Education.

means that youngsters can develop creative talents in any one of several fields. On the basic level, the student is exposed to classes and lab activities that stimulate questioning, analysis, and problem solving. This orientation can then serve as the springboard to individual research in the area of the student's choice.

Independent Thought and Research

What specific program can direct students toward thinking and working individually? A program that prepares students to carry out independent research should at the same time significantly increase their level of creative productivity. While learning to do research, students are also learning to

- set up hypotheses
- design an original experiment
- develop laboratory techniques
- evaluate data
- cooperate with other students
- develop confidence in their ability to work independently

In the ninth year, the entering youngster selects either biology, chemistry, or physics. A student starting with biology takes a course whose core content is mandated by, but goes significantly beyond, the New York State Regents' syllabus. All three courses of study call for motivation of the students and movement toward higher-level thinking skills. For example, lessons about "cell division" are a legendary part of basic biology courses. Describing what happens when one cell forms two is not seen, however, as the end point of the intellectual process that the course seeks to develop. Students are encouraged to question why it is that a cell is stimulated to begin the process of division. Is there any stimulus that can relate the division of an amoeba to the division of, for example, the skin cell in a higher organism? Why do cells divide at all? By asking questions of this sort, the student is taken beyond the basic curriculum, both in content and in process.

The Socratic method of teaching is used, with emphasis on recognizing problems and offering hypotheses. Because each course meets for 10 periods per week (5 "double" periods), many open-ended laboratory experiments can be performed. For example, students might be asked to devise a procedure to determine the difference between organic and inorganic catalysts. "Ongoing" experiments are also used. For example, a series of experiments with slime mold led students, after a group of simple preliminary investigations, to go on to design and carry out their own original experiments with this organism. Pre- and postlab discussions allow students to plan their lab time properly and to analyze and critique their results. Both classroom and laboratory activities are designed to teach students to identify a problem, offer a hypothesis, plan an experimental design, do data analysis, test hypotheses, and come to an appropriate conclusion.

Their work is done within a basic Piagetian framework of guiding students from the concrete or recall level to the higher or formal/operational level. More simply put, the teacher tries to ask fewer and fewer "what" questions and more and more "whys." Assignments do not merely review the material of the day. Homework, which stresses the recognition and analysis of problems, is given after a concept is developed in class. This procedure discourages clinging to disjointed facts and allows students to focus on the processes of science.

From the incoming pool of students, several classes of honors science are formed. Students are selected for honors classes based on entrance examination scores and expressed interest in being a part of the program. The approach in the honors classes includes all of the techniques described and then goes a step further. Less time is spent on class discussion per se, and students are given more lab experiences of an open-ended, ongoing nature. In addition, content is enriched in scope, and more opportunity is provided to learn the process of scientific thinking by reading research articles. Students are also given guidance in how to read a body of related articles and then formulate a research question suggested by the literature. These classes are also scheduled for 10 periods per week.

In the 10th year, those students who took biology then select chemistry or physics as their "basic science." The orientation in the physical science department continues to develop and refine problem-solving and creative-thinking skills.

Students With a Bent for Science

Selected students are then recommended for a biology research class in their junior year. What criteria are used in the selection process? Our experience has indicated that students who are creatively gifted in science have the following characteristics. They are

- strongly and sincerely motivated toward learning and achieving in science
- able to work well independently in the laboratory, the library, and the classroom
- curious, seeking explanations
- very much interested in getting answers to questions suggested by their work and by their teachers
- askers of many questions
- stimulated by problem-solving approaches to learning
- good at identifying significant problems in a mass of information
- readily able to induce, deduce, and make connections between related ideas

- often able to see different approaches or come up with "offbeat" ideas
- full of creativity and achievement, which extend to many other areas
- able to relate well to their peers and to their elders
- often able to establish long-term goals

This year, about 100 students are enrolled in the 11th-grade biology research class. Of these, 40 do their research within our building in a specially equipped projects room. Students are individually guided by a teacher or a team of teachers as to the advisability and practicality of their research proposals. The remaining 60 students have been helped to secure positions in local hospital and university laboratories under the direct tutelage of faculty mentors, who maintain close communication with a member of the Bronx Science biology department. This process affords the students the benefit of close contact and work with professionals in the field. It has been our experience that most research scientists are very willing to help young people solve science problems.

These professionals are often pleased and surprised by the knowledge, interest, and ability shown by the youngsters. For example, one student in introductory honors biology began to explore the literature about a particular snail and became fascinated with its neurobiology. In her junior year, she became a research assistant at The American Museum of Natural History and extended her interest to the lab. Now, six years later, she is still working with the same senior scientist, her career orientation clearly established. Another youngster did his research at home, virtually "under the bed"—that is, without sophisticated equipment or lab space—but with much determination to see a project through to completion. This kind of dedication can lead to a lifetime goal.

Students in the research phase of the program do extensive library work to learn how to read scientific papers and how to abstract relevant information. They learn how to use *Biological Abstracts, Index Medicus,* and various specialized journals. Procedural details for each project are discussed with each student individually. Periodic progress reports are expected, and a full written report is required at the end of the year.

Helping Students Grow

Bronx Science provides reinforcement and rewards in many ways. A departmental student publication, the *Journal of Biology,* publishes many individual research papers. Students are encouraged to enter local and national contests, where they are often among the prizewinners, although, in a very real sense, all who see a problem through to completion are winners. It must be stressed, however, that winning awards is the "by-product": The true prize is something all participating youngsters will have for life—a questioning, creative approach to learning. Not every child enters or wins contests, but all who pass through our program, we hope, are enriched.

What qualities would be desirable in teachers involved in implementing such a program? Our experience shows that they should be
- open, flexible individuals capable of stimulating higher-level thought and questioning
- tolerant of diverse approaches to solving a given problem
- capable of critiquing student research papers
- experienced in teaching gifted students
- strong in subject matter
- philosophically committed to the inquiry method of teaching/learning and to the value of student research
- capable of identifying and motivating underachievers who might benefit from the research program

These observations strongly coincide with a study by Jack A. Chambers (1973), in which science professionals were asked to identify the characteristics in their teachers that the scientists felt had stimulated their creativity. The traits most often selected were that the teachers were well-prepared for class, taught in an informal manner, and accepted disagreement. Their students "viewed the facilitating teachers as more often personally interested in teaching and in their students and as having a high level of commitment to their field. The students' image of these teachers was of a hard-driving, dynamic individual who was very intellectually demanding of students" (p. 330).

Students and staff have been overwhelmingly enthusiastic in their reactions to the program. It is not uncommon to see two students standing in the hallway arguing about such questions as the relative merits of different techniques for culturing protozoa or for objectively measuring mating behavior in fruit flies. Teachers freely share information and approaches to subject matter with one another. Our department offices and laboratories have become a place where students and staff can work together as colleagues. Not every Bronx Science graduate goes on to a career in science or mathematics. Over the years, this figure has stayed at about the 60- to 65-percent mark. Many students have stated that our program has helped direct their lives, not only in terms of subject matter, but also, more significantly, in fostering enthusiasm for, and a creative approach toward, the entire learning experience.

The almost legendary mural above the main entrance to the Bronx High School of Science has woven among its tiles a quotation from John Dewey— "Each great advance in science comes from a new audacity of the imagination." Dewey's statement exemplifies the spirit of our program to develop our students' creative potential. Imaginative programs, coupled with talented students, take the learning process that giant step forward where youngsters become not only consumers but also the producers of knowledge.

References

Chambers, Jack A. (1973). College teachers: Their effect on creativity of students. *Journal of Educational Psychology, 65*, 326–334.

Cox, Catherine M. (1926). *The early mental traits of three hundred geniuses.* In Lewis M. Terman (Ed.), *Genetic studies of genius* (Vol. 2, pp. 11–842). Stanford, CA: Stanford University Press.

Gould, Stephen Jay. (1981). *The mismeasure of man.* New York: W. W. Norton.

Kopelman, Milton, Galasso, Vincent G., and Strom, Pearl. (1977, Spring). A model program for the development of creativity in science. *Gifted Child Quarterly, 21*(1), 80–84.

Kopelman, Milton. (1988). *A brief history of the Bronx High School of Science.* Unpublished manuscript.

Taffel, Alexander. (1987, September). Fifty years of developing the gifted in science and mathematics. *Roeper Review, 10*(1), 21–24.

Witty, Paul A. (1958). Who are the gifted? In N. B. Henry (Ed.), *Education of the gifted, Part I* (pp. 41–63). (Fifty-eighth Yearbook of the National Society for the Study of Education, Part II.) Chicago: University of Chicago Press.

Problems in the Development of Programs and Science Curriculums for Students Gifted/Talented in Areas Other Than Science*

Irving S. Sato

For several decades, the need for differentiated education for gifted/talented students‡ has been documented in numerous sources. Milton J. Gold (1965/1982) stated in *Education of the Intellectually Gifted,*

> Education of the gifted . . . must be seen in its proper perspective as simply an extension of the doctrine of individual differences . . . The theme of self-realization epitomizes education in a democracy because of the value placed on the individual human being. To the extent that school programs are truly adapted to individual differences, they contribute to the self-realization of each student. (pp. 1–2)

Other sources echoing this need through the years include S. P. Marland, Jr.'s, *Report to the Congress* (1971) and James J. Gallagher's surveys of

---

*The author gratefully acknowledges the editors for their verification of extended sections of this paper.

‡ The term *gifted/talented student,* as used in this paper, refers to students who excel consistently, or show the potential to do so, academically, creatively, kinesthetically, and/or psychosocially.

U.S. education for the gifted/talented (1983), research that summarizes the results of national surveys conducted about a decade apart.

Subscribing to this need for educational changes for the gifted/talented, this paper presents some key considerations for developing science curriculums for students whose special interests and/or abilities lie in areas other than science. Within this group are students who generally elect a laboratory course in high school science to satisfy requirements for graduation; the course is usually high school biology. These gifted students include those who incline to the social sciences, humanities, and the arts. Their destinations, in terms of careers (lifeworks) may be law, history, sociology, or psychology; sometimes they choose teaching in schools or universities; sometimes they turn to business. Others enter careers in writing, journalism, music, painting, dance, and the varieties of graphic and theatrical fields. Those inclined to mathematics, often erroneously considered to be a subset of science, may turn to physics, computer science, or engineering.

In any event, the students with whom this paper is concerned are in the target group of the 3 to 5 percent with exceptional talents and interests, so estimated in Marland's *Report to Congress* (1971). This gifted group includes some with interests in science and mathematics. To be sure, there are some gifted/talented students so extraordinary that they excel consistently in all academic areas and in some nonacademic ones as well. Eventually, as these polymaths select areas for concentration, some will incline to the sciences.

Before examining what would be considered appropriate science curriculums for students gifted/talented in areas other than science, one should study the conditions essential for the development of schooling and education of the gifted/talented in general. The most vital conditions for ensuring substantive programs for the gifted are (a) effective administrative arrangements, (b) necessary curricular and instructional changes, and (c) planned staff development.

Administrative Arrangements in the Education of the Gifted/Talented

Certain administrative arrangements are necessary to guarantee three basic types of experiences in educational settings for the gifted/talented. These students need opportunities, on a regular basis, some of the time, to interact with one another, to mix with students of varying abilities, and to work independently.

Different writers classify many program alternatives for gifted/talented students in various ways. However, these program options fall basically into three categories when classified in terms of their primary mode of delivery:

enrichment, acceleration, and special groupings. Enrichment extends learning environments in such ways as to offer substantive opportunity for inquiry and discovery. Acceleration involves grade skipping and/or speeding up coverage of subject matter (beyond what would be done in ordinary situations). Special grouping could include cluster grouping (placing some gifted/talented pupils into settings with students of varying abilities), special classes (scheduling only gifted/talented students into a special section), and independent study, which, for science, often includes research and the solution of unknowns. All alternatives in programs available are administrative arrangements for organizational purposes and do not guarantee in themselves significant or substantive experiences for gifted/talented students. Certain curricular and instructional changes must take place in these settings to ensure vital learning experiences.

Curricular and Instructional Changes in the Education of the Gifted/Talented

Curriculum can be viewed generically as being composed of organized sets of purposeful experiences in education: These not only include schooling, but also the learning environments and experiences at home and in the community that help students become all that their potentials allow them to be. Addressing himself specifically to the construct of curriculum in the school, Ralph W. Tyler pointed out (1949) four basic questions that must be faced in developing any curriculum and plan of instruction. They are

1. What educational purposes should the school seek to attain?

2. What educational experiences can be provided that are likely to attain these purposes?

3. How can these experiences be effectively organized?

4. How can we determine whether these purposes are being attained?

In other words, an effective curriculum has clearly stated goals and objectives for the learner, learning experiences organized to attain these purposes with continuity and sequence, and evaluation to assess the extent to which the aims have been realized.

What should a curriculum for the gifted/talented be? What features would distinguish this curriculum from that for students of other abilities? To serve the gifted/talented most effectively, curriculums must be appropriately differentiated, articulated from kindergarten through grade 12, sequential in content to be assimilated and in skills to be acquired, substantive in subject matter, and linked in a recognizable construct to the regular curriculum.

219

What constitutes appropriate differentiation for the gifted/talented? Differentiation is a construct of our educational responses to the characteristics of such students. Some general traits that often make gifted/talented students different from their age peers have been delineated by Ruth A. Martinson (1974). Because many of the educational needs that result from these unique characteristics are inadequately addressed by general education, the need for a specialized kind of education for the gifted/talented has arisen. But differentiation does not exclude interaction with other students; indeed, it often means the opportunity to interact freely to enrich one's learning.

Because gifted/talented persons are usually advanced, sophisticated, and mature, appropriate ways educators respond to these characteristics would involve the gifted/talented in experiences with content (subject matter) and processes (strategies for learning), which are complex, intricate, and challenging. The gifted/talented are then more likely to create outcomes and results that reflect their successful assimilation of these intended contents and processes; the products of the gifted/talented are also likely to be advanced, sophisticated, and mature. When translated into desired outcomes in terms of differentiated program goals for the gifted/talented, what would these necessary educational adjustments be? How would they differ from goals in general or regular education?

Goals for Programs for the Gifted/Talented

The key question school districts must address is "What are the essential knowledges, attitudes, and skills that gifted/talented students should acquire through schooling and education?" In the nineteenth century, Herbert Spencer (1854/1966) first raised a similar question: "What knowledge is of most worth?" Recognizing that teachers have too many demands made upon them and that expectations must, therefore, be realistic in light of available time and resources, school systems must concentrate on the absolutely essential elements of education and recognize that teachers may not have the time and/or the resources for the important-but-not-required or the "nice-to-do."

In response to the question posed by Spencer, program goals for the gifted/talented must address six areas where the regular curriculum is inadequate and which are absolutely essential for the gifted/talented. Although listed here separately, a district can readily combine two or more areas when generating its goals. The areas are

1. self-concept and affective needs related to pupils' giftedness (what it means to be gifted, ways to deal productively with giftedness, etc.)

2. interrelationships not only with the gifted but also with children and adults of varying abilities and society as a whole (e.g., finding personal satisfaction in sharing one's gifts with others)

3. courage and skills to become self-directed, happy, and productive gifted/talented individuals

4. advanced thinking skills (related especially to critical and creative thinking)

5. creative production (shifting from emphasis on regurgitation to greater utilization of the known as a springboard to generate exciting personal breakthroughs)

6. intricate, complex, abstract generalizations and diverse product development to increase comfort and versatility in self-expression

Several school districts nationwide follow this goal in construct and emphasis in their programs for the gifted/talented. For example, the Amarillo (Texas) Independent School District's goals are so structured: There, gifted/talented students

• develop healthy self-concepts relative to their giftedness and interact effectively with other gifted students, peers, and society

• use in-depth contents to employ advanced critical and creative thinking skills and generate complex and intricate products appropriate to their giftedness

• acquire the necessary advanced thinking and self-directed learning skills to become independent creative producers

Actually, these goals are the bases underlying *all* curriculums for the gifted/talented in a school system. Indeed, these goals should underlie everything the gifted experience in schooling and education, in and out of school and in interaction with *all* whom they meet in the normal course of events. Therefore, all objectives for all courses, units, lessons, indeed, all of schooling should emanate from and reflect specific ways by which these goals could be carried out at various grade levels in the different disciplines.

To assure that these goals are addressed through the curriculums for the gifted/talented, school districts select from research and literature some guidelines for the teachers and support staff who will be working with these students. In other words, local districts now determine which principles to follow to effect differentiation and, at the same time, to relate directly to how their staff will keep the specific goals defined for these students from contravening the superordinate goals of schooling in an open society. The National/State Leadership Training Institute on the Gifted and Talented Curriculum Council has developed one set of principles of differentiated curriculums. (See figure 1 below from Sandra N. Kaplan's [1979] manual, where she discusses these principles extensively.) Philip H. Phenix's *Realms of Meaning* (1964/1986) also offers some principles that can be readily adapted for the gifted/talented.

Figure 1
Principles of a Differentiated Curriculum
for the Gifted/Talented*

- Present content that is related to broad-based issues, themes, or problems.
- Integrate multiple disciplines into the area of study.
- Present comprehensive, related, and mutually reinforcing experiences within an area of study.
- Allow for the in-depth learning of a self-selected topic within the area of study.
- Develop independent or self-directed study skills.
- Develop productive, complex, abstract, and/or higher-level thinking skills.
- Focus on open-ended tasks.
- Develop research skills and methods.
- Integrate basic skills and higher-level thinking skills into the curriculum.
- Encourage the development of products that challenge existing ideas and produce "new" ideas.
- Encourage the development of products that use new techniques, materials, and forms.
- Encourage the development of self-understanding, i.e., recognizing and using one's abilities, becoming self-directed, appreciating likenesses and differences between oneself and others.
- Evaluate student outcomes by using appropriate and specific criteria through self-appraisal, criterion references, and/or standardized instruments.

For optimal function of any program for the gifted/talented, the identified students must have appropriate educational settings with appropriate curriculums, instructional practices, and teaching materials. However, the staff of the school program designed for these students must receive training in order to work with these students effectively. Administrators, supervisors (i.e., building principals and subject area chairs), and classroom teachers must understand the special educational needs of these students, provide the suitable educational climate for them, and know how to modify and/or adapt curriculum and/or instruction suitably.

---

*Developed by the National/State Leadership Training Institute on the Gifted and Talented Curriculum Council (James J. Gallagher, Sandra N. Kaplan, A. Harry Passow, Joseph S. Renzulli, Irving S. Sato, Dorothy Sisk, Janice Wickless). From Sandra N. Kaplan (1979), *Inservice Training Manual: Activities for Developing Curriculum for the Gifted/Talented* (p. 5). Los Angeles: National/State Leadership Training Institute on the Gifted and Talented. Reprinted by permission.

Staff Development in the Education of the Gifted/Talented

Depending upon the assessed needs of the staff in a particular school system, the major purpose(s) of the training might be (a) to heighten awareness, (b) to motivate staff members to change attitudes and adapt alternate values on what must be taught and how it might be taught most effectively, and/or (c) to build upon what staff members are already doing successfully in teaching and guidance.

In the sciences, because the rate of scientific and technological growth and change is phenomenal, continuing staff development becomes critical. Mandatory, too, are changes in curriculums that bring to the fore recent findings and methodologies. In a sense, then, teachers and supervisors of science must undergo continuous retraining and must plan (and execute) continuous curriculum revisions.

Many times, cooperation between school districts and various community agencies results in interesting inservice linkages. For instance, in the summer of 1987, the Los Angeles Unified School District worked with the University of Southern California to offer "Advancing Science with Advanced Placement" (ASAP), an intensive, federally funded program made up of a two-week summer workshop with monthly follow-up meetings during the school year. For many years, the Los Angeles Department of Water and Power has provided the city's schools with grade-level science-oriented materials, workshops, demonstrations, speakers, and exhibits on water, energy, conservation, and safety.

Even if the ideal setting is present in terms of program prototype(s) and substantive curriculums, very little of significance happens without a staff with appropriate attitudes and necessary skills in teaching the gifted/talented. Staff development for those who work with the gifted/talented, then, might be viewed as organized sets of experiences that change educators so that they relate effectively to gifted/talented students.

The Gifted/Talented in Science and in Other Areas

What is the distinction between students who are gifted/talented in science and those whose skills lie in areas other than science? Many times both categories of students are in the upper levels of academic achievement. In order to serve the two groups we must first locate their members. Martinson (1974), Joseph A. Platow (1984), and Gallagher (1985) describe various ways to find gifted/talented students with and without special science aptitudes as well as indicate the advantages and disadvantages of the methods used.

Paul F. Brandwein (1955/1981), in attempting to resolve problems in the identification of the gifted, developed a program in science in which self-selection or self-identification was the guiding principle. In essence, the three-year program of science (biology, chemistry, physics) offered an alternative to students in their sophomore year at Forest Hills High School (New York City): Students with an interest or career plans in science could enter—without regard to IQ or prior achievement—a program offering opportunity for advanced work in science and for research on originative science problems. Approximately 10 percent of the student body taking science (some 200 students on the average) elected the program. (Not all continued through the three years or persevered through the research required—approximately 15–20 percent turned to other areas. In the 10 years of the program where assessment was possible, the school placed approximately the same number of finalists in the Westinghouse Science Talent Search as did two select schools of science (Brandwein, p. 47, and in this volume).

Thus, Brandwein does not accept a distinction between the gifted/talented in science and the gifted/talented in other areas. He presents the thesis that all students should be given equal access to a program in a school without prior testing, stressing that "it is by their work we come to know them" (Preface, 1981 edition, p. xi). Brandwein continues,

> It seems to us that at present it is not necessary for a teacher to depend on tests which seek to identify students with high-level ability in science in the early grades. If qualified teachers were to furnish sufficient opportunities in science to all students, those with high-level ability would come forth and identify themselves. (p. 24)

The concept of selection through performance is also advocated by Joseph S. Renzulli, Sally M. Reis, and Linda H. Smith (1981), who recommend a "revolving door policy" so that gifted students may have a choice in their work. Michael A. Wallach (1985) makes the essential point that giftedness in an area may be "field-specific"—that is, *work* in a specific field tends to bring forth the trait of giftedness; predictive tests are uneventful.

For these reasons, among others, it is important to seek clues to the identification of the gifted/talented through their *work*. Naturally, this work is offered mainly or even only in well-contrived, appropriate curriculum and instruction. Therefore, the search for provisions for the gifted/talented in science or in other areas should take place through opportunities for the kinds of experiences described in the foregoing pages—particularly those in figure 1 (page 222).

The curriculum, especially as expressed in course work and instructional practices, that most schools offer students gifted/talented in science should be different from that provided for students with special skills in other areas.
• Those gifted/talented in areas other than science generally follow elementary school course work in science (assuming some exposure there) with

middle school general science, which is often a sequence of three years (frequently lasting two to three periods a week) in life, earth and space, and physical science. In most instances, those gifted in the performing and kinesthetic arts end their course work in biology, as do other gifted/talented college-bound aiming at careers other than science. Both groups use biology to "satisfy" their laboratory science requirement if only *one* such course is a prerequisite for graduation or for college entrance.

• Those science gifted/talented who may choose science as a career generally go on to take chemistry and physics. In small high schools, the latter courses may be provided in alternate years. Small schools may not make special provisions for the gifted/talented in science; the available arrangements for gifted/talented individuals are generally in the form of pullout project work.

• A number of schools, particularly those in or near college and university towns or government or corporate laboratories, offer opportunity for originative work. (See elsewhere in this volume, particularly A. Harry Passow's "School-University, Laboratory, and Museum Cooperation in Identifying and Nurturing Potential Scientists.")

• In most circumstances, the gifted/talented in science may identify a talent in science through their activity in a field-specific area (i.e., biology, chemistry, geology, or physics). In addition to required course work, they may select themselves through the Science Talent Search, conducted in a good number of high schools throughout the nation, by submitting reports of individual investigations together with certain records and recommendations by teachers. Or students may enter State Science Talent Searches now offered by a majority of states. (See Brandwein's papers in this volume.) Or the gifted/talented in science who have selected a high school science major may identify themselves as having *talent* in science through their work in a college or university (Humphreys, 1985).

Although the changes in science curriculum examined in this paper make provisions for opportunities in science instruction in the first three categories indicated above, the practice common in the United States is still typically the last. However, the evidence appears to indicate that a rich program in science, one that makes provisions not only for courses but also for individual projects and originative work, often stimulates students to choose a lifework in science and engineering. This paper notes that new provisions in curriculums are in the offing: These may provide a kind of science more fulfilling for those who do not incline to careers in science than that generally available.

Curricular and Instructional Changes in Science

Phenix in *Realms of Meaning* (1964/1986) firmly states that the ultimate goal of any course of study should be to fulfill meanings in life for children. He elaborates:

... an understanding of the fundamental patterns of meaning enables the educator to make a successful attack on the various sources of frustration in learning, such as fragmentation, surfeit, and transience of knowledge, by showing what kinds of knowledge are required for full understanding and how the essential elements may be distinguished from the unessential ones in the selection of instructional materials. In this fashion the curriculum may become a means for the realization of the distinctively human potentialities. (p. 49)

Phenix then proceeds to organize the major disciplines and other important areas of living into six categories, or "realms," with which educators must deal directly if they are to help students fathom the significant meanings of life and living. The physical sciences, life sciences, and social sciences (where empirical studies are required) are in the realm he labels "empirics"; this realm concerns itself " ... with material truth expressed in the general laws and theories of the actual world as studied in the *natural* and *social sciences"* (p. 26, italics mine). Thus, the science educator who is searching for ways to work more effectively with the gifted/talented student in areas other than science would ask a variation of the question posed earlier: *What essential knowledge, attitudes, and skills characteristic of the sciences should nonscience gifted/talented students assimilate into their lives before they graduate?*

Goals in Science Curriculum and Instruction

In answer to the question above, certain areas in science must be successfully addressed between kindergarten and grade 12 for students gifted/talented in areas other than science. The College Entrance Examination Board in *Academic Preparation in Science* (1986) has suggested a framework of academic outcomes for students:

1. *Gathering Scientific Information:* The skills to gather scientific information through laboratory, field, and library work
2. *Approaching Scientific Questions Experimentally:* [The possession of] sufficient familiarity with laboratory and field work to ask appropriate scientific questions and to recognize what is involved in experimental approaches to the solutions of such questions
3. *Organizing and Communicating Results:* The ability to organize and communicate the results obtained by observation and experimentation and [the ability] to interpret data presented in tabular and graphic form
4. *Drawing Conclusions:* The ability to draw conclusions or make inferences from data, observation, and experimentation, and to apply mathematical relationships to scientific problems
5. *Recognizing the Role of Observation and Experimentation in Theories:* The ability to recognize the role of observation and experimentation in the development of scientific theories

6. *[Understanding] Fundamental Concepts:* [The ability to] understand in some depth the unifying concepts of the life and physical sciences (from pp. 20–23)*

Relating these suggested student outcomes in science to the suggested program goal areas for the gifted/talented discussed earlier (figure 1, page 222), science educators can arrive at appropriate science goals. For example, the following science goal incorporates items 3. through 6. of the goal areas for the generally gifted/talented listed on pages 220–221, while simultaneously dealing with the first four outcomes for the science-gifted listed above. The College Board suggested (1986) that students gifted/talented in areas other than science investigate through experimentation self-selected scientific questions; gather essential information through related laboratory, field, and library work; and communicate in an original manner the results drawn through appropriate inferences and conclusions.

Guidelines for Science Educators
To Differentiate Curriculum and Instruction
for the Gifted/Talented in Areas
Other Than Science

By examining the criteria the Biological Sciences Curriculum Study staff has followed since 1958 to plan, conduct, and implement new projects, science educators can use the insights of a respected science curriculum group as guidelines for their own efforts in this area. The implications are also useful for sciences other than biology. Joseph McInerney (1986/1987), quoting Arnold B. Grobman's earlier framework, asks

Criterion 1: How well does the information in question illustrate the basic, enduring principles of biology?
Criterion 2: Do teachers and administrators perceive the proposed material as useful and important?
Criterion 3: What is the relationship between the proposed curriculum materials and the prevailing context of general education?

The essential thrust of the question—Criterion 3—concerns itself with the phrase "the prevailing context of general education" in relation to that for the gifted/talented. To those unfamiliar with the contexts that prevailed in critical curricular development prior to 1983, it might seem as if the curricular contexts now in use are new. This criterion cannot be fully understood unless

---

*Reprinted with permission from *Academic preparation in science: Teaching for transition from high school to college,* © 1986 by College Entrance Examination Board, New York.

current curriculum developers recognize that their predecessors facing similar problems and solutions in similar contexts had, in the past, mounted concentrated and wide-ranging efforts affecting curriculum and instruction in science and mathematics.

In briefest review, this "crisis" in curriculum and instruction was a general one affecting schooling in all areas for all students. The crisis did *not* arise with the marvel of Sputnik; it was most likely a response to concentrated attacks on schooling by the postsecondary education community, especially as defined by Arthur Bestor (1955). Indeed, committees of concerned educators and scientists had been meeting to consider curricular and instructional changes in science and mathematics since the early 1900s. Nonetheless, Bestor was a central figure in a group of critics who pressed for the return of "standards"—meaning generally what is covered by college preparatory curriculum. Essentially, this group called for the education of an elite. But the Soviet Sputnik made the point as no book, argument, or supplication could. When it circled the globe in 1957, it accelerated both action and reaction already under way. But why do we need to wait for a serious "crisis" nationally or internationally to improve teaching in any area, particularly in science? We don't; there was no failure of learned groups to call into question policy and practice (curriculum and instruction) in science teaching over the decades.

In 1958, shortly after it was organized, the Biological Sciences Curriculum Study (BSCS) invited Paul deHart Hurd to undertake a historical—and critical—study of secondary school curriculums in biology. His study took him into the wider field of science curriculum. An attempt here to summarize Hurd's remarkable compilation of studies and the analysis and synthesis he brings to bear on them would have to fail. In his study, the point and counterpoint of numerous probes of reputable scholars are sequenced in the five decades between 1900 and 1950, ending with "The Crisis in Science Education and a Reappraisal, 1950–1960." While Hurd emphasizes trends in biological education, he reports significant recommendations and trends in science education generally.

One is taken by the fecundity of the contributions of students of curriculum about which Hurd's study reports, as well as by the contrasting, sometimes contradictory, views of teaching and learning in the area of science these thinkers offer. Their recommendations are stated with confidence, and we may be sure all the commissions Hurd summarizes wished to advance the causes of learners, teachers, and schools. But, while the science espoused was based on scientific methodologies that were by their very nature *evidential,* one may note, at the same time, that not all the studies of curriculum and instruction offered similar assurances of validity.

Nonetheless, Hurd's study is of central importance in understanding the role of social events in the formative years of science curriculum and instruction. Within the compass of this discussion, to avoid reinventing the wheel, we must content ourselves with a brief analysis of the curricular thrusts and parries. We begin with the seminal work of the early 1950s before placing it into the larger course of events: As the powerful redirection of science curriculum and instruction at midcentury extended to the 1960s; then, innovation gave way to the languishing times of the 1970s and 1980s; now, there appears to be an effort at rejuvenation.

For example, one of the early studies was organized in 1952, when Harvard University President James B. Conant, aware of the coming crisis in science education, called upon a national group of 50 outstanding science teachers and 25 science supervisors to discuss the grave state of affairs. Conant, steeped in the philosophy of science and himself a brilliant chemist, had wide access to the body of scientific study. Indeed, he served as an early adviser to the Office of Scientific Research and Development in the 1940s and 1950s. In the discussions he conducted with the teachers assembled in 1952 and 1953, he emphasized that the cumulative body of science consisted of a series of conceptual schemes based upon preliminary observation and experiment (inquiry), leading to further observation and experiment, leading to newer conceptual schemes. With this position, he expressed not only his own thinking but also that of prominent theorists in science and science education (Conant, 1947). As we shall see, this conceptual approach became the core of a construct for science curriculum and instruction: *The conceptual approach remains the heart of modern science curriculum as inquiry does of modern science instruction.*

The consensus, reached by the teachers during two weeks of contemplation and discussion of newer curricular and instructional strategies, was as follows:

1. The curriculums in general science, biology, chemistry, and physics were based rigidly in the textbook, and the method of instruction in class was based in the lecture.

2. Sessions in the laboratory were based upon exercises to "verify conclusions already stated in the textbook." Except in a few high schools, there was little if any independent study.

3. Very few (most said "no") opportunities were given for independent study as well as opportunity for individual research by gifted students (particularly those inclined to a career in science). In short, there was little attempt at *differentiation* of opportunities for the students in science. The vast majority of the exceptions constituted those schools stimulated to enter the Science Talent Search.

4. Teachers who were adequately trained in science were in serious demand; they were, however, generally unavailable.

5. The curriculum in science was based upon fragmented topics. There seemed to be no coherent philosophical or curricular base (summarized by Watson, Brandwein, and Rosen, 1953).

The Conant group thus described the nature of the situation, defining the needs of science instruction in 1953.

Conant had planned to pursue his investigation into reform in science, but other events intervened. In 1953, he was appointed U.S. high commissioner for

West Germany; then, ambassador. When he returned in 1957, the committees devoted to science reform had begun their work; thus, he took on the important role of analyzing general education, turning his attention to the reform of American schooling. Especially in such works as *The American High School Today* (1959), he established a structure for comprehensive high schools organized to serve the diverse needs, interests, and abilities of the young.

The Organization of New Curriculums— Reform Groups in Science

In 1952–1953, the federal government had not yet been stimulated by the Soviet hardware, which was to go into orbit in some five years. Nonetheless, reacting to the obvious need for change in curriculum and instruction in science and mathematics, the community of scientists began to intervene in studies that eventually had considerable impact on teaching in science.

Major committees and commissions, including federal and state groups, began to organize. Some prominent ones included the following: In 1952, Jerrold Zaccharias of the Massachusetts Institute of Technology convened informally the Physical Science Study Committee (PSSC); after its reorganization (1955–1957), the School Mathematics Study Group (SMSG) was formed. Next came the Chemical Bond Project (1958–1959); the Biological Science Study Committee (BSSC) (1958–1959); and the Chemical Education Materials Study (CHEMS) (1959–1960). Between 1958 and 1961, Congress passed the National Defense Education Act, providing some $95,000,000 to $100,000,000 in the early 1960s (this would be about $376,000,000 to $395,000,000 in 1988) for science education, as well as explicit acknowledgment of the relation of education to national defense. The initial funds for organization were supplied by the National Science Foundation, by the Ford and Sloan Foundations, and by a number of industrial corporations. But when Sputnik orbited, a torrent of additional funds flowed from the federal government and the private sector as well.*

In the succeeding years, 1960–1965, a remarkable series of unusual textbooks, laboratory, research and project guides, and laboratory materials

---

*Some estimate that the funds available for curriculum and instructional development over the 8–10 years of activity by all groups during the 1950s and early 1960s were in the range of $80,000,000 to $100,000,000 (in 1988 dollars, about $316,000,000 to $395,000,000), possibly more. These funds also subsidized numerous institutes given over to the reeducation of teachers. On this subject, see Grobman (*The Changing Classroom,* 1969), an exceedingly important account that is highly applicable today. In that work, Grobman, who was director of BSCS, gives a detailed account of the preparation of curricular and instructional materials in biological science that could furnish a paradigm for curricular development in all the sciences.

specially designed for new approaches in inquiry (i.e., observation and experimentation) were developed for students at all presecondary levels. New kinds of films and filmstrips to accompany the textbooks were also produced as well as materials (pamphlets, specially designed laboratory equipment, and guides for using all equipment). Further, the various groups developing these new, imaginative curricular and instructional materials also conducted intensive, wide-ranging, short- and long-term residential institutes for teachers nationally. Many of these meetings were held as a full summer program as well as during the year. Teachers were able not only to gain familiarity with the curriculums and the instructional materials but also to practice using the variety of instructional approaches they had met.

The BSCS approached the task by developing three programs accompanied by new kinds of texts, laboratory materials, teaching guides, and instructional materials for the classroom: the first, a "general approach" to the concepts of biology; the second, a "molecular approach," which emphasized contemporary researches in molecular biology; the third, an ecological treatment. The curriculum materials, developed mostly for high school students by PSSC, CHEMS, BSCS, and SMSG, as well as by groups preparing materials in earth science and astronomy, were not designed especially for gifted/ talented young but were aimed at the schooling and education of *all* students.* So too were the three curriculums intended largely for the elementary school—Elementary Science Study (ESS), the Science Curriculum Improvement Study (SCIS), and Science, A Process Approach (SAPA), the latter contribution from the American Association for the Advancement of Science.

It is also well to emphasize that research projects and research problems were available even in the late 1950s and early 1960s for gifted science students. There was thus a clear application of what is now known as "differentiated education" for both the science-gifted as well as the gifted in areas other than science. For example, BSCS had a special committee that prepared materials specifically for the gifted (Grobman, 1969).

Much could be written about the curricular innovations, but suffice it to say that their footholds were these:

- *A conceptual approach to curriculum*
- *An inquiry approach to instruction* (with laboratory activity based in innovative investigations)
- *A differentiated approach* in investigation for group laboratory work and individual probes (research) on problems selected by the students themselves

---

*SMSG works in both elementary and high schools. In the opinion of many science educators and scientists, all these materials could, with revision and updating of content for free differentiation by schools and school districts, be useful for heterogeneous student populations now.

These curriculums were designed primarily to organize content on a conceptual basis, using the disciplines of science as overall structures to increase interest of all students by emphasizing inquiry in the processes of scientific study. The curriculums did not emphasize social issues arising out of then-existing technology. On this, see also Hurd (1961).

Diverging Curricular Efforts

During the 1970s, mindful of the preceding de-emphasis on the social and cultural consequences of science, the BSCS produced relevant materials on population, natural resources, and drugs. The Educational Development Corporation, under the leadership of Zaccharias (see above), prepared curriculums linking science and social studies. The Educational Development Corporation also produced curriculums and films attending to the special needs and interests of minority groups in science and mathematics. In 1970, the first edition of Harvard Project Physics (by Gerald Holton, Fletcher Watson, and F. James Rutherford) was published. It developed a new program in physics that, with the help of experimental films and other accompanying materials, emphasized the cultural implications of the field.

In addition, the National Education Association and National Science Teachers Association proposed in 1959 in their project on academically talented students the following guidelines for curriculum development in science:

- The content should be developed in the light of contemporary scientific thinking and theorizing. As often as possible, pupils should glimpse the frontiers of scientific investigation. . . .
- Content areas should be developed in depth. To this end, a few well-chosen areas should be studied. . . .
- The overall science program for the talented should be planned carefully to assure continuity of the program from elementary school through college. . . .
- Science content should be developed in such a way that pupils have many opportunities to work with science materials, equipment, and apparatus. . . .
- The choice and development of content areas should be planned to reveal the relationships among the sciences. . . .
- The science program should be planned for the early and increasing application of mathematics to the precise formulation of scientific laws and principles. . . .
- The teaching of science should be concentrated increasingly on the development of the concepts, principles, broad generalizations, and great issues of science rather than on the accumulation of unrelated facts. . . .
- Full advantage should be taken of the science resources of the community for enriching and supplementing the science program for talented students. . . .
- Science should be taught in such a way as to reveal the influence of science in such *other areas of culture as politics, economics, world outlook, standard of living, and the influence of these areas in the development of science. Recognition of the social responsibility of science should be developed.* [Italics mine] (pp. 14–25)

In the ensuing decades, the science curriculums developed in 1958–1962 had considerable influence on publishers' preparation of textbooks and other instructional materials; in many cases, the materials produced by the various curricular study groups were revised over three editions. The chemistry curriculum was so organized that a second edition could be published without the supervision of CHEMS; PSSC and BSCS continued a certain collaboration with the publishers. However, in the 1970s the influence of these innovative programs began to decline. And in 1983, NSTA published its account of their waning.

Recent Developments

In a thumbnail sketch, the findings published in the NSTA yearbook (Brown and Butts, 1983) *Science Teaching: A Profession Speaks* conclude that "a wide variety of writings and reports, current projects, and research converges in a characterization of *current science* as plagued by 10 common, recurring problems" (p. 4 [Italics mine]):

1. *The textbook is the curriculum.*
2. *Goals are narrowly defined.*
3. *The lecture is the major form of instruction, with laboratories for verification.*
4. *Success is evaluated in traditional ways.*
5. *Science appears removed from the world outside the classroom.*
6. *A shortage of science and mathematics teachers has led to the widespread use of unqualified and underqualified teachers.*
7. *The outdated curriculum neglects the needs and interests of most students.*
8. *Current science instruction ignores new information about how people learn science.*
9. *Supplies, equipment, and other resource materials are severely limited or obsolete in most science classrooms and laboratories.*
10. *Science content in the elementary schools is nearly nonexistent.* (pp. 5–11, with supporting statements)

The preceding brief history of the enormous effort and time, which took the energies of several hundred leading scientists, teachers, and leaders of the teaching community between 1957 and 1965, as well as those educators who attempted revisions in curriculums during the 1970s, should be compared with the ten deliberate conclusions offered by the 1983 NSTA yearbook. These conclusions come some 30 years after Conant called his group together and some 25 years after the incredible thrust of the reform movement in science schooling and education discussed in this paper (pages 227–233). Readers who have studied Hurd's analysis will note certain familiar oscillations in the current recommendations for curricular reform with those of the past.

Below are some features of the situation that may have led to the debacle of the curricular reform movements of 1955–1965. The 1983 NSTA yearbook points to

- *Federal, state, and local responsibilities.* In the U. S., *delivery* of precollege education is clearly a state responsibility* . . . [And research] is an appropriate sphere of activity for federal agencies. (p. 13)
- *Federal initiatives.* It is the National Science Foundation [NSF], however, that has the authorization to maintain the health of science and the science education which supplies scientists. As described in Public Law 507 . . . : "The Foundation is authorized and directed to initiate and support . . . science education programs at all levels. . . . " The existence of the present crisis and the variety of Congressional initiatives are clear evidence that NSF's policy-making body, the National Science Board, has been negligent over a long period of time. . . . (p. 14)
- *NSTA Initiatives.* NSTA has been involved, in a variety of ways, in addressing the crisis in science education. The NSTA position statement "Science-Technology-Society: Science Education for the 1980s" (Appendix, pp. 109–112) has received widespread acceptance. The officers and staff of NSTA are working with the media to inform the public . . . *We,* [i.e., science educators] not others, are setting the agenda for science education in the 1980s and beyond. (p. 17)

Moreover, this and subsequent yearbooks recommend two major reforms: *First,* NSTA called for a National Laboratory for Science Education (1983, pp. 101–108). Considering that a boom-and-bust cycle—both in education in science and in education for the gifted/talented—does indeed exist, NSTA proposed the following ameliorative responses: (a) A K–12 experimental school and (b) "think tanks" of scholars to study developments in science education. The school might serve as an "early warning system" to prevent deterioration of curriculum and instruction in science programs (for all students). The National Laboratory would also have arms for research, development, and communications. *Second,* the 1983 yearbook offered a position statement—*Science-Technology-Society: Science Education for the 1980s*—which includes curricular and instructional positions (pp. 109–112). These positions have been further explicated in the 1984 and 1985 yearbooks.

Since the publication of the 1983 yearbook, NSTA has begun its program of correction with three publications: *Redesigning Science and Technology Education* (Bybee, Carlson, and McCormack, 1984), *Science Technology and Society* (Bybee, 1985), and *Focus on Excellence: Science Teaching and Career Awareness* (Penick, 1987).

The yearbooks detail the kinds of necessary projects now being undertaken to achieve the goals that need to be met in renovating and reconstructing science curriculum and instruction for the 21st century. But those goals,

---

*(Nonetheless, in the past 20 years, federal support has been reduced in the funding of teachers' salaries and in preservice and inservice education.)

which have yet to be fulfilled, invite considerable discussion and participation. Suggestions from these publications and others, as well as ideas coming out of various conferences, are just beginning to take effect. Yet at this writing, the "back-to basics" movement is still having a profound—and adverse—effect on science. After all, this movement of the middle 1970s and early 1980s has been around for some time. Its view of schooling and education runs counter to that of programs melding the newer science with its impact on interdependent technologies and their incredible, not fully comprehended, influence on social changes. Merely to use the title "Science, Technology, and Society" (STS) promises a new direction, but its offhand use as a rubric for the profound shift in policy suggested by the philosophical, curricular, instructional, and administrative changes true STS requires slights the gravity of parallel changes in schooling, education, and, concomitantly, society.

Focus on Society

Daniel Bell's analysis in *The Coming Post-Industrial Society: A Venture into Social Forecasting* (1973) has implications relevant to those changes STS infers and implies for the direction of philosophical, curricular, instructional, and administrative directions. Bell's work could serve as a primer for developers of a policy for schooling and education in the coming decades. Thus, the NSTA statements of the 1980s may be considered in light of how the networks of science impinge on those of society: These statements offer new conceptualizations of the relationships between science and society.

Moreover, Bell's philosophy, as reinterpreted in the yearbooks, helps to frame a new responsibility in the education of science teachers: Not only do teachers need grounding in the science they will teach, but also they need to understand the particular technocratic society that will be affected by the science they teach and by the students they prepare for life in it. Specific examples of concept and content may be taken from the 1984 and 1985 NSTA yearbooks proposing curricular and instructional changes through STS. This writer runs the risk of inviting comparisons by emphasizing certain papers as examples of the curricular and instructional changes that will be required but does so in the hope that workers in curriculum will examine the entire texts of these two yearbooks. Their references are also rich in pertinent material.

Thus, please note Rodger W. Bybee's "Global Problems and Science Education Policy" (1984, pp. 60–75)—particularly the ranking of problems in tables 1 and 2 (pp. 65–67). Scientists and engineers considered 12 societal problems related to science and technology as follows: population growth, war technology, world hunger, food resources, air quality, atmosphere, water resources, land use, energy shortages, hazardous substances, human health and disease, extinction of plant and animal species, mineral resources, and the problems posed by nuclear reactors. It is clear that examination of the impact of these problems would, in turn, pose problems not only for teaching in the varieties of communities in the United States but also for the developers of curricular and instructional materials. So, too, would the issues raised in Part III of the 1984 yearbook "Redesigning the Essential Components of Science and Technology Education" (pp. 149–244), as well as Part IV "An Agenda for Action" (pp. 245–248). All these pieces offer concepts central to the curricular movement embodied in STS.

In highly useful discussions for curricular constructs, authors writing in the 1985 yearbook emphasize

a basic theme (Part I)
rationale and goals (Part II)
a new curriculum emphasis (Part III)
new instructional strategies (Part IV)
managing change (Part V)

In short, both the 1984 and 1985 yearbooks are basic to adopting current curricular reforms both for the citizens who will become our experts in science and technology and for the citizens who will apply those changes with insight and foresight—our change-agents in society. These books should also be useful as guides not only to those who are elected to make the laws but also to those who live by them.

Indeed, projects in some 20 states (and more to come) center on curriculums generally titled *STS* or some variation upon it. Essentially, the samples of curricular materials already developed are intended to prepare all students, both those gifted/talented in science and in other areas, to inquire into, probe, analyze, and synthesize the impact of *science* and *technology* on life and living and on *society* generally.

One should note, however, that curricular efforts to address the "Science of Society" in other times did not prevail. For example, there is Jerome Bruner's *Man: A Course of Study* (1966, pp. 93–101). The course curriculum and instructional materials engendered considerable enthusiasm and, in some quarters, concern and active antagonism. Because social problems addressed through the methods of the psychologist, and intended for the elementary school, did not—for many parents and some school districts— seem appropriate for children, the instructional program did not survive its first edition. Lawrence Senesch's (1973) attempt to introduce the principles of economics (treated from the viewpoint of some social scientists) floundered in its later editions. The impact of science and technology on society is of considerable, if not critical, significance to the future course of our civilization. Is the prevailing curricular context favorable today?

To help gifted/talented students in areas other than science, as well as the others who will also be responsible for the future of civilization, educators are required to modify the curricula *in all disciplines,* not only in science. This modification must be accomplished systematically to ensure curriculums—including the sciences—which are appropriately differentiated, articulated kindergarten through grade 12, sequential in content to be assimilated and skills to be acquired, substantive in subject matter, and linked meaningfully to the regular curriculum.

Even as this is written, F. James Rutherford, of the American Association for the Advancement of Science, heads the ongoing *Education for a Changing Future* (originally called *Project 2061*). (See "By Way of Beginning" in this volume.) The deliberations of this project's collaborating scientists and educators will surely have a further impact on the thrust in policy, philosophy, curriculum, instruction, and administration of schooling and education.

Principles for Differentiating Science Curriculums for Students Gifted/Talented in Areas Other Than Science

Some 30 years after the early intervention in 1952–1965, this writer has formulated some suggested guidelines for curriculum selection, development, and/or modification for gifted/talented students in areas other than science. For those young, science courses and units should

1. be consistent with the district's program goals for gifted/talented students

2. be organized around themes, problems, or issues in interdisciplinary settings

3. emphasize enduring generalizations, concepts, principles, and theories, instead of only facts and statistics, to effect more lasting understanding of life meanings

4. be basically open-ended and offer gifted/talented students in areas other than science opportunities to probe intensively in detail into some self-selected science fields

5. lead toward mathematical and scientific literacy by assisting the gifted/talented in areas other than science to become comfortable with the symbols and language of mathematics and science

6. emphasize independent inquiry and research including problem finding, hypothesizing, problem solving, and data collection, analysis, and validation

7. include deliberate ways for the gifted/talented in areas other than science to develop and apply inductive and deductive thinking, dialectical reasoning, analogical thinking, decision making, and self-evaluation skills

8. encourage gifted/talented students in areas other than science to extend knowledge and to experience personally novel insights and other exciting scientific breakthroughs

9. create situations where the gifted/talented students in areas other than science can experience working as a scientist does

The foregoing, of course, is a composite set of guidelines. All appropriately differentiated science curriculums for students gifted/talented in areas other than science will not satisfy all these criteria all the time. Rather, separate units will incorporate different guidelines. If, however, a given unit is truly differentiated, it will certainly embody several of these items simultaneously. Note that not only do these guidelines relate to the principles of differentiation (figure 1—page 222) and the science recommendations for the gifted/talented above, but they also respond to the unique characteristics of gifted/talented students. Ultimately, these guidelines become ways by which educators can ensure that they are actively helping gifted/talented in areas other than science attain the stated differentiated science goals of a school system.

An Articulated, Differentiated Science Curriculum for the Gifted/Talented in Areas Other Than Science

Desirably, a school district should develop an articulated science curriculum, kindergarten through grade 12, which, as a continuum of courses, guarantees that the science curriculum guidelines in the preceding section are met. *Improving or developing this type of curriculum takes a long-term commitment of at least five to seven years, because it involves training the writers and then composing, modifying, piloting, field-testing, revising, and then implementing the curriculums.* Indeed, it may well be that by the time the curriculum is finished, the context and content of science (as catalyzed by the computer and the artifices of the information age) may have begun to change radically. This writer (1987) has described one way in which curriculum has been developed for various disciplines (including science) for the gifted/talented through his piloted and field-tested C^3 Model—Comprehensive Curriculum Coordination (figure 2, below).

Obviously, while this articulated, differentiated curriculum is under development, the gifted/talented students in areas other than science cannot be expected to take science courses without some appropriate changes. Two ways by which some short-term changes might be effected would be (a) by training science teachers working with gifted/talented in areas other than

Figure 2
The C³ Model
Comprehensive Curriculum Coordination

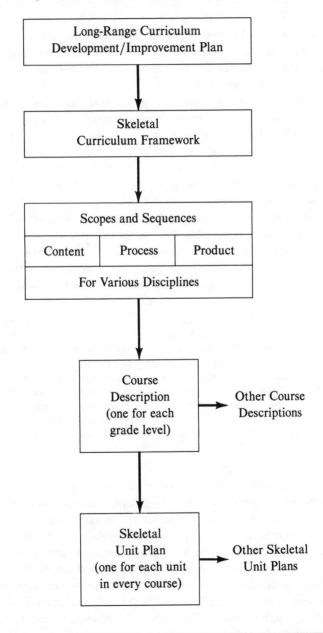

science to modify curriculum and instruction, and (b) by selecting and using more suitable materials, some already published and some being devised in new computer-driven techniques. The latter also have a promise for adaption to the needs of individualized instruction.

First, science teachers should learn to apply the science curriculum guidelines (pages 237–238) above to alter their teaching objectives and experiences to differentiate them for gifted/talented learners. The teachers should try to emphasize more open-ended inquiry, self-directed learning, and primary and multisource approaches while dealing with the relatively stable concepts of the sciences.

Next, school systems should select published materials or prepare specially devised materials for the science courses for the gifted/talented in areas other than science, with specific criteria that parallel the science curriculum guidelines above. Most of the instructional materials developed by BSCS, PSSC, and CHEMS are still available for reference. So are the instructional materials of the reform curriculums of the 1970s. In fact, most of the courses (amended by other publications and brought up-to-date by mimeographed material prepared by selected teachers) are still in use.

However, it is lamentable that, as is customary in response to cultural crises, society often sweeps useful work aside. For example, many of the excellent curricular materials prepared for the elementary school during the 1960s were unused even then; they are certainly ignored now.

Shymansky, Kyle, and Alport looked at research studies reporting on the effects of the activity-oriented, inquiry-based elementary programs of Elementary School Science (ESS), Science Curriculum Improvement Study (SCIS), and Science—A Process Approach (SAPA) in the performance areas of student achievement, attitude, process skills, related skills, creativity, and Piagetian tasks. They state, "Our quantitative synthesis of the research clearly shows that students in these programs achieved more, liked science more, and improved their skills more than did students in traditional, textbook-based classrooms." ... But many teachers never used these projects even in their heyday, and even fewer are using them now. (p. 61, NSTA 1983 yearbook)

Disuse is also the fate of many of the elementary science curricular and instructional materials designed by groups of scientists and educators who collaborated in their development during that early period of reconstruction of science curriculums. And a number of publishers worked in tandem with the educators in the preparation of the instructional materials devised during the reform movements of the 1950s, 1960s, and 1970s, developing high school and elementary school instructional materials based upon the conceptual and inquiry approach created before and after Sputnik. A number of laboratory approaches also form a *base* for the selection and development of curriculums for today.

Wider Views

To return to a discussion of programs for the gifted: Further attention to "instructional" schema (as distinguished from schema of "content"), such as those developed in the methods of self-selection (Brandwein, 1955/1981), the revolving door model (Renzulli, Reis, Smith, 1981), and the "individual" instructional device and the "Unit-Pak" by which it is implemented (William Georgiades and seven researchers, 1979), may provide clues to constructs that permit the gifted in areas other than science and the gifted in science to live and work together. All these models permit well-developed, even rigorous courses of study for the gifted/talented whether the students are committed or not yet convinced that science, social science, or any other area of intellection is appealing as a lifework. What is available to them is a personal choice, through long-term originative research in science, intensive work beyond the requirements of the course; this work may include (as these models offer) individual, self-directed research.

Thus, if a district developed its science curriculum by using the C^3 Model (figure 2, page 239), it would have a major theme, problem, or issue (with accompanying overarching concepts, which apply from kindergarten to grade 12). Above each major theme, problem, or issue, the district would in turn design a science scope and sequence, which would itself have themes, problems, or issues, which would again apply to concepts for each science course at all primary and secondary levels. (And these themes, problems, issues, and concepts would developmentally relate to the original overarching ones.) The district would then create course descriptions based upon its scope and sequence. Finally, the district would have units, which would include objectives for the learner and learning experiences related to the program goals for the gifted/talented, as well as integrating the principles of differentiation with the science curriculum guidelines above.

Of course, in the final analysis, it is the teacher who really decides whether or not the students gifted/talented in areas other than science will benefit from suitable science experiences. It is the teacher who ultimately determines the climate and environment in which learning will occur in the classroom setting, and it is s/he who must know how to use the curriculum wisely. This teacher must personally believe in self-directed learning and scientific inquiry, be knowledgeable about, comfortable in (and, if possible, have a passion for) the sciences, and view science as a group of disciplines with broad bases and with numerous possibilities for integration with other fields. Such a teacher can create passion for the sciences in students gifted in other areas and thus unlock new possibilities.

In Sum

1. This discussion is concerned with certain essential similarities and differences in the curricular and instructional strategies and tactics affecting students gifted/talented in areas other than science, as well as those who aspire to a lifework in science. The essential difference seems to be that those inclined to science are, while still in high school, offered opportunities for individual research or investigation into problems without known solutions.

2. The recommendations of commissions and bodies of educators concerned with the schooling and education of the gifted/talented are briefly sketched.

3. A review of early curricular thrusts developed by scientists and educators in the 1950–1965 period is made; the return in emphasis to the "basics" (and away from the 1950–1965 movement) of the 1970s and 1980s is described.

4. The cycle is now being closed by a return of curriculum movements to an emphasis on science and technology to meet the requirements of the postindustrial society. Thus, with a melding of the impact of science and technology on society, new curricular and instructional directions are beginning to appear in STS programs.

5. Because the problems in fashioning the new curriculums promise to be different, a discussion of the principles followed by developers of plans *of* instruction (curriculum) and plans *for* instruction (teaching) are discussed.

6. It seems that curriculum may still be based on the construct of conceptual schemes, while instruction may still be based in the inquiry approach. The new element of the 1980s curricular movement seems to be an emphasis on science and technology related to significant social problems.

References

Bell, Daniel. (1973). *The coming post-industrial society: A venture into social forecasting.* New York: Basic Books.

Bestor, Arthur. (1955). *The restoration of learning.* New York: Alfred A. Knopf.

Brandwein, Paul F. (1955). *The gifted student as future scientist: The high school student and his commitment to science.* New York: Harcourt Brace. (1981 reprint, with a new preface [Los Angeles: National/State Leadership Training Institute on the Gifted and Talented])

Bruner, Jerome. (1966). *Toward a theory of instruction.* Cambridge, MA: Harvard University Press.

California State Department of Education. (1979). *Principles, objectives, and curriculums for programs in the education of gifted and talented pupils: Kindergarten through grade twelve* (rev. ed.). Sacramento: Author.

College Entrance Examination Board. (1986). *Academic preparation in science.* Princeton: Princeton University Press.

Conant, James B. (1947). *On understanding science.* New Haven: Yale University Press.

Programs

Conant, James B. (1952). *Modern science and modern man.* New Haven: Yale University Press.

Conant, James B. (1959). *The American high school today.* New York: McGraw-Hill.

Gallagher, James J. (1985). *Teaching the gifted child* (3rd ed.). Boston: Allyn and Bacon.

Gallagher, James J., Weiss, Patricia, Oglesby, Krista, and Thomas, Tim. (1983). *The status of gifted/talented education: United States survey of needs, practices, and policies.* Los Angeles: National/State Leadership Training Institute on the Gifted and Talented.

Georgiades, William, et al. (1979). *Take five: A methodology for the humane school.* Unpublished manuscript.

Gold, Milton J. (1982). *Education of the gifted/talented.* Los Angeles: National/State Leadership Training Institute on the Gifted and Talented. (Reprint of 1965 edition)

Grobman, Arnold B. (1969). *The changing classroom: The role of the Biological Sciences Curriculum Study.* Garden City, NY: Doubleday.

Humphreys, Lloyd G. (1985). A conceptualization of intellectual giftedness. In Frances Degen Horowitz and Marion O'Brien (Eds.), *The gifted and talented: Developmental perspectives* (pp. 331–360). Washington, DC: American Psychological Association.

Hurd, Paul DeHart. (1961). *Biological education in American secondary schools 1890–1960.* Washington, DC: American Institute of Biological Science.

Kaplan, Sandra N. (1979). *Inservice training manual: Activities for developing curriculum for the gifted/talented.* Los Angeles: National/State Leadership Training Institute on the Gifted and Talented.

Kaplan, Sandra N., Passow, A. Harry, Phenix, Philip H., Reis, Sally M., Renzulli, Joseph S., Sato, Irving S., Smith, Linda H., Torrance, E. Paul, and Ward, Virgil S. (1982). *Curricula for the gifted—Selected proceedings of the First National Conference on Curricula for the Gifted/Talented.* Los Angeles: National/State Leadership Training Institute on the Gifted and Talented.

Marland, S. P., Jr. (1971). *Education of the gifted and talented. Report to the Congress of the United States by the U.S. Commissioner of Education* (Vol. 1); *Background papers* (Vol 2). Washington, DC: U.S. Government Printing Office.

Martinson, Ruth A. (1974). *The identification of the gifted and talented.* Los Angeles: National/State Leadership Training Institute on the Gifted and Talented.

McInerney, Joseph D. (1986/1987, December-January). Curriculum development at the Biological Sciences Curriculum Study. *Educational Leadership, 44*(4), 24–28.

National Education Association and NSTA. (1959). *Science for the academically talented student in the secondary school.* Washington, DC: Author.

NSTA. (1983). *Science teaching: A profession speaks.* Faith K. Brown and David P. Butts (Eds). (Yearbook.) Washington, DC: Author.

NSTA. (1984). *Redesigning science and technology education.* Rodger W. Bybee, Janet Carlson, and Alan J. McCormack (Eds.). (Yearbook.) Washington, DC: Author.

NSTA. (1985). *Science Technology Society.* Rodger W. Bybee (Ed.). (Yearbook.) Washington, DC: Author.

NSTA. (1987). *Focus on excellence: Science teaching and career awareness.* John E. Penick (Ed.). Washington, DC: Author.

Phenix, Philip H. (1986). *Realms of meaning: A philosophy of the curriculum for general education.* Los Angeles: National/State Leadership Training Institute on the Gifted and Talented. (Original work published 1964)

Platow, Joseph A. (1984). *A handbook for identifying the gifted/talented.* Los Angeles: National/State Leadership Training Institute on the Gifted and Talented.

Renzulli, Joseph S., Reis, Sally M., and Smith, Linda H. (1981). *The revolving door identification model.* Mansfield Center, CT: Creative Learning Press.

Rutherford, F. James. (1985—). *Education for a changing future.* Washington, DC: American Association for the Advancement of Science. (Originally called *Project 2061: Understanding science and technology for living in a changing world*)

Sato, Irving S. (1987). The C^3 model: Resolving critical curricular issues through Comprehensive Curriculum Coordination. *The Journal of the Gifted, 11*(2), 92–115.

Senesch, Lawrence. (1973). *Our working world.* New York: Science Research Associates.

Shymansky, James A., Kyle, William C., Jr., and Alport, Jennifer M. (1982, November-December). How effective were the hands-on science programs of yesterday? *Science and Children, 20*(3), 14–15.

Spencer, Herbert. (1966). What knowledge is of most worth? *Herbert Spencer on education* (pp. 121–159). New York: Teachers College Press. (Original work published 1854)

Tyler, Ralph W. (1949). *Basic principles of curriculum and instruction.* Chicago: University of Chicago Press.

Wallach, Michael A. (1985). Creativity testing and giftedness. In Frances Degen Horowitz and Marion O'Brien (Eds.), *The gifted and talented: Developmental perspectives* (pp. 99–123). Washington, DC: American Psychological Association.

Watson, Fletcher, Brandwein, Paul F., and Rosen, Sidney. (1953). *Critical years ahead in science teaching.* New York: Carnegie Corporation. (A report to James B. Conant)

School, University, Laboratory, and Museum Cooperation in Identifying and Nurturing Potential Scientists

A. Harry Passow

Cooperation between colleges and universities and schools in the identification and development of potential scientists is a widespread phenomenon which has quite a long history. Some postsecondary educational institutions have been making opportunities available to precollege students with high ability for as long as 30 to 40 years or even more. These are special programs for such students, separate from arrangements made for individual students to enter college or university early and/or be accelerated through a regular undergraduate or graduate program.

The *1988 Directory of Student Science Training Programs for High Ability Precollege Students* (Science Service, 1987) lists hundreds of universities and colleges, as well as a number of centers and laboratories, which offer programs for high-ability elementary and secondary school students. These special science programs are designed "to provide talented high school students with educational opportunities beyond those normally available in courses or laboratory work in the students' schools" (p. iv). This is accomplished by bringing these outstanding precollege students "into contact with the instructional staff, research personnel, and general resources of colleges, universities, and research institutions" (p. iv).

The programs generally are of three types:

• **Research** involves students as junior associates on a research team or principal investigators of a problem of appropriate difficulty, under the direct supervision of an experienced scientist. This type of activity is generally of most benefit to students with a strong background in science who can work on a single problem for the duration of the project.

• **Courses** provide students with academic experiences not available in their high schools for some reason (too high level, too few students interested, lack of qualified teachers, budget, etc.). These offerings may be specially designed for the program, regular early college courses, advanced placement courses, regular high school courses, and/or remedial courses.

• **Research and Courses Combined** offer instruction in conjunction with the opportunity to do some sort of research (Science Service, p. iv).

These college/university programs may be offered during the academic year, during the summer, or year-round. Some programs are available only for local area students who commute to and from the institution, others are primarily residential, and still others cater to both commuters and residential students from outside the immediate area. Some provide for elementary as well as secondary school students, although the majority tend to include mainly the former. Students pay no fees to participate in some programs while others require tuition, as well as room, board, and travel. Scholarships are available for some programs. The number of students enrolled varies from a handful to as many as several hundred.

Students qualify and are selected for such programs in a variety of ways. Usually a high grade point average (GPA) in high school is a major requirement. Often some combination of standardized tests, evidence of interest, recommendations from high school teachers and/or counselors, or high Scholastic Aptitude Test (SAT) or American College Test (ACT) scores are also required. Some programs are specifically aimed at recruiting minority students and are restricted to qualified students who come from particular racial or ethnic groups.

Thus, each year thousands of precollege high-ability students are chosen for and participate in a tremendous variety of science and mathematics programs on college and university campuses. In addition, many institutions of higher education provide staff and resources for programs held within elementary and high schools rather than on the college campus.

A survey of 260 institutions by Janet E. Lieberman (1985) found three major approaches or strategies for cooperative or collaborative efforts between schools and colleges and universities. The oldest and probably the most popular is early admission of academically able students to college. These programs usually admit and integrate students to regular classes with little or no restructuring. A second pattern involves cooperation between faculties of the colleges or universities and schools with the ultimate aim "to improve teacher training and professional growth and performance" (p. 2). The third pattern involves institutional restructuring aimed to change articulation patterns through new high school to college structures; to loosen the rigidity of the typical 12 grades, then college, process; and to unify the sequence of education.

Some Illustrative Programs on College or University Campuses

Some of the variety and range of postsecondary campus-based programs for precollege students of high ability can be seen in the following examples. This list is representative rather than exhaustive or qualitative. Programs have been chosen to exemplify diversity of such offerings, not necessarily because those described are particularly outstanding—although many are.

Alabama State University and University of Alabama at Birmingham Minority High School Student Research Apprentice Programs. The Alabama State program is limited to 5 to 8 juniors and seniors; that in the Birmingham program enrolls 12 to 18 12th graders. The former provides a research program; the latter focuses on research in biomedical science. The Alabama State program seeks out minority students with "outstanding grades; interest in a health research career; aptitude for and demonstrated interest in biomedical sciences; strong recommendations by science teachers" (p. 1). The University of Alabama also admits minority students able to commute daily to the Birmingham campus if they are interested in biomedical research careers, have aptitude for and demonstrated interest in science and math, good language skills, and outstanding high school grades (Science Service). The Alabama State University program lasts eight weeks each summer; the Birmingham program, ten weeks. Neither program charges any tuition, and students are paid minimum wages while enrolled.

The College Studies for the Gifted at Fort Hays State University *(Kansas).* This cooperative program involving Fort Hays State University and a number of school districts aims to provide advanced educational opportunities for intellectually, dramatically, artistically, and musically gifted precollege students. The program makes it possible for students ages 10–18, selected on the basis of IQ/scholastic aptitude scores, achievement scores, and screening forms in the music, drama, and art areas, to choose from 80 standard college courses. The College Studies for the Gifted attempts to serve the entire state, many of whose students come from rural and agricultural areas. Because of the long travel distances, during the academic year most students spend only one or two hours each day at the University in classes with regular university students taught by regular faculty. The rest of their time is spent in their regular senior high or junior high classrooms. Participants also attend afternoon, evening, and Saturday classes; Telenet and media classes through the

Public Broadcasting System; classes at four Fort Hays State University off-campus centers; and summer residential programs at the University. The University also offers six workshops each year in special interest areas, each of which is taught by regular faculty members and by guest experts, sometimes on videotape. Past workshop topics have included water resources and pollution problems, bioethics, computers, and laser research. Special afternoon and evening seminars are offered on current issues and topics of interest. The program, a cooperative effort to educate young gifted students in an appropriate academic environment, provides them—sometimes for the first time—with access to high-level, challenging learning experiences. The students can generate "dual credit" for their studies—credit at their school as well as in college.

Columbia University Science Honor Program *(New York)*. For at least 40 years, every Saturday morning, 350–500 outstanding 11th and 12th grade students (a few 10th graders are also included) gather on the Columbia University campus for a multidisciplinary program focusing on the physical and life sciences and mathematics. Students qualify for the program on the basis of high scores on a special examination, but they are permitted to take the exam only after being nominated by their high schools, which submit GPAs, aptitude exam scores, and other supportive data. The program consists of courses and lectures by university faculty, laboratory and research experiences using the university facilities, and seminars with outstanding professors, including some Nobel Prize winners. The program runs throughout the academic year. There are no tuition or other charges for the student participants.

Johns Hopkins University Center for the Advancement of Academically Talented Youth *(Maryland)*. Since 1972, Johns Hopkins University's Study of Mathematically Precocious Youth has provided opportunities for "radical acceleration," beginning first in mathematics and moving to the sciences and the humanities. Its Center for the Advancement of Academically Talented Youth—located at college sites other than Johns Hopkins University in Baltimore—now offers courses in science, mathematics, and the humanities at high school through university levels for almost 3,000 students, ages 12–16. Students qualify for the residential programs at ages 12–13 by scoring over 500 on the SAT-Mathematics or a combined score of over 930 on the SAT-Mathematics and SAT-Verbal. The courses, for which tuition is charged, are each three weeks long. Students are offered standard 1-year high school courses in biology, chemistry, or physics in 3 weeks—5 hours per day for 15 days (Stanley and Stanley, 1986).

Long Island University Center for Gifted Youth *(New York)*. Long Island University offers Saturday classes and a summer program on its three campuses for gifted students in kindergarten through grade nine. Included among the offerings are courses titled: Explorations in Science, Forensic

Science, Marine Biology, Robotics, Chemistry, Computer Programming, Optics and Lasers, Biology, The Nature of Life, and Electricity. The purpose of the program is to augment local programs by providing children with superior intellectual ability learning experiences that deepen and extend intellectual interests, as well as develop the skills of independent learning. Students attend for a three-hour block on Saturday mornings and during a two- to four-week summer program. The Long Island University Center represents a program in which the university provides only facilities and resources (e.g., laboratories, classrooms, and computers) for the tuition-paying students.

The University for Youth (University of Denver) *(Colorado)*. This program is one of several offered by the University of Denver. During each quarter of the academic year and for six weeks during the summer, the University of Denver provides a program for gifted and talented students from preschool through junior high school. Graduate students from various university departments serve as instructors in course offerings which include computer science, physical sciences, and mathematics. Participants are selected on the basis of having already been identified for a program for the gifted and talented by their school or, if there is no such program, recommendations of teachers and the principal. During the academic year, as part of the regular school day, selected students from the Denver Public Schools are transported to the university campus for classes in the areas of science, mathematics, and computers. The University also offers the **Early Experience Program,** designed for academically gifted high school seniors and highly qualified juniors who wish to enrich their secondary school experience with university-level courses. The Early Experience Program enables students to take one or more University of Denver courses while still pursuing their high school diplomas. The students receive college credit at the University of Denver or other cooperating institutions of higher education.

Laboratories, Museums, and Research Centers as Resources for Precollege Science Education

Resources for identifying and nurturing potential scientists also appear in laboratories, museums, and research centers, many of which are actively engaged in educational efforts. Almost since they were founded, the eight Department of Energy National Laboratories have involved university faculty and students in their work for both educational and research purposes. More recently, the laboratories have extended their educational activities to the precollege level.

In addition to programs aimed directly at precollege students, the national laboratories also have a number of programs aimed at providing opportunities for teachers to upgrade their knowledge about science and scientific developments as well as improve their teaching skills. These programs are based on the assumption that teachers play a key role in the identification and development of potential scientists.

Argonne National Laboratory *(Illinois)*. The Argonne National Laboratory, for example, offers several precollege science and engineering programs aimed at "encouraging highly talented and motivated students to embark on a science/engineering career" (Thomas, 1985, p. 67). Its **High School Sophomore Research Apprenticeship Program** is a six-week program which enrolls 40 students, a majority of whom are from minority groups. The **Argonne National Laboratory Precollege Program** enrolls some 65 graduating high school seniors who have chosen to pursue science and engineering careers. The students work with a scientist or team of scientists on a research project and, at the end of the summer, must prepare a paper outlining their research activities. In addition, Argonne engages in several cooperative programs, such as the **Northwestern University Midwest Talent Search Program,** which involves about 60 youngsters, including several minority students, in a summer program, and an **Early Identification of Science/Math Talented Students** in cooperation with the University of Wisconsin—Madison; this program involves 40 students from grades 2 to 5. Argonne also participates in the consortium of industries and national laboratories' program titled **Tomorrow's Scientists and Tomorrow's Managers,** the purpose of which is to encourage students to explore careers in science and mathematics. Argonne National Laboratory's **Explorer Club** makes it possible for students to be introduced to scientists and scientific principles.

Battelle Pacific Northwest Laboratories *(Washington)*. Battelle has created its **Sharing Science with Schools** program aimed at discussing current research and development being conducted at Battelle with students in area high schools. The presentations, offering advanced scientific information about such topics as biotechnology, robotics, and food irradiation, are intended to be incorporated into each school's science curriculum.

Center for Excellence in Education *(Virginia)*. The Center for Excellence in Education's annual Research Science Institute, a six-week summer residential program, is aimed at fostering the development of the intellectual and practical skills of gifted and talented youth. The program emphasizes advanced theory and research in applied mathematics, the sciences, and engineering. Sixty-seven scholars (50 from the U.S. and 17 from abroad, all in their last year of high school) are selected on the basis of demonstrated academic excellence in mathematics, the sciences, and the verbal arts, and potential for continued superior performance. The Research Science Institute combines on-campus course work in scientific theory with off-campus work in scientific research, under the direction of research scientists, mathematicians, and engineers. On-campus lectures offered during the first two weeks are followed by a three-week research internship and a final week during which the interns make oral/written presentations. The morning lectures explore important areas of scientific theory and research in courses such as Research Frontiers in Physical Science, Research Frontiers in Biological Science, and Frontiers in Pure and Applied Mathematics. The afternoon courses, which provide training in technical skills needed to do research, have titles such as Computer Skills for Scientific Research and Quantitative Research Methods. Other activities include field trips to research centers and museums and guest lectures by distinguished scientists, humanists, and researchers. The internships focus on explicitly defined, narrowly focused research problems in applied mathematics, biology, physical science, or engineering, with students matched with scientists on the basis of mutual research interests.

Chicago Museum of Science and Industry *(Illinois).* While not organized to provide only for the potential scientist, Chicago's Museum of Science and Industry's Education Department offers a variety of programs for children, youth, adults, and families, many of which contribute to the development of potential scientists. The museum provides a wide range of classes, workshops, and lectures and sponsors a museum science club and several science fairs. Its Kresge Library makes available a rich source of science and science education materials. The Museum's youth section contains a comprehensive collection of books, magazines, films, videocassettes, audiocassettes, filmstrips, computers and computer software, and science kits. Its Seabury Laboratory is the site for classes and workshops, the science clubs, and science demonstration development programs. The Museum also offers an extensive program of courses, workshops, and services for teachers.

Fermi National Accelerator Laboratory *(Illinois).* Fermilab provides several programs for precollege students. They may attend 1 of the 3 10-lecture sessions of Saturday morning physics offered each year, each making up a survey course in particle physics. Each two-hour Saturday morning lecture is followed by a tour of some Fermilab facility to let students see what experiments are like. A second program recruits high school students from inner city schools who are brought to the laboratory to work with a supervisor.

Lawrence Livermore National Laboratory *(California).* Among the activities of the Lawrence Livermore National Laboratory are the following:
• 25 scientists from Lawrence Livermore National Laboratory have taught science to some 9,000 fourth and fifth grades in a program called **Lawrence Livermore Elementary School Science Study of Nature.**
• Part-time jobs for high school students, first created in 1971, were followed by full-time summer jobs in order to recruit more minority students.
• Lawrence Livermore National Laboratory scientists make presentations to high school classes about future careers in science and engineering.
• Scientists serve as role models at Oakland High School, a magnet school in the Oakland area.

• Through a cooperative arrangement among the Livermore Laboratory, the schools, and the business community, a career exposure program for students interested in science has been initiated.

• A program to teach basic science to junior high school students was begun in 1984.

• Lawrence Livermore National Laboratory loans equipment, including computers and word processors, to schools.

• A Visiting Science program is available to schools. (Perry, 1985, p. 69)

Los Alamos National Laboratory *(New Mexico)*. Among several programs offered by the Los Alamos National Laboratory to encourage interest and study in science are ones called **Science Beginnings** and **Careers in Science and Engineering**. The target audience for **Science Beginnings** is children in grades four to six, and the program operates through visits to the Los Alamos National Laboratory Museum with its state-of-the-art hands-on exhibits and through demonstrations at elementary schools in the seven surrounding counties. Close to 10,000 junior high students a year are reached through its **Careers in Science and Engineering Program.**

In Conclusion

Cooperative efforts between schools and colleges and universities; research centers; laboratories; museums and similar agencies and institutions make an important contribution to the identification and nurturance of youngsters with potential for becoming scientists and mathematicians. Such cooperative efforts, which extend learning opportunities beyond those normally available in schools in a number of ways, provide students early access to

• more advanced content—knowledge, concepts, processes—thus facilitating curricular acceleration and/or enrichment by enabling students to participate in college-level learning experiences

• skilled, knowledgeable instructional staff and researchers who are often persons likely to be involved with the constantly changing frontiers of knowledge in their disciplines

• contact with persons who are engaged in "doing science" rather than "teaching about" science, so that students can learn the nature of science and scientific research from practitioners

• resources—laboratories, equipment, and operations—which are far richer and more authentic than those found in most schools

• mentors and role models, enabling the young to be involved with science practitioners and researchers engaged in doing scientific research, thus fulfilling an important function for the development of talent potential

• opportunities to be challenged and to be engaged in learning experiences which are relevant and appropriate

- more chances to interact with other capable youngsters, often resulting in mutual stimulation to be creative and productive

In sum, these cooperative efforts are highly significant in the identification and development of scientific potential. They are part of the overall process in talent development, supplementing—not replacing—the learning opportunities provided within the school itself. Whether students are able to profit from additional personnel, materials, programs, and resources depends, to a great extent, on the kinds and quality of experiences provided by their schools.

References

Lieberman, Janet E. (1985). An overview: Collaborative programs "new articulation models." In *Conference on collaborative programs: 1985* (pp. 1–8). New York: City University of New York.

Perry, Manuel. (1985). Programs for high school students. In Frank M. Vivio (Ed.), *A national resource to meet a national need: The role of national laboratories in precollege education* (pp. 69–70). Argonne, IL: Argonne National Laboratory.

Science Service, Inc. (1987). *1988 Directory of student science training programs for high ability precollege students.* Washington, DC: Author.

Stanley, Julian C., and Stanley, Barbara S. K. (1986, March). High school biology, chemistry, or physics learned well in three weeks. *Journal of Research in Science Teaching, 21*(3), 250–263.

Thomas, Juanita R. (1985). Programs for high school students. In Frank M. Vivio (Ed.), *A national resource to meet a national need: The role of national laboratories in precollege education* (pp. 67–68). Argonne, IL: Argonne National Laboratory.

Identifying and Nurturing Future Scientists in Other Nations

Pinchas Tamir

Although there is general recognition that the education of gifted children poses special problems, in most countries these problems have not been considered top national priorities for at least two major reasons. First, since World War II, the major educational effort of practically all nations has been to make formal schooling available to all and, as much as possible, to offer "equal educational opportunities"* to all children. Because many of the newcomers to formal education have come from poor and culturally disadvantaged families, the needs of these children have, justly, received top

---

*For further discussion of the complex nature of this aim, please note in this volume "Science Talent: In an Ecology of Achievement" and "Equality, Equity, and Entity: Opening Science's Gifts for Children."

Acknowledgments: The author would like to thank the following individuals for the help they offered through personal communication: David Cohen, Macquarie University, Australia; John B. Holbrook, University of Hong Kong, Hong Kong; Peter Fensham, Monash University, Australia; Myriam Kresilchick, University of São Paulo, Brazil; Wim Nijhof, University of Twente, the Netherlands; M. M. Ryan, Department of Education, Wellington, New Zealand; Chhotan Singh, National Council of Educational Research and Training, New Delhi, India; and Ross S. Wenn, Gifted Children Task Force, Ministry of Education, Melbourne, Australia.

priority; most money and activities are, therefore, directed toward compensatory education. Second, in many countries, especially in western Europe, special treatment of the gifted is looked upon as suspect—as an attempt to undermine egalitarian society and reestablish frameworks that offer an advantageous education to the upper classes and other elite groups.

In preparing this chapter, reports from responsible observers have been included without interjecting comment on the political, economic, or social practices in the countries whose programs and practices are described. It seems desirable to be aware of the growth, or lack of growth, in various countries' national efforts for the gifted in science, since these efforts express a judgment on the importance of the programs in relation to the national welfare.

As shown by a recent survey (Mitchell and Williams, 1987), Belgium, Denmark, Greece, Italy, Norway, Portugal, Spain, Sweden, and Switzerland have no special programs for the gifted. In contrast, other countries, while sharing the concern for providing "equal educational opportunities," find it desirable to pay special attention to the education of the gifted. Bruce M. Mitchell and William G. Williams (1987) write that "it is no accident that South Africa, Israel, and Taiwan—three nations that are facing immediate external and internal threats—have developed extensive educational programs for the gifted and talented students" (p. 534). For various reasons, these countries need to ensure the best use of their available resources and seek to do so through the use of special programs. As one writer put it, "We Israelis of all people cannot afford the luxury of neglecting our gifted children" (Milgram, 1979, p. 13).

Five Questions to Thirty Nations

Very little information is available in the literature about the approaches used in different countries to the education of the gifted in general and to the nurturing of science talent in particular. Hence, as soon as I received the invitation to write this chapter, I mailed a brief letter to close to 30 colleagues in different countries, asking 5 questions:

1. What special actions have been taken in your country in relation to the identification and education of the gifted in science?

2. Are there special schools or classes for the gifted in science?

3. Are there special training programs for teachers who instruct the gifted?

4. Are there special science curriculums for the gifted?

5. Are there offerings of special informal (out-of-school) learning experiences for the gifted?

The answers I received serve as one source of data. I also approached several colleagues at international conferences and obtained some direct information from them. In addition, during a recent trip to South Africa I visited some educational centers for the gifted. Finally, as an Israeli concerned with the education of the gifted in science, I have firsthand information on many of the available programs in my own land.

Since my information on different countries' programs varies in nature and detail, I cannot report it in a uniform framework. Rather, I present a description for each country on the basis of available information, which is admittedly scattered and incomplete. Naturally, I tell the Israeli story in greater detail than any other. I have divided the material geographically, roughly by continent, and within each division I present some brief comments on specific countries, followed by more detailed reports if information permits.

North and South America*

Education in Canada is the responsibility of the provinces; each has a different program for the gifted. In Ontario, for example, a special bill in effect since 1981 requires all boards of education to provide programs for exceptional students. Unfortunately, I have no information about special programs in science.

In most Latin American countries governments do not support special programs for the gifted; however, private organizations often do. In Brazil, for example, organizations sponsor a number of annual activities.

Brazil

In Brazil special attention is given to the needs of the gifted in some training projects for teachers. In addition, the following events take place:

• an annual contest called Scientists for the Future (the winners present their projects at the meeting of the Brazilian Society for the Advancement of Science; the winning project receives a special award)

• an annual congress where young scientists in the state of São Paulo present individual project reports and discuss their projects with scientists

• a number of science fairs distributed throughout the country, the largest taking place in the states of Rio Grande do Sul and Amazonas

---

*Because most of this volume is devoted to discussion of the educational system of the United States, I do not attempt a description of attitudes and policies prevalent in the U.S. in this paper.

Europe*

In spite of a general feeling in western European countries that gifted students can thrive in regular programs, several countries offer special programs for the gifted.

In Great Britain students are grouped by ability in some subjects in secondary schools. In addition, scholarships are available that enable selected students to attend certain schools.

In France special classes for the gifted have been established in middle schools since 1979.

In the Netherlands the government is supporting a series of research studies aimed at providing data on the basis of which a national policy related to the gifted will be formulated.

In West Germany the government supports experimental programs for the gifted in mathematics and literature, as well as national contests in music, drama, and science. However, the teachers' union is strongly opposed to special programs for the gifted. A private organization, the German Association for the Gifted, supports extracurricular activities such as clubs and camps.

As mentioned earlier, many European countries, including Denmark, Norway, Sweden, Belgium, Switzerland, Greece, Italy, Spain, and Portugal have no special provisions for the gifted.

The Soviet Union offers a wide variety of approaches and programs that encourage students to develop their talents, despite a lack of consensus concerning the desirability of having special schools and classes. The special provisions are described in detail below.

Great Britain

In Great Britain a report from the Department of Education and Science (1967) states that because the nation's primary education "at its best is better adapted than any other" program to meet the needs of gifted children, separating the gifted into special classes is not required (p. 307). As to secondary schools, despite the recent trend away from selective grammar schools, aptitudes and abilities are evaluated and students are grouped according to their ability in particular subjects, including science. Consequently, students with special science abilities study science at an advanced level.

The Department of Education and Science has special funds available for scholarships for students to attend selected schools, as well as money earmarked for educators to do research projects related to the gifted. The

---

*The decision to group the Soviet Union, which spans both Europe and Asia, with European nations in this paper was an arbitrary one made for convenience.

National Association for the Gifted sponsors "explorer" clubs, which meet on Sundays and during the summer holidays.

The Soviet Union

In the absence of personal communications concerning education in the Soviet Union, I have relied on sources in the literature for information. My account draws heavily on an excellent piece by John Dunstan (1983) entitled "Attitudes to Provision for Gifted Children: The Case of the USSR."

Soviet science and technology has had an impressive reputation since the 1950s, when the successful launch of Sputnik stunned the world. However, the United States reacted to the early Soviet space achievements by intensifying the competition between the two countries for scientific and technological dominance. For this and other reasons, the education of highly trained specialists became a matter of highest priority in the U.S.S.R. (Dunstan).

In debates preceding the introduction of a new school law in 1958, Khrushchev advocated "schools for gifted children" in the sciences (DeWitt, 1961, p. 18). However, because of criticism from politicians, psychologists, and educators, who argued that such schools "would encourage snobbery, egoism, anticollectivism, and other undesirable social evils," Khrushchev did not insist upon their inclusion in the law (Brickman, 1979, p. 319). Nevertheless, schools for mathematics and science sponsored by the universities and the U.S.S.R. Academy of Sciences were established. Of these, the Moscow School of Mathematics and Physics and the Novosibirsk School of Mathematics and Physics are the best known (Brickman, 1979).

According to Dunstan, "the minority viewpoint of 1958, which was predominantly concerned with ensuring the replenishment of the Soviet scientific leadership, turned into a majority stance . . . by the middle of the next decade" (p. 292).

One example, Dunstan writes, is that

> from 1962–63 we see the emergence of the *fiziko-matematicheskie shkoly* or FMShs, boarding schools specializing in mathematics and science (usually physics), attached to and partly staffed by a higher educational institution, and offering along with the general subjects a very high level of specialized teaching to their rigorously selected 14- to 16-year-old entrants (about 2 percent of those initially interested). Including the four original FMShs in Moscow, Leningrad, Kiev, and Novosibirsk, by 1970 there were some eleven such schools over the whole country. Their purpose was unashamedly to produce a scientific and technological elite. (p. 296)

By 1970, Dunstan continues, mathematics and science boarding schools made up 0.02 percent of the "general day schools with the senior course" (p. 298).

Another development in the 1960s was that many general (that is, not technical) schools introduced advanced vocational programs in the "senior forms" in computer programming and other fields. These schools, known as "mass schools with a special profile," (p. 297) made up 1.21 percent of the general day schools in 1970.

A third development was the introduction of elective courses, many of which catered to the talented students, in the general schools (Dunstan, 1983).

Finally, a variety of extracurricular opportunities became available to talented Soviet students. These included correspondence and summer schools sponsored by universities, as well as "study circles" offering a wide range of activities, which were meant to "help enrich the curriculum and identify the gifted in such academic fields as physics, astronomy, technology, biology, literature, history and others" (Brickman, 1979, p. 321).

Dunstan (1983) reports that "the long-standing Moscow Mathematics Olympiad (academic contest) was put on an all-Russian basis in 1961 and an all-Union one in 1967, now covering physics and chemistry also" (p. 298).

It is interesting to note that there is still strong opposition to the provision of special education for the gifted in the U.S.S.R. This opposition is stronger among educational theorists and psychologists than among teachers and scientists. However, the majority in each of these groups still supports the special mathematics and science schools. In addition, Dunstan reports that between 1958 and 1973 support for special provisions for gifted children was expressed by 87 percent of the scientists, 70 percent of the teachers, and 59 percent of the educationists and psychologists in written materials later evaluated as part of a study of Soviet attitudes. Moreover, support for special provisions for the gifted in the area of science has been much stronger than in other areas such as, for example, foreign languages.

Africa

My information on Africa is confined to the Republic of South Africa, which I recently visited.

South Africa

Although most of this commentary is based on direct experience, I have also referred to an article by A. L. Behr (1983).

The Republic of South Africa pays special attention to the education of the gifted. A number of educational authorities as well as private individuals and organizations have implemented special programs for gifted students. Following is a brief description of some of the programs that bear on science education.

The most conspicuous to a visitor are the extracurricular private centers for the gifted. Eight such centers were established in the large urban areas in the past decade. They offer enrichment programs in various subjects, including sciences (e.g., computers, astronomy, paleontology, electronics, petrochemistry, and geology). Children are selected on the basis of questionnaires filled out by their parents, interviews of parents and children, recommendations of the schools, and results of IQ and creativity tests. The range of activities offered by these centers is very broad and caters well to the abilities and interests of the children selected. The instructors consider collaboration with parents desirable.

In the Cape Province, special program committees function in various regions. These committees compile enrichment materials and keep them in resource centers, where they are made available to the schools. In many schools gifted students are pulled out of the mainstream classes for certain periods of the day for special enrichment classes. In large schools gifted children may be permanently grouped for the study of particular subjects such as science. Courses focusing on teaching gifted students are compulsory at some teachers' colleges and are also offered at certain universities. Special inservice education is offered as well.

Similar activities are offered in other South African provinces. In Natal, enrichment and extension programs are offered within the schools. These include clubs, societies, excursions, and outdoor activities. Some schools organize regular meetings for selected students so that enriched courses in subjects such as mathematics and biology may be followed. In 1981 an adviser on talented children was appointed and 18 pilot schools were selected for more systematic implementation of programs for the gifted.

The Department of Education and Training, which is responsible for the education of the black population, established a Saturday school for gifted students in 1982 at Soweto Teachers' College. This school offers enrichment programs in a variety of subjects, including the natural sciences, to several hundred gifted students. It is sponsored by a private banking firm.

Unfortunately, the vast majority of regular schools in the country are unable to provide sufficiently for the needs of the gifted. Although the number and variety of activities available to the gifted in South Africa are quite impressive, the percentage of the gifted—black or white—who actually enjoy and participate in these activities is rather low. An estimated 5 percent of the gifted population is at present being served (Meyer, 1983).

Australia and New Zealand

Australia and New Zealand, although neighbors with much in common, have in recent years taken different approaches to serving their gifted populations.

Australia

Although education in Australia is the responsibility of the different states, the national government supports certain programs; for example, it pays each state to employ resource teachers to initiate and support enrichment programs for the gifted. In some cities, gifted science students spend half a day each week away from their regular classes to participate in special programs that focus on scientific investigation. In addition, a special four-year program for outstanding science students is offered at Melbourne High School.

The state of Victoria has shown a far greater commitment to providing special programs for the gifted than any other state; consequently, the rest of this discussion is devoted to describing the practices there. In Victoria a special Gifted Children Task Force coordinates various activities aimed at the development and nurturing of "children with special abilities"—a description preferred to "gifted" because it focuses on the child's ability rather than on placing a label on the child. The work of this task force is based on this philosophy: Schools should provide early and continued opportunities to recognize and foster special abilities, and children with special abilities need opportunities to develop purposeful skills at a pace appropriate for them and to learn to apply their advanced skills in a range of situations.

Three kinds of activities have been offered for students with special abilities: extension, enrichment, and acceleration. These children receive support through a number of means, such as close home-school partnerships, counseling services, and community support. Their teachers receive preservice and inservice education, as well as curriculum materials that take into account the students' special needs.

Some of these special programs operate in schools, while others occur in other settings. For example, in 1986, 11 cluster groups involving more than 2,000 students and 200 teachers studied a variety of enrichment materials. Two additional groups worked in collaboration with postsecondary institutions. Students who wish to carry out individual projects are provided with expert mentors, and, during school vacations, enrichment study camps have been conducted in several locations. The Gifted Children Task Force evaluates all of these activities and publishes relevant materials, such as a teachers' guide on how to deal with students with special abilities and booklets describing programs available for these students.

New Zealand

New Zealand's education policy has been based for many years on the premise that all students should have access to schooling of high quality. Although all students follow the same curriculum until the end of the 11th year of school, the New Zealand Department of Education believes that the syllabus in each subject is sufficiently flexible to enable teachers to cater to the special needs of small groups or individuals within their classes and that a special talent in science would thus be recognized and fostered. In spite of the official position, three private organizations actively support talented students.

Asia

In most of the east Asian countries, competition is fierce among students, as they move from one level of schooling to the next, to get access to better schools. This contest is essential because quite often the kind of school one attends is a factor in determining future education and opportunities for careers. Consequently, most students are grouped and schooled by their "ability" (which is measured in various ways). Hence, countries like Hong Kong, Singapore, and Japan (which is discussed in detail below) have no special programs for gifted students.

In some countries, such as the Philippines, where there is a severe shortage of qualified science teachers, special day and residential schools with modern equipment and highly qualified teachers cater to the gifted students. In these countries there is a large gap between the science education offered to the gifted and that provided to the rest of the students.

Korea has established special schools for gifted science students.

Taiwan operates a comprehensive plan for the gifted that includes special programs, curriculums, and classes in elementary and junior high schools. (There are 100 such special classes.) In addition, there are particular requirements for teachers of the gifted and even extra pay for these teachers.

India's approach centers on a search to identify gifted secondary students through the use of examinations.

Israel's educational programs, planned and present, and philosophy toward education are discussed in detail at the end of this section.

Japan

According to my colleagues, Japan, like Hong Kong and Singapore, has no special programs for gifted students. This may be explained, on the one hand, by the effect of a longtime tradition and, on the other, by the special role that schools play in Japanese society.

Tetsuya Kobayashi (1976) describes the tradition as follows:

Japan is almost the only nation in the world whose population—exceeding over 100 million—still consists of a single national group with a common historical heritage, a single language and relative homogeneity in culture and belief. There is no doubt that such unity contributed to the swift modernization of Japan and assisted during the process in creating certain features of its national tradition in education, such as centralization of control and uniformity in the content of education. (p. 173)

This tradition of uniformity is not conducive to the development of special programs for selected groups of students. Yet Japanese students have consistently exhibited a high level of achievement in science (Comber and Keeves, 1973). This high level of achievement owes much to the initiative of the government as well as to the concern of industry.

However, as Kobayashi observes, the great progress and impressive achievements of the Japanese educational system since World War II would not have produced the results we now see without the support of the people at large, to what has been called their "zeal for education." Almost all Japanese parents have been eager to give their children education beyond the compulsory age, and most Japanese are as concerned with education as with problems of health and living. There are at least two explanations for these attitudes:

• In Japanese society, education has long been considered an effective way of climbing up the social ladder.

• The rapid economic and industrial development that followed World War II greatly changed the structure of the society and increased the demand for qualified personnel in many fields.

Consequently, the "zeal for education" has resulted in fierce competition for places in schools. In the public school system, a student has to go through at least *four* selection procedures before completing a university education, that is, entering elementary school, moving to middle school, going to high school, and matriculating at a university. At each stage, but especially in the last two stages, the kind of school to which one is admitted determines the opportunity to have access to a better school at the succeeding stage. This large-scale screening mechanism results in a de facto provision of quality high-level education to the most talented students, who have been repeatedly selected for and admitted to the better schools. Japan has succeeded in offering high-level science education to all its students, thereby increasing the country's pool of future scientists.

There is a wide range of out-of-school programs and resources that contribute to the high quality of science education in Japan. These include educational television programs—some of which are at a very sophisticated level (Troost, 1984)—as well as science magazines, science kits, natural history museums, zoos, and botanical gardens. There are about 50 centers for science education in the country. Their primary concern is the inservice education of teachers, but some of the centers set up science rooms for children where interested students can work on individual projects, and sometimes the centers hold exhibitions of children's work (Takemura, 1984).

Inventiveness and creativity are further encouraged through annual science invention prizes, including the Prime Minister Prize and the Encouragement Prize of the Minister of Education (Education Council, 1985).

In Hiroshima there is a flourishing Scientists in School program in which a wide variety of scientists come to schools, participate in lessons, and engage in team teaching with the schools' teachers.

However, regardless of the nature of the out-of-school activities, the tremendous emphasis upon entrance examinations and the required preparation for them lead many gifted students to pass up the opportunity to attend the activities, and if they do attend them, to be too tired to participate actively or benefit fully from them (Imahori, 1982).

The Philippines

In the Philippines the Department of Education, Culture, and Sports (DECS) has issued to all public and private schools the following guidelines:

1. Identification of gifted and talented children should begin as early as at the kindergarten level.

2. Special grouping can start in the second grade. Some strategies that may be adopted are special classes, partial integration, acceleration through skipping or telescoping grades, or through early entry into high school.

3. Curriculum enrichment may be accomplished through additional reading assignments, independent research, and supplementary studies.

4. Teachers for the gifted should have training in the education of the gifted.

5. Periodic evaluation of the program should be undertaken to determine needed improvements.

The Department of Science and Technology (DOST) regularly offers science scholarships at the undergraduate and graduate levels to students with marked intellectual gifts toward science careers.

Philippine Science High School, a chartered institution, serves as the country's principal training ground in the sciences at the secondary level. The school has recently been expanded to include a university. (The expanded school, now called Philippine Science High School and University, is supported by funding from DOST and DECS.) Admission to Philippine Science High School is selective and is based on a two-step, examination-based screening process: the Scholastic Ability Test and the Science and Math Aptitude Test.

Standards for the recruitment and retention of the staff of Philippine Science High School are kept high, since the quality of education at the school depends greatly on the quality of the faculty. Demonstration teaching is required of all applicants. Academic qualifications and teaching experience are also considered. At least once a year the school holds a forum for teachers on giftedness.

Public science high schools under the supervision of the DECS, such as the Manila Science High School and Quezon City Science High School, are also available to students gifted in science and mathematics. There are 25 other science high schools distributed throughout the country. Several high schools have special classes for the gifted and offer elective courses in math and science.

Training programs in special education for teachers are offered during the summer at the universities.

The curriculums used at Philippine Science High School and Manila Science High School were patterned after that of the Bronx High School of Science in New York City. The curriculums offered are slanted toward the sciences and mathematics; however, languages, the social sciences, and the humanities are also stressed to provide well-rounded schooling that addresses all aspects of the students' academic and social development.

At Philippine Science High School students are required to do hands-on research on a topic of their choice, with a faculty member acting as an adviser. As in other science high schools, students meet in science clubs, attend seminars and forums, view showings of scientific films, and participate in local and international science fairs and competitions.

India

Since 1963 India has sponsored a special national talent search, operated by the National Council of Educational Research and Training, in which 750 scholarships are awarded every year to talented students for work in natural sciences, social sciences, medicine, and engineering. Candidates are selected at the end of the 10th year of school, on the basis of state and national examinations. Every scholarship winner intending to study basic sciences either at the undergraduate or at the postgraduate stage is required to attend a compulsory summer school. In this way, India selects and nurtures its future scientists, doctors, and engineers.

Israel

Israel's strong commitment to nurturing talent is no doubt in part a result of the Jewish people's traditional respect for learning and for learned people, as well as a response to its needs as a small, newly developing nation with poor natural resources, hostile neighbors, facing constant threats to its security and economy. Before I detail the various programs aimed at the gifted, it is worth emphasizing that Israel's first and most urgent educational priority by far has always been to offer "equal educational opportunities" to all Israeli citizens. Hence, the major efforts of the educational system since the establishment of Israel as an independent state in 1948 have focused on compensatory education for the culturally disadvantaged. Yet, without denying in any way the importance of remediation, Israel recognizes that gifted students are also entitled to opportunities to develop their intellectual potential and, at the same time, that the country cannot afford to neglect their talent and thereby lose their potential contributions. It is possible to cater to the gifted without investing great sums of money or diverting funds and other resources from the education of the rest of the students.

In 1973 the Department for the Education of the Gifted was established in the Ministry of Education. Its task has been to coordinate various activities already in operation and to initiate and promote new ones. Although there are neither special schools nor special classes for gifted students in elementary schools, measures to identify such students are taken in the lower grades. These measures include teachers' recommendations and psychometric tests. Children with IQ scores higher than 140 are directed to after-school classes held in institutions of higher education, to special classes offered during regular school hours for half a day each week, and to special summer camps.

At the secondary level, Israel has special schools for gifted students in music and the arts and plans similar ones for gifted science students. Meanwhile, a variety of procedures that help meet the needs of the gifted operate within the framework of the general educational system in the junior and senior high schools.

In the junior high schools (grades seven through nine), all students follow the same syllabus; however, in certain subjects, including mathematics, English, and in some schools, science, students are grouped, as they are in England, into "settings," so that high-ability students face a more challenging course.

Senior high schools often divide their students by ability and interest. Students who elect to (and are judged able to) specialize in physics or biology, for example, study together. Although such classes do not admit only the gifted, the screening out of students who cannot cope with the requirements of the matriculation examination creates selective classes in which conditions are quite favorable for the gifted.

There are further provisions for gifted high school students. Any student whose final grade at the end of the 10th grade is *B* or higher in a particular science subject may, in lieu of the matriculation examination in the subject, carry out an individual research project under the supervision of a research scientist. In the natural sciences, this project requires the student to do hands-on research, write a comprehensive report (including review of pertinent literature), and take an oral examination administered by a scientist who serves as an external examiner. Many talented students have chosen to carry out such individual research, which gives them the opportunity to develop and exercise their intellectual potential and at the same time helps pave the way for a future science-related career.

In certain subjects, such as biology, the high school curriculum as well as the structure and demands of the matriculation examination provide opportunities and incentives for gifted students to develop and apply their talents. All high school biology students in grades 11 and 12 study an inquiry-oriented curriculum that is based heavily on weekly investigative laboratories, the analysis of research papers, invitations to inquiry, and class and small-group discussions; individual student projects in ecology are also required. All of these activities are reflected in the matriculation examination, which comprises a high-level, research-oriented paper-and-pencil test (60 percent of the total score) and a highly demanding, inquiry-oriented practical test (40 percent of the total score). I have described these examinations in greater detail in other works (Tamir, 1974, 1985). But what I would like to emphasize here is that this kind of framework, while suitable for average students, is at the same time open enough to enable the more talented students to develop and apply their talents and reach very high levels of performance indeed.

In recent years a new avenue has been made available to high school students. In an attempt to promote teachers' autonomy, the Ministry of Education now encourages teachers to design their own special programs in any area that they choose. Every student can choose one such special program as a substitute for a regular matriculation examination. Within this new framework one science teacher, for example, offered a unique research course to selected highly motivated students. The course required that these 12th-grade students spend one day each week doing individual research projects in the national agricultural research institute, under the supervision of local researchers. In addition, the students met twice a week with their teacher, who structured the classes on the basis of her students' research activities. This teacher's success offers a model framework for other special science-related programs.

Israeli research institutions and universities for many years have been offering summer science camps as well as year-round science clubs for interested students. Many gifted students have taken advantage of these offerings.

Another successful program, supported by special funds, has been helping gifted students from culturally and economically disadvantaged homes. These students are identified at the age of 11, at the end of the sixth year, which is their last year of elementary school. Often the families of these children live in rural areas, far away from prestigious, selective city high schools from which most students continue to further education. Many such students enroll in special classes in one of these high schools, and some are also offered boarding services. Special instructional strategies are used to help these children develop their potential and overcome any academic deficiencies so they can pass the matriculation examinations and continue their education in postsecondary institutions.

Finally, a special project instituted by the Israeli Defense Force offers a number of gifted individuals the opportunity to participate in an accelerated program leading to a B.Sc. degree in the natural sciences. These individuals ultimately serve in the army as researchers.

In Summary

Because the information in this article is sketchy and incomplete, the conclusions and generalizations here must be regarded as tentative.

The opportunities available to the gifted students in a given country appear to depend on three major factors, namely,
- the official provisions made by the country's government
- the nature of the educational system and the curriculum
- the meaning assigned to equality of educational opportunities

Official Provisions

Review of the individual reports reveals four patterns:
- no official governmental provisions, as in Hong Kong, New Zealand, and many countries in western Europe
- no official provisions by the government, but some by private organizations, as in several countries in Latin America
- governmental provisions made only on the basis of students' prior achievement, as in India's talent search program
- governmental provisions supplemented by ones from other public institutions (such as universities) to identify potential and nurture it, as in Israel, Taiwan, and the Soviet Union.

The Nature of the Educational System and the Curriculum

The actual opportunities open to gifted students may depend on the general approach of the countries' schools. New Zealand's schools, for example, seem to provide for the special needs of individual students, including the gifted. There is ample evidence that the "open classroom" often found in British primary schools and characterized by an informal atmosphere, mixed age groups, a stimulating environment, and the free use of space and materials by the students (Weber, 1971) offers excellent opportunities for identifying and nurturing talented students. (The question remains, however, to what extent individual teachers take full advantage of these opportunities.)

The problem appears to be more complex in secondary schools, where the educational approach and the curriculum are of singular importance. For example, the introduction of college-level courses in science in the high school sometimes deflects the intention of providing high-quality challenging learning experiences—for many such courses place strong emphasis on the memorization of vocabulary and concepts and devote little attention to the nature of science and an investigative approach. On the other hand, Israel's inquiry-oriented high school biology curriculum, which is designed for the general student population, illustrates an approach that provides opportunities for significant intellectual development for all students.

The Meaning Assigned to Equality of Educational Opportunities

Obviously, the pattern of provisions adopted by a country depends on available resources and, in many cases, the ideology of the country. In certain developed countries, there appears to be a conviction that any special provision for the gifted deviates from the principle of "equal educational opportunity" by offering special privileges to children who are already lucky to possess natural advantages. Those holding this position commonly believe that the gifted will "make it" anyway and that any available resources should be directed to the weaker student populations, which typically lag far behind.

An alternative policy, held by countries like Israel as well as by both individuals and organizations in many other countries, assigns a quite different meaning to the principle of "equal educational opportunities." In this view, the principle is understood to mean giving free access to full educational opportunity, and its goal is to provide each student the opportunity to develop fully his or her potential. The need for special provisions for the gifted is supported by the conviction held by many people that a number of gifted children underachieve in regular classrooms or develop negative attitudes toward school and learning and that, ultimately, these students' potential to make a significant contribution to society may be lost.

It may be commonly observed in developed nations worldwide that special opportunities are offered to the young who display abilities in the arts (particularly music, painting, and writing) and in the kinaesthetic arts (particularly dance, gymnastics, and individual and team sports). In certain countries this policy also covers children with high-level mathematical abilities. Now, in the light of the requirements of societies' increasing dependence upon science and technology, some countries are extending special opportunities for laboratory work to students with high ability in the arts of investigation (i.e., scientific research).

Such work often goes beyond regular course requirements; the student undertakes the kind of research practiced by scientists in the field, most often guided by a teaching scientist or research scientist. Thus, the young scientist-to-be forges an early relationship with a mentor. For, in the end, such a special educational opportunity afforded during schooling is amply returned to society by the recipient's lifetime of contributions.

References

Behr, A. L. (1983). Educating the gifted: Some psycho-didactical considerations. In A. L. Behr (Ed.), *Teaching the gifted child* (pp. 107–134). Durban, South Africa: Yesten University Press. (Publication series of the South African Association for the Assessment of Education, No. 17.)

Brickman, William W. (1979). Educational provisions for the gifted and talented in other countries. In A. Harry Passow (Ed.), *The gifted and talented: Their education and development, Part I* (pp. 308–329). (Seventy-eighth Yearbook of the National Society for the Study of Education.) Chicago: University of Chicago Press.

Comber, L. C., and Keeves, J. P. (1973). *Science education in 19 countries.* New York: Wiley.

Department of Education and Science. (1967). *Children and their primary schools: A report of the Central Advisory Council for Education* (Vol. 1). London: Her Majesty's Stationery Office. (Often referred to as the Plowden Report)

DeWitt, Nicholas. (1961). *Educational and professional employment in the USSR.* Washington, DC: U.S. Government Printing Office.

Dunstan, John. (1983). Attitudes to provision for gifted children: The case of the USSR. In Bruce M. Shore, Françoys Gagné, Serge Larivée, Ronald H. Tali, and Richard E. Tremblay (Eds.), *Face to face with giftedness* (pp. 290–327). Monroe, NY: Trillium Press. (World Council for Gifted and Talented Children)

Education Council, Japan Committee for Economic Development. (1985). A proposition from businessmen for educational reform in pursuit of creativity, diversity, and internationality. In *Discussions on educational reform in Japan.* (Reference Reading Series, No. 15) (pp. 35–42). Tokyo: Foreign Press Center.

Imahori, K. (1982). Out-of-school science activities in Japan at the primary and secondary school level. *The Journal of Science Education in Japan, 6*(2), 47–58.

Kobayashi, Tetsuya. (1976). *Society, school and progress in Japan.* Oxford: Pergamon Press.

Meyer, P. S. (1983). The education of gifted children and manpower situation in South Africa. In A. L. Behr (Ed.), *Teaching the gifted child* (pp. 17–33). Durban, South Africa: Yesten University Press. (Publication series of the South African Association for the Assessment of Education, No. 17.)

Milgram, R. M. (1979). Gifted children in Israel: Theory, practice and research. *School Psychology International, 1*(3), 10–13.

Mitchell, Bruce M., and Williams, William G. (1987, March). Education of the gifted and the talented in the world community. *Phi Delta Kappan, 68*(7), 531–534.

271

Takemura, S. (1984). *Study on the position of science education in Japan.* Hiroshima: Hiroshima University, Faculty of Education.

Tamir, Pinchas. (1974, Spring). An inquiry-oriented laboratory examination. *Journal of Educational Measurement, 11*(1), 5–16.

Tamir, Pinchas. (1985, January). The Israeli "Bagrut" in biology revisited. *Journal of Research in Science Teaching, 22*(1), 31–40.

Troost, K. M. (1984, January). What accounts for Japan's success in science education? *Educational Leadership, 42*(5), 29.

Weber, Lillian. (1971). Infant schools in England. In Lee C. Deighton (Ed.), *The encyclopedia of education* (Vol. 5, pp. 68–74). New York: Macmillan and Free Press.

Apprenticeship to Well-Ordered Empiricism: Teaching Arts of Investigation in Science

Paul F. Brandwein
Evelyn Morholt
Sigmund Abeles

Permit us the strategy of gliding into the concept of well-ordered empiricism through observation of the young as they learn to *think* and *do* science.* We begin with the first meeting of teacher and class.

The First Day
We begin the first day of our meeting in biology class. We go out into the world. For science seeks to explain how the world works.‡

---

*On the concepts undergirding the curriculum and modes of instruction expressed here, please see the companion paper "Science Talent: In an Ecology of Achievement" in this volume.

‡A field trip is possible in all environments and with students of all ages. For example, a tree on a city street will serve; so will those ubiquitous plants called weeds growing in a crack in a sidewalk.

We propose a simple task: Students are to find a green plant and to write a brief description in answer to the question, "Where is it growing?" The answers are various, but all agree that the green plant is growing in soil.

And, "Where is soil to be found?"

On earth, to be sure.

We probe further, "What does a green plant need to grow?"

The answers are clear and show a grasp of information: good soil, minerals, warm temperature, and, of course, light.

And, "Where does the green plant get its light?"

All affirm: From the sun.

And, "Where is the sun?"

A bit of hesitation but again the information is forthcoming—93 million miles away. (Some convert the distance quickly into about 149,000,000 kilometers.)

Well, then, the question again, "*Where does a green plant grow?*"

It is gratifying, no, satisfying, to find a sudden urgent and happy light in the eyes of most; smiles and waves of hands affirm the discovery. (Eureka! Insight!) *The green plant grows in the solar system!* Of course. The explanations come thick and fast, some students wondering why they hadn't grasped such a simple idea before. Soon, some contribute other facts: The solar system is, after all, held in interaction with other stars and celestial bodies by the force of gravitation. And, of course, animals depend for their lives on the green plant. The class has arrived at one grand, significant explanation, a *concept,* of a way the world works: The environment of an organism is the universe. A concept is a form of orderly explanation, a result of well-ordered empiricism.

Second, students have already begun to grasp the purpose of a study of science, indeed one purpose of the year's work: to seek explanations for their observations of purported *fact* and purported *theory.* As we shall see, the young are to learn that facts are not isolated; they fit in interrelationship to other facts (Miller, 1987). Thus, one characteristic of a fact—in the area we assign to science—is that it is an index of a fuller explanation, to come, perhaps, in greater maturity: A scientific fact is one related to the world accessible to the scientist. In time, the students are to learn that the entire world is not accessible to science: The probes of scientists are limited by their need for hard-tried and hard-won verifiability of their findings. *Fields of study not accessible to the testable methods of scientists are thus simply not in the area of science.* Further, the students are to learn that frequently the verification of a finding is not solely in the hands of the finder but is the collaborative activity of those working in the same field.

Third, the students will find that a *concept* generally directs the planning of an investigation. The knowledge that a green plant derives its primary energy from the sun and its matter from the soil and air must enter into the substance of students' investigation, namely the form of their observations and their experiment. But we anticipate.

Obviously, we have begun our work with what most call the "processes of science." However, in this first day, both process and content have been shown to be interrelated: *How* something is known, a process, cannot, in the end, be separated from *what* is to be known, that special kind of content we call a concept. Indeed, a process, *observation* of a plant's growth, combined with critical thought, has led to a *concept:* A green plant is interdependent with the sun.

The function of a *concept* needs further clarification. A concept may be "true" or "false," but it is *thought* to be true at the time it is used. Thus, at one time, the concept of a flat earth was held to be true. A concept may also take the form of a hypothesis, theory, principle, or law—or just an informed idea, but an idea that deserves the kind of groping that is the beginning of science. Exploration (a form of observation) proved the concept of a flat earth to be false, so this concept was replaced with a new one: The earth is a sphere. The *thought* of the concept of the earth as a sphere then directed further *doing,* that is, exploration. In brief, a concept precedes exploration, or, if you will, experimentation. Thought before action. Thus, in certain aspects of the future laboratory work our students were to do, the concept—a green plant is dependent on the sun's energy—was central to their work; it was not, could not, be disregarded. When young seedlings were deprived of the sun's light, the students found the darkness not conducive to their plants' growth. They were beginning to learn the meaning of well-ordered empiricism: Begin with a concept.

Well-Ordered Versus Disordered Empiricism

But of course there is more to well-ordered empiricism than just the bare idea of doing science. In his Bampton Lectures, James B. Conant (1952) took occasion to clarify the meaning of empiricism and to distinguish between its well-ordered and disordered forms. One can, for example, determine what is wrong with an electric light by systematic, well-ordered empiricism, by testing one limited working hypothesis after another. Or one may do so by helter-skelter disordered empiricism, by repeating the same trial again and again to no purpose.

Thus, one may hypothesize: The plug is disconnected, so see to it first. Then, if the light still does not operate, one may change the bulb, etc. It would, for example, be *disordered* empiricism and wasted, if not fruitless, activity, if one went about the task by assuming that something was wrong with the plug, then immediately proceeding to wire it anew, without making the first tests in an ordered, systematic way. Perhaps, the only thing "wrong" was that the plug was disconnected or the bulb was not screwed in properly.

In a trial-and-error method there is obviously a difference between well-ordered and disordered empiricism.

Conant emphasizes, "What is often defined as the scientific method is nothing more or less than an approximate description of a well-ordered systematized empirical inquiry" (p. 46). And, Conant continues, "Any philosopher who happens to read these lectures will note that throughout I am using 'empirical' and 'empiricism' in a sense other than that to which he is accustomed; I am using these words as they are commonly employed by scientists" (note, p. 46). Conant goes on to state a significant concept in the scientist's mode of inquiry:

> Now, systematized or well-ordered empirical inquiries are one element in the advance of science; *the other element is the use of new concepts, new conceptual schemes that serve as working hypotheses on a grand scale. Only by the introduction of a theoretical element can the degree of empiricism be reduced* [italics ours]. (p. 47)

Thus, without the concept of the *germ theory of disease,* one might have tried, and indeed "physicians" of the early periods in medical history did try, disordered empirical approaches, such as driving the devil or evil spirit out of the skull (perhaps by drilling a hole there) or by letting blood. For example, Hippocrates could not know that bacteria caused disease. Centuries elapsed before the experiments of Louis Pasteur and Robert Koch; even their work did not fasten the concept in the practice of physicians of the time. The concept became part of the culture only after much education, and it is still not part of common knowledge in underdeveloped countries. (The reader will think of other examples of helter-skelter empiricism.) With the formation of the germ theory, a systematized, well-ordered empiricism was possible. The concepts of the atom, of space, of photosynthesis, and of energy also resulted in well-ordered empiricism. Modern science is, one may say, the interaction of concepts (theoretical notions) and experimentation (well-ordered empiricism).

The activities of a given scientist may thus include "hands-on" work, but, if so, it remains directed by the cumulative knowledge of *prior research*—a primary tool of all scientists. Possibly it is the function of a class discussion, a piece in a book about the subject under investigation, a part of a useful text, a laboratory lesson, or, when necessary, a good lecture, to add to the cumulative knowledge that is antecedent to "hands-on" investigation.

We hear now of "hands-on science." From the viewpoint of well-ordered empiricism, science begins with a concept, an understanding, a hypothesis, a carefully devised design for hands-on activity. This means "brains-on" *before* hands-on. For the very young, hands-on activity at the beginning of an attempt to understand the way things work may well be part of the "concrete operation" urged by Bärbel Inhelder and Jean Piaget (1958). But often hands-on activity remains only play—a good thing in itself. For the young, *play* becomes *science* when the hands-on activity is in the service of a search for meaning. Thus, the very young should be allowed to play with magnets or paper airplanes. But the child's hands-on activity doesn't lead to science unless it leads to a search for the properties of magnets or, in the case of paper planes, certain principles of flight. Nonetheless, as students go about their work, they should be encouraged to use a concept (brains-on) to design an investigation (hands-on) in planned laboratory work in search of meaning. Then, brains-on once again, as students record (write or enter into a computer) their observations in the form of a systematic assertion.

Work in the Service of Knowing

To learn to originate, in science at least, the young should have opportunity to innovate, to discover, that is, to do an original piece of work: an investigation or experiment (on their level, to be sure). This principle implies both time and motivation within an encouraging school atmosphere. These, in turn, imply a teaching method that recognizes pupils' strengths and abilities, as well as their weaknesses and lack of experience in the arts of investigation, that is, in the arts of well-ordered empiricism. Originative investigation means constant labor, a scholar's attitude, and work—above all—work. And first, this task requires an opportunity to learn something about well-ordered empiricism.

Our strategies and tactics are to describe in fair detail an environment in which this happens. We follow this course:

1. In teaching, we offer circumstances that allow students to select themselves for immersion in an augmenting environment based in apprenticeship to well-ordered empiricism. This augmenting environment follows a channeling environment that both motivates students to pursue a course of personal development and persuades them that they might like to try science as a possible lifework.

2. As we shall see, in the channeling environment, students' interests are given free play—more or less. They are given the opportunity to probe the lives of scientists so as to recognize what scientists, as artisans, are about. As students proceed, they learn to determine what is a *fact* (acceptable data). The young come to learn how to discern, distinguish, and evaluate variables; how to weigh gathered evidence (facts) for support of hypotheses, theories, concepts, principles; and how and where to get knowledge and skills. They begin their apprenticeship in a certain form of behavior epitomized by Jacob Bronowski's classic statement in *Science and Human Values:* "We OUGHT to act into act in such a way that what IS true can be verified to be so" (1956, p. 74).

3. After six months to a year within this channeling environment, the young are free to determine whether they wish to have a try at the augment-

ing environment. They express their wish by a first act: They find a problem of their own and express a wish to *do,* to learn something of "the art of investigation" (Beveridge, 1957) by submitting a *plan of work* aimed at solving the problem they have uncovered. Thus, they come face-to-face with the first thrust of a personal apprenticeship to well-ordered empiricism. For original investigations, as original as they can be in high school, may well be apprenticeships that enable the young to determine whether to have a try at becoming scientists.

Our model for doing an investigation is best expressed by Bronowski (1978):

> In summary, first, science is not an independent, value-free, dissociated activity which can be carried on apart from the rest of human life, because, second, it is, on the contrary, the expression in a very precise form of the species-specific human behavior which centers on *making plans*. Third, *there is no distinction between scientific strategies and human strategies in guiding our long-term attack on how to live and how to look at the world*. Science is a world view based on the notion that we can *plan* by understanding. Fourth, science is distinguished from magical views by the fact that it refuses to acknowledge a division between two kinds of logic. There is only one logic: it works the same way in all forms of conduct, and it is not carried out by any kind of formula but by an active view of how you apply the logic of *long-term planning strategies to the conduct of the whole of your life* [italics ours]. (p. 17)*

We call our model, stemming from Bronowski's thesis, a *planning model for research;* that is, it is a plan for well-ordered empiricism. It is the foothold idea for the apprentice scientist, whose work is to be carefully planned to meet a scientist's specific objective on the apprentice's level: to add to the store of knowledge, either through verification of existing knowledge or through addition of new knowledge.

Environments for Apprenticeship

Thus, we shall proceed along these lines. We shall probe

- *Channeling Environments:* emphasis on certain behaviors
- *A Nexus between Channeling and Augmenting Environments:* processes of individual research
- *Augmenting Environments:* arts of investigation—the well-ordered experiment learned through personal and usually individual experience

In a sense, then, we shall be describing a "Teaching Model in Schooling of the Talented in Science: Potential Through Identified Performance." But it

---

* © 1978 Columbia University Press. Used by permission.

is necessary to contemplate that equal access to opportunity in *schooling* is not the same as equal access to opportunity in *education.** Recall, educational environments include prior growth and development at home and with family and in a given community; these environments also include opportunities devised by parents and teachers for learning in the host of environments available in out-of-school experiences, particularly during the early years of childhood, as well as those during the high school years. Thus, the individual young who take on the channeling-and-augmenting environments will not have had equal access to *educational* opportunity but, in the work we describe, will be given equal access to opportunities devised in *schooling*.

We intend to validate the model based on 10 years of extended study in a school accepting a heterogeneous population living in the school's district; entrance tests are not required or permitted. The program, involving special adaptations of instruction in empiricism, has been described in initial detail by various studies (Pressey, 1955, and Brandwein, 1955, 1981, 1986). In supporting the nature of a differentiated curriculum, Pressey remarked, " ... in proportion as they [students] are very able, and especially as they have special talents, special adaptations of the usual curriculum are likely to be desirable" (p. 124). After this introduction, we are now able to describe the strategy and tactics by which students proceed from a modest experience in well-ordered empiricism to the time when they go on in their own unique ways to *plan* and *do* an *experiment*.

Problem Solving Versus Problem Doing

Suppose now, the 9th or 10th grade students have already been to the laboratory. They have, for example, read or found that green plants use carbon dioxide in making food for energy and growth, and, as a by-product, that green plants store the excess in leaves, stems, or roots.

They do simple "labs": They find, for example, that exhaled air differs from inhaled air and that the former turns bromthymol blue to yellow-green. Inorganic acids turn phenolphthalein from pink to colorless; inorganic bases turn it from colorless to pink.

TEACHER. How is it that all of you have gotten similar results? How do you explain your finding identical facts?
STUDENTS *(Different explanations are offered).*
ONE. We used similar substances!
ANOTHER. We used the same methods!

---

*For a further discussion of essential differences between the two, schooling and education, see in this volume the prologue, "By Way of Beginning," "Science Talent: In an Ecology of Achievement," and Brandwein, 1981.

A THIRD. The conditions for doing the "experiment" were the same; we had the same room temperature and the substances came from one bottle. We followed the same directions. So we found the "same facts."

In short, they were *doing* an "experiment" whose results were not to be a surprise; the equipment was laid out to assure a common finding.

The Next Two Weeks

To reemphasize, we have placed quotation marks around that maligned term "experiment." The students were not, in fact, doing an experiment. They were not engaged in *solving* an unknown; they were *doing* a problem, an exercise whose solution was known in advance; they were finding the knowledge designed for them by a teacher. This is not to denigrate the function of the oft-criticized "labs" as "cookbook experiments," nor for that matter, the much-denigrated textbooks that may affect and interact with the "lab" experience. When textbooks and labs are properly used, together with critical class discussion, they do offer students experience with the commonalities in the objects and events of the world necessary for their probes. Acquaintance with such commonalities furnishes the young with the beginnings of a skill: the recognition of *discrepant events* that may be subjected to investigation. Critical thinking begins with the ability to identify a discrepant event; hence, the importance of reading as a "brains-on" activity and the "lab" as "hands-on" activity as a combined base for further exploration. Indeed, students could find the "solution" to their probes in the "labs" in most textbooks, high school or college. But shortly, these students will engage in problem solving, a different kettle of fish.

To emphasize, *problem solving* is different from *problem doing*. However, problem doing in what we normally call the "lab" is quite valuable—even apart from learning skills. Students can attain their first approximation of the meaning of a "fact." They find, essentially, that *a fact has its credential outside of its believer.* Thus, inorganic bases turn colorless phenolphthalein pink not because you, we, or I say so, but because the results have been verified by countless qualified observers. $KClO_3$ is written thus because its formula has been similarly verified. In the same way, $F=ma$ remains a valid statement; the speed of light is a constant in $E=mc^2$; the mammalian heart generally has four chambers, etc., etc., etc.

All these statements of fact have their credentials in the detailed description of well-ordered empiricism, resulting in the discovery of a new and meaningful statement assessing the facts, the validation of the facts, and the concepts embodied in the statement. This is an exceedingly important finding for all the young, whether as citizens in general or as scientists-to-be in particular. The experience needs constant affirmation: Facts have credentials, that is, credibility, and those credentials—their confirmation and affirmation—exist by means of well-ordered empiricism. The believer may attest to the statement as "fact," but the discovery may not have been his or hers.

Do Facts Change?

There are several ways to press a finding to this question. One young teacher we observed last year collected old textbooks—some from as far back as 1916. She told us of one simple way of uncovering observations on the question: *Are facts affirmed by scientists forever true?* After students had studied the intricacies of the brain's function for a while, she asked whether anyone would like to find out what the thinking was on the substance acetylcholine earlier in the century. Several students said they would be happy to report the next day. For their "data base," the teacher gave different students textbooks from 1916, 1920, 1925, 1930, and asked them to include the facts they found in the books in their reports. As one student said,

> The author must have known what he was writing about; he's a Ph.D. in zoology. But the textbook he wrote in 1916 talks [sic] about neurons only, not neurohumors. Then I found out that *neurohumor* is of recent use, and *acetylcholine* didn't come into common use until later, I suppose.

Facts change as new discoveries are made. "Facts" are made falsifiable in the light of fresh inquiry, newer knowledge, and newer discovery. Well-ordered empiricism is not based on the permanence of a fact—if only because new variables may enter the picture.

Variables as Conditions of Experiment and Amendments of a "Fact"

Early on, after the first day's acquaintance with an organism outdoors, each student *grows* an organism. Purpose: to relate growth of a living thing to its environment. All students grow a plant from the "beginning." They grow tiny chunks of potato in *light* and in *darkness;* at the temperature of a *refrigerator* and of the *room;* in *moist* soil and in *dry* soil; *with* an eye (the bud) and *without* one. They learn that the *conditions* of the environment and the *conditions* of the material must be *controlled* to secure valid data. Thus, a potato at *room temperature,* in the *light,* in *moist* soil, *with a bud* is likely to produce a green potato shoot. No bud (a variable), but all other variables present, no shoot. In darkness, a spindly etiolated shoot. In dry soil, a sickly attempt at a shoot, soon dead (etc, etc.). They learn that the *variables* in the environment of the experiment must be precisely *controlled;* that is, a well-ordered empirical approach demands that the different variables remain precisely similar, if validity is to be attained. The control of the variables amends the precision of the statement of the fact observed. Students thus begin to get their first "smell" of the meaning of the terms *true* or *valid* as the scientist uses them. The sequence in the teaching of the meaning of a *fact* is but one indication of the teaching devices used in the channeling environment designed to develop certain aptitudes, that is, essential behaviors, toward one end—a respect for well-ordered empiricism.

Soon, students will immerse themselves in the augmenting environment in which well-ordered empiricism will be required. Such essential behaviors as accuracy in measurement, precision in language, honest reporting, assessment of permissible error, stating credit for prior discovery, relentless examination of the literature, etc., etc. will come forth, as students probe on their own or in teams. There are, however, still other essential behaviors—that is, knowledges, attitudes, and skills—that students should develop before they are let loose in an apprenticeship to empiricism, before they proceed to an originative work on their own. The most important is a taste, even a preference, for a certain useful *independence*.

An Indispensable Behavior: Independence and Independence Training

It is clear that a scientist *is,* within the graceful limits of the necessary cooperative aspects of science, *an independent thinker.* Even within a team, different scientists remain independent researchers who have agreed to uncover a piece of knowledge or skill necessary to the shared team effort. Such has been our own experience in scientific research, and such are the behaviors we try to foster in our students—gifted or not. Gifted students seem particularly to thrive and exult in a certain independence. Obviously, they are not yet independent, but they begin to get their first taste of the meaning of freedom.

At Forest Hills High School in New York City, we began our work in independence training when we talked to the general science classes in our "feeder" schools (eighth grade and middle schools). Or, we deferred our efforts to the time the students entered high school. We suggested that students who might want to have a try at a career in science join a special class in biology (Biology H). Usually we began with two classes of 30–36 students who applied for entry into the program for which they "selected" themselves by registering for Biology H. All were accepted—no written tests, no searches of past record at this point—self-selection!

We did not deny anyone entry into the program by use of a test as a measure of selection. In our view, a heterogeneous school in an open society allows access to equal opportunity: Any program is available to all students who wish to avail themselves of it. The test—if it may be called a test—is appraisal by students of their achievement in the work. As we shall see, during the 10-year study (1944 through 1954), ingress to and exit from the program were permitted.

Neither should our use of experimental and control groups be misunderstood (Brandwein, 1955, pp. 31–32). The control group did not use a placebo program; the control group was offered the same opportunities as the experimental. The control was simply one that consisted of fairly matched students to the experimental; the latter was one that, in our observation, seemed more dedicated to the completion of an experiment than did the control.

Equal opportunity for the young in a democracy does not now and did not then, in our view, permit an experiment that denies opportunity to one group and permits it to another. In this, we hold the position expressed by Julian C. Stanley and Camilla Persson Benbow (1986):

> We were rather sure that the smorgasbord of accelerative educational opportunities we planned to offer the "experimental" subjects in the study were much more likely to help than to harm them. Therefore, it would be inadvisable to withhold such opportunities from a portion of the subjects (probably half of them), who in a controlled experiment would be assigned randomly to a "control" group. (p. 369)

(See also Brandwein, 1986, or Brandwein, 1955/1981, for certain strategies useful in a school accommodating a heterogeneous population.)

Early in the course, we presented the students with the notion that the study of biology had changed somewhat; it had become an amalgam of the study of living things (biology) and the study of nonliving things (chemistry, geology, a bit of physics, and, of course, mathematics, among others). Biology has become a study of living things adapting to a physical environment and one of living things utilizing nonliving things (substances) in their physiology. There were decent biology textbooks, but at that time none was a satisfactory amalgam of the newer aspects of biological science. (Even in 1947–1956, biology was no longer a discrete science.)

We offer the students a proposition. We shall use a biology textbook as a springboard for our study. But there is much else to learn beyond the textbook. Would they be willing to collaborate in the task of building the course of study? If only because teachers and students need to know what they must prepare in advance, we propose the following strategy. For the first two weeks of the course, the students may examine a selected textbook and find one "unit" they could study and learn on their own; they may do this with friends. At the end of the two weeks, we propose that all meet in class to build the outline of the course. Meanwhile, they will be developing certain interesting skills that all must have to study and learn in any science—let alone biology. At the same time, students become acquainted with two teachers in biology who will become their mentors; these two are to be available for questions and resolution of problems or conflicts, upon request for an appointed time, before or after school, during study or lunch periods, or in time "hollowed out."

In the first two weeks, then, (even as they plan the course), the young begin to learn how to use library, laboratory, textbooks, scientific data bases, and other resources. They continue to learn from others; they learn from all manner of contributors, including the scientific workers in the field they are probing.

Further Probes

Early on, we probe further into the internal environment of the organism. (We have found it easiest to work with plant organisms at the beginning of a year's work in biology; animals require a bit of handling.) We begin by examining a green plant's cells; elodea is a good source. Again, students confirm one another's observations, reinforcing their understanding of facts as *confirmed observation embraced within a concept.*

There is no lecture on the plant cell; the laboratory, textbooks, and library are first resources. Early on, the students begin to enter upon the habit of taking responsibility for enhancing their observations with readings of prior discovery: A good textbook serves one such function, for it is, in a sense, a history of prior discovery. Their readings and discussions modify the textbooks' contributions—even the college textbook students will eventually use.

Young potato plants growing in water in the classroom serve as raw materials for preparing slides of plant cells, including upper and lower epidermis of young conducting cells (xylem and phloem cells in the stem). Roots and rootlets are visible in the water; they, too, are subject to study under the microscope. And where does the potato plant, the bud, the "eye," get its energy for growth? The answer is plain: Potatoes are good sources of starch, easily converted to sugar useful for the cells' energy. (Here, a laboratory on the conversion of starch solution to sugar by the cell enzyme *diastase.*) And the first time students do an investigation involving substances—or one involving *any* element of danger—the students are required to read, study, and understand a statement on safety procedures.

Thus, the first week is spent on plant growth and cell studies; the second week is planned for animal cell studies. Laboratory work almost every day—with 15 minutes at the end for questions and suggestions for preparation for the next day. Being gifted students, most read, apprehend, do, and follow instructions easily; the work on cells advances rapidly. Meanwhile, a few are anticipating the planning of the course. They ask, "When?" They are impatient.

At the end of the first week, a problem is raised by the teacher regarding some interesting stone objects from the Petrified Forest in Arizona. For all the world they look like stone parts of trunks of trees, stems, and roots.

Indeed, there are ways of cutting bits of them very, very thin and preparing them for microscopic examination.

And this is the way the thin sheets of stone-plant look under the microscope (a slide is placed, illuminated, and focused on the screen). The clear image of plant cells—walls, nucleus, etc.—is obvious. A statement is to be found on the board, "You may read further about petrified plants in these botany books in the library. Or in your encyclopedia." A few hands are already calling for attention; information is at hand. The teacher suggests, "Let's wait until everybody has had a chance to look into the problem; let's discuss it next Monday. Perhaps you'll want to write a few hundred words on your findings."

Next Monday, the period begins with the question, "What have you found is the meaning of the slide of petrified material we observed Friday?" All do *not* raise their hands. "How many of you are not prepared?" A few raise their hands. A short statement by the teacher follows along these lines, "You are aware, I think, that scientists must be ever-learners, and they must be independent learners. Next time you are not prepared, you are obliged to give me a note at the beginning of class, thus—'I'm not prepared today'—and explain your reason." And the stage is set for that most perdurable of developed aptitudes: *independence in learning.*

Nonetheless, the discussion proceeds. Students report (factually, thoughtfully, and in decent English): The objects are clearly cells. Therefore, these petrified plants—called fossils or fossil remains—were once alive. Now, they are preserved. There is general agreement. (Some offer an estimate of age.)

Further Advances in the Learning of Process Skills

But the objectives of the experience are clear in the teacher's mind: Students are not only to advance in knowledge (of *content*) but also in methodology (of *process*). The teacher asks a question:

Is it fair to put your findings thus?

IF these are indeed cells, as they appear to be,

THEN, the petrified trunks, or parts of organisms, should be parts of organisms that lived in the past.

Some students recognize the statement to be in the form of a hypothesis (the *IF/THEN* form). Some are ready to define the hypothesis by the clichéd "educated guess," but, of course, it is more than that. A few questions call forth the notion that a *hypothesis* is indeed *part of a process; it is a plan for further experiment.*

Suppose the *IF* portion of the statement to be "true"; the *THEN* part follows as a statement requiring evidence. To make the statement evidential, the scientist is obliged to find living organisms with similar forms of cells or to find chemical equivalents to which living and dead cells conform. To repeat, a hypothesis is a plan for *work*, not a substitute for *conclusions*. It is a plan for gathering a multitude of *observations*. The aim of all these is, in the end, an orderly explanation of the phenomenon, a new and meaningful statement, possibly a *theory*.

A theory, recall, is an explanation of a considerable number of independent, interrelated observations (e.g., the atomic theory). And the students have many such observations ready at hand on which to base a first theory: observations of at least eight plants—onion, elodea, *Tradescantia*, strips of potato epidermis, xylem and phloem cells, root hairs of potato, stem cells of the clear weed, *Impatiens*. Within the coming week the students will also have examined a multitude of cells of animals, including mammalian (their own cheek cells and cells of other animals) and amphibian tissue cells. The theory, then: *Organisms are composed of units called cells.**

Thus, the students begin to appreciate certain first concepts of the nature of scientific processes, to be modified as learning proceeds. Thus, the young discover, in general, that a hypothesis is a grand design for planning an investigation and that a theory not only explains the multitude of facts gathered through observation but also offers a design for directing further work. Thus, the cell theory directs research on the nature of the cell, on the manner cells maintain the internal environment, on the cell's function in heredity, in homeostasis.

The students have reached the objective of the first two weeks of the course in biology: a *melding of content and process* derived through *observation, experiment, and a search through the cumulative knowledge* that are the remarkable resources of scientists. The first two weeks are, therefore, a first and smallish introduction to the kind of research, or *plan for discovery*, which is in the repertoire of the scientist. The rest is to follow. The first steps are being taken.

A Further Thrust Toward the Skills and Privileges of Independence

In our experience, in the second week, after the labs on cells have been completed, students begin again their eager, much-appreciated nagging: *When are we going to build the course?*

---

*Of course, some students have already read that Matthew Schleiden and Theodore Schwann had developed this "cell theory," after many years of work. This in the 1830s. The students also learn that the two scientists had worked separately (Schleiden on plants and Schwann on animals), but their work was published in a related time span; thus, they are jointly credited with the theory.

TEACHER. How many are ready? *(Not all raise hands.)* Clearly not all are ready. Let me make a suggestion. Why don't you group yourselves into six groups? *(The classroom is so organized that this is possible; it has movable chairs and tables.)* I'll finish up my own plans. *(On the unusual occasions when the discussion becomes heated, the teacher flicks the lights on and off—a signal: "light, not heat.")*

Toward the end of the class, a short discussion leads to the provision that some groups will meet at different homes to "finish" their plans.

Almost always (in our experience), students are ready to plan the course early, at the beginning of the second week. A committee of students lists all the topics and page numbers their classmates are willing to study on their own. The suggestions vary from "the book's too easy" to "that topic is hard." The final plan after two days of discussion (a number of plans are telescoped) is something like this.

First, we may conclude that one student or another will mention or submit topics she/he can study. The net result—the entire textbook (an instructional device) might, if the central concepts were dispersed among the volunteers (who offer to study and report on them), be covered in sequence by different volunteers during the year. But the teachers in the group have agreed to offer a different tactic. They are not willing to be left out—not only because some concepts are inadequately treated in the text but also because

• new researches not readily available in our library need to be discussed
• laboratory investigation must be included
• an individual piece of research, which all will be required to do, will need to be planned
• time must be left for recreation, for play, for other subjects

Finally, the class agrees upon the topics that may safely be left to the students on their own. These generally include

• classification of organisms (with specimens provided in the laboratory)
• general body anatomy and physiology (with the options of discussions in class on special aspects—i.e., enzymes, hormones)

The students obligate themselves to study in groups that they organize. (Note that the activities described above are now called "cooperative learning"; on this subject, see Roger T. Johnson and David W. Johnson's paper in this volume.)

The teacher may then undertake to probe the concepts involved in the course titled *Organisms: Internal and External Environments—Past, Present, Future.* It includes, in rough progression,

1. interaction of cells in maintaining the structure and function of organisms (chemistry and physiology of cells, organs, tissues, homeostasis, and health generally)

2. interaction of cells in maintaining the heredity of the organism (genetics mainly)

3. interaction of organisms with environments, past and present (ecology of present environments [ecosystems] and ecology of past environments [evolution])

4. future of environments (including use of resources, space, etc.)

Some Sticking Points

Clearly, the class's proposal that it is to study certain topics on its own and the general willingness of students to do so is a *test* of willingness to undertake *independent study*.

All is not smooth going; nothing ever is, is it, in plans that involve the human condition? Other students have learned of "what's going on," and they "want in." So do certain of their parents. However, not all participating students are "sold on the plan": They hadn't counted on being responsible for work on their own; they prefer the lecture as a mode of instruction; they are mindful of the end-of-the-year state examination in biology. (In New York State, this is the Regents examination.) These students "want out"; so do their parents.

The class discusses the meaning of independent study, its values and its hazards. The teacher helps by suggesting, in answer to a question on testing, that students will be given the opportunity to test themselves at certain planned periods. Further, in our experience, the classes have done splendidly on the state examination. (Evidence is offered.) In the end, approximately 4 to 8 students (a total for both classes) ask for transfer to the "regular" classes; about 10 students, sponsored by parents—some quite insistently—seek entry. The department chair, in discussion with the teachers in the department and the principal, has foreseen such problems, and in the first two weeks the transfers are made. (In the practice of the science department, planning for the first two weeks of the semester included concepts taught by all science teachers in their special courses; therefore, students who transferred in the first two weeks were not penalized by inadequate preparation in subject matter.)

The requirement of independent study is not the only reason why some students are uncomfortable with the class they originally "selected for themselves." The 5 of 13 teachers (in the various sciences) who have volunteered to participate in the program are committed to reducing the lecture as a mode of instruction. In fact, in this course, the "lecture" will be reduced from some 70 percent to 30 percent, so that students may devote time to independent study in and out of class. This, we have called the maturity-independence model. In essence, the transfer of knowledge from teacher to student is reduced gradually so that the transaction in learning is not of *teacher to learner* but is turned to a certain independence: the *learner as learner* and the *teacher as mentor* and guide. That is, the teacher acts as a *guide to* not as a *guardian of* the archives. After all, the students are gifted. Recall, we are engaged in probing whether they *wish to turn their gifts to a talent in science.* And the study of science and the work of the scientist require the habits of independent study.

Figure 1
Changing Initiatives

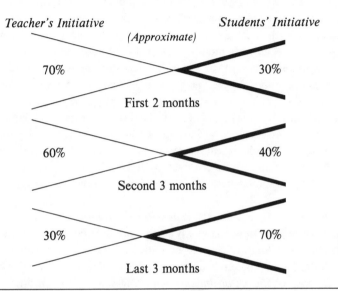

Teacher's Initiative		Students' Initiative
	(Approximate)	
70%	First 2 months	30%
60%	Second 3 months	40%
30%	Last 3 months	70%

Teacher's Initiative
Includes lectures, visiting speakers, demonstrations, literary and note-taking skills; teaching, discussion, and seminar skills; conferences; individual testing; additional laboratories; teaching uses of texts and reference papers, etc.

Students' Initiative
Includes individual study, cooperative learning, library work, requests for extended lectures, laboratories, projects, seminars, extensions of topics, etc.

Thus, the teacher turns from a presenter to a questioner, persisting in a certain strategy, always probing: *What do you know?* Statements beginning with *what, why, how, when, where* ask about substantive content matters—not *who* will tell me (or us). (Students are expected to be prepared.)

How do you know it? (Students are expected to know the sources they quote or those they have studied.)

How well do you know it? (Students are expected to offer evidence or reasons for their statements.)

Is it true? (Students are expected to give some evidence of verification.)

Now, all this is not rigid. Teaching, we realize, is not only a plan for combining two loves—the young and their concepts and processes of thought and action with certain knowledges, values, and skills that will come out of the interaction that is the essence of teaching. Teaching also requires the practice of extreme patience, of kindness, of compassion. Students do not easily learn that a given teacher won't always be with them; they not only need to learn how to learn but also to understand the significance of independence in learning, an essential life-affirming behavior. They are put at ease: If they find they are not comfortable with the approach, they may change class again at the end of the semester. And, if they have done the work as diligently as possible, we promise that the lowest grade will be a *B*. Only one to three students leave at the end of the semester; the rest find it easier and easier to accommodate the "new" mode of teaching-learning. Besides, all students may take a given test a second time until they learn how to accommodate themselves to the arts of study. And the mentors are at hand to help them encompass these arts. Further, we are committed to the Rogerian principles of "psychological safety and psychological freedom" (Rogers, 1961): There is an absence of threat. Mistakes are not "punished"; there is freedom to disagree; there is also freedom to leave the course of work proposed.

This, then, is a picture of the *channeling environment*. In the last months, students will be introduced slowly to the *augmenting environment*. In the latter, a talent for science is to be indicated through *performance* in research, that is, in problem finding and problem solving. A first objective: the *development of the art of finding compelling problems for study.*

Augmenting Environments: Discovering Problems for Research

Students are, however, *not* required to wait one year before they are permitted to engage in individual or group research. The channeling environment is rich enough to stimulate some students to probe individual—sometimes astonishingly unique—problems long before that. For some students, the process has begun, in fact, before the first day; they bring certain research problems to the work at hand. Early on, we recognized that the habits of well-ordered empiricism are best taught in a time-honored and effective manner; they are to be experienced personally in individual work. And individual research is central to the augmenting environment. *Learning* an art of investigation means *doing* an investigation. In 1947, we had begun visits to the elementary schools that sent their students to Forest Hills High School. (This is relatively easy to do in a school with a known population of elementary and junior high school students destined for the school.) We explained our program to students and teachers and, in fact, developed a program of visits to our science classrooms by teachers and students in the feeder schools. Certain students became interested in protist (then protozoan) studies. They came to the school in the 9th or 10th grades, prepared and eager to conduct their own "experiments." However, most problems directed toward possible individual or group experimentation came to the fore almost with the first day.

A Nexus Between Channeling and Augmenting Environments: Certain Processes of Individual Investigation

We have noted a critical ability in our samples of those who turn their talents to science; they are powerful conceptualizers. (Of course, such is also often characteristic of the young generally gifted in the habits of intellection.) In fact, one had to guard against their easy knack of "putting together" observations and facts into theoretical assertions—often ingeniously contrived.

One instance will illustrate at once the ease with which problems are defined by these students, as well as the opportunities and difficulties of seeking solutions. We had established a curricular structure centered in concepts and, therefore, in concept seeking and concept forming; we set aside one period a week—or whenever students were ready for such a discussion—on "individual research problems." It was understood, however, that, once a student brought a problem to the attention of the class, it was his or hers to *solve*—if a solution proved possible. The group's function is, then, that of discussion.

To illustrate the process of initial discussion by the class followed by individual work: One of the students *(R)* put a problem before the group couched in the language of the hypothesis—it becomes a habit. (Note the *IF-THEN* construct.)

IF the sun's light—and thus light generally—is essential to the growth of organisms,

THEN the growth of trees situated on street corners near street lights may be affected.

Wonderful! *R* was asked whether she wished to discuss the problem in class or whether she wished to proceed on her own—to make the problem available for her individual research. In any case, she would be required to present a "plan of research." No, *R* said, she would be pleased to have help; she said she was "swamped in variables."

The class responded. First, she would need to know the names of the trees she planned to study. (*R* responded by holding up "Key to Plants' Identification" and explaining how she had begun to use it.) Then, different members of the class suggested the following variables:

1. She would need to compare the same species. (*R* had chosen a frequently planted tree—*Tilia michauxii,* a common horticultural variety of the Linden.)

2. She would need to compare trees of similar trunk diameter, height, and branching.

3. She would need to compare trees on corners with similar geographic locations. (She chose the southeast corners to compare with northwest ones.)

4. She would need to measure light intensity at given periods—morning, afternoon, evening. (*R* agreed that, for safety reasons, she would do her evening measurements in the company of an adult.)

Other variables were width of leaves, change in color, and differing moisture content of soil. Even in this first attempt, the students had begun to reduce the empirical aspects of their research and to nurture an ability to design a *well-ordered empiricism;* they had *determined the concept* involved in order to form a hypothesis, which, in turn, directed their search for variables. (Recall that *disordered empiricism* is trial-and-error, somewhat like poking about in the haze; well-ordered empiricism begins with a theoretical underpinning, a concept.) The concept in this case: An organism is the result of the interaction of its heredity and environment.

After two weeks of pondering the problem, *R* came to the conclusion that she would prefer to engage in laboratory experimentation where *she* could control the variables. She came up with another problem.* Note another aspect of well-ordered empiricism: the use of a device (in this case, the laboratory) to control variables.

Other Projects: Other Problems

The point that needs making is that gifted students with a bent toward science (we call them "science-prone") are quick to find problems. After all, the structure of the course puts them in the way of wide reading of college texts, the basic high school biology book, and technical magazines—*Scientific Monthly, Scientific American,* and so forth were the wont of many of them—and they also paid wide attention to newspaper, magazine, and TV accounts. Then, there was the science and mathematics journal they had organized; the monthly seminars and yearly science congresses; the science fairs they attended; and the like. Some of them had begun attending the open meetings at the Museum of Natural History, seminars of the New York Academy of Sciences, etc. Thus, *finding* problems was not difficult; finding problems *accessible* to work on and *fundable* by the high school laboratory was. Some students, however, were able to work in nearby college and

---

*Her new study brought to successful conclusion: "The Influence of the Antibiotic Terramycin on Reproduction of *Paramecium caudatum.*"

hospital laboratories (with college mentors approved by their parents and high school adviser-mentors).

However, not everyone in the class elected to do research; approximately 50 percent of the two classes, each comprising 36 or so students in any given year, found the inclination and time—and possessed the *predisposing factors,* particularly that of *persistence*—to pursue the problem to the end in honorable success or honorable failure. (For further elaboration, consult "Science Talent: In an Ecology of Achievement" in this volume.) All students at Forest Hills High School were required to take a year of laboratory science and almost all elected biology. Approximately 40 percent of our enrolled students selected preparation in the commercial area and in other so-called nonacademic careers. About 60 percent elected college-preparatory programs, but all could, if they wished, change programs. About half of the college-preparatory group selected themselves for immersion in more rigorous study in a variety of subject matters. However, to clarify, we regularly found that in any year approximately 70–80 students selected themselves for the "science-prone" group.

In any event, a spate of problem finding resulted from our practice of insisting on reading (where applicable) the scientific literature, sharing the information in class, and regular discussions of "problem seeking" by those who wished to proceed with their own research. The science talented found plenty of bona fide subjects for research. For example, here is a list of problems presented in one class (similar in number and interest to other years):

• Is there a variation in background radiation in the Flushing region (an area covering two square miles)?

• What precision is there in the order of flowering of four selected species of plants on selected dates in spring, early summer, early autumn, and late autumn, in specially situated regions in Forest Hills?

• Is it possible to design a machine to perform a multipartite operation (a game of cards) previously done only by humans?

• Is the "rising" of *Wolffia* commensurate with its increase in number of buds?

• What is the rate of population growth in comparative inoculations of *Paramecium caudatum* and *P. aurelia* in like culture media?

And, in the form of topics, these:

• comparison of growth of sessile organisms *Vorticella* in different environments

• problems in the photography of crystalline structures

• the ultraviolet photosensitization of *Tribolium confusum*

• the harmony of Mendel's results: an inquiry into statistical measures

• the uses of computers to harmonize music

• the content of galls on the cherry *(Prunus serotina)*

- the embryology of a snail *(Physa)* in relation to variation in light, temperature, and isotonic media
- meteor showers in August: seeking new constellations

Some 30 problems were commonly generated each year by students. Most problems and topics were derived "independently," stimulated, aided, and abetted by reading and discussions in class. Some projects were "uncovered" by readings suggested by the teaching mentors the students had chosen. Some topics were stimulated by parents who were scientists or in the out-of-school seminars the students attended. (On this see A. Harry Passow's "School, University, Laboratory, and Museum Cooperation in Identifying and Nurturing Potential Scientists" in this volume.)

It is well to consider preliminary and interim reports of research problems by two students, samples of the distinctly individual researches carried through during the 10 years of our study (Brandwein, 1955/1981). We may note that each report is, in its own way, an index of the first requirement in self-expression and conduct of the "apprentice scientist": well-ordered empiricism as an index of the critical thinking that is the habit of the scientist. It is also to be noted that each piece of research was monitored by a scientist knowledgeable in the specific area of the research.

Michael Fried and his Tribolium confusum

Preliminary Report: An Ultraviolet Photosensitization in Para-aminobenzoic and Pantothenic acid fed to *Tribolium confusum*

I had read (see references)* that when mice were fed buckwheat and were placed in a strong light they died, while mice lacking either the light, the buckwheat, or both thrived. I tried to duplicate the results on insects. I worked with the confused flour beetle *Tribolium confusum.*

The effect in mice can be duplicated on the flour beetle. I am reasonably certain that:

1. The ultraviolet rays of the light, acting with an agent (or agents) in the buckwheat, seem to cause the reaction known as photosensitization.

2. When pantothenic or para-aminobenzoic acids (in a concentration of 5 percent and higher) are added to the diet of the flour beetles, the photosensitizing effect does not occur.

It may be that the photosensitizing reactions are caused by the conversion of either (or both) pantothenic or para-aminobenzoic acid (PABA), both of which

---

*Notes and references appended to original paper not included here to conserve space. Paper, later enlarged by some 1,500 words, was part of student's submission to the Westinghouse Science Talent Search.

are needed by the cells to synthesize antimetabolitic structural analogues. The cells of *Tribolium* seize upon these structural analogues but cannot utilize them; the cells thus suffer from a deficiency of these vitamins. Death may be the result.

Michael Fried, Senior
Forest Hills High School

Michael Fried did not sail smoothly through his investigation: He had certain problems to solve. For example,

1. He cared for two colonies of *Tribolium,* one in the school laboratory, one at home. This, a precaution against catastrophe.

2. He spent considerable time in a laboratory (over a period of four months) developing the range of dosages of ultraviolet light and pantothenic and para-aminobenzoic acid until he obtained optimum concentrations. This, under the supervision of a scientist in the laboratory.

3. He had his data reviewed by biochemists and physicians at a nearby university and hospital.

4. His mentor—and the scientist who monitored his work—thought his procedure and technique elegant. His experimentation proceeded over a year with constant improvement in method.

Carl Koenig and His Machine Performing Multipartite Operations

Interim Report: The Design of an Electrical Machine Performing a Multipartite Operation of a Mechanical and Mathematical Nature Previously Done Only by Man*

The question of machines replacing man has been asked again and again during the last two centuries. As of now, machines have been developed that can completely replace man in certain jobs concerning mechanical operations and arithmetical computations. I was interested in knowing if a machine could be designed that could completely replace man in a more complex, multipartite job.

---

*(Of course, Carl meant *human* throughout.)

Full description, diagrams, tables, and references originally appended are not included here to conserve space.

Perhaps it will be useful for the reader to know that, if the Search is a valid test of finding those who might succeed in science, then—as compared to entrants at two select schools—students at Forest Hills, including *R,* Michael Fried, and Carl Koenig, performed favorably during the 10-year period of our study. (See table 1 in "Science Talent: In an Ecology of Achievement" in this volume. However, the reader of that paper will also note that during this period, not all schools, some with highly able students, entered the Science Talent Search.)

All jobs involve one or more of the following factors: mechanical operations, arithmetical computations, observations, reasoning, and the element of chance. I wanted to see if it was possible to design a machine that could perform a job that contained all five factors. Since the dealer in the game of blackjack has a job that involves all five factors, I decided to try to design a machine which would take the dealer's place.

First, I unified into a definite set of rules the various ways of playing blackjack. Then, after arranging the controls of the machine for convenience of operation and prevention of confusion, I prepared an instruction sheet to tell how to operate the machine. Because of its multiphase operation and its resulting complexity, the machine was considered as 17 individual units, each being a block of cognate functions. All the functions of each unit were tabulated, and then wiring diagrams of each unit were developed. A method was developed for the machine to decide mathematically what the dealer would decide. The designing of the machine was more than just a problem in designing circuits. It was also a problem in unification, tabulation, and classification of information, problem analysis, and probability.

My machine has the possibility of taking over the jobs of man. It could be assumed, however, that I am inferring that machines will replace man himself. If a machine can reason, it is following the reasoning that man told it to follow and not reasoning by itself.

Carl Koenig, Senior
Forest Hills High School

We were, and are, persuaded that, in these and several hundred other investigations and experiments, we could glean evidence of the kind of *questing* and *persistence* of scientists-in-embryo. We find a kind of active seeking and finding of a *channeling environment* and a further probing and seeking that thrust the student into an *augmenting environment*. We see the beginnings of a personal transformation into the attitudes, aptitudes, and behaviors we recognize as those of the maturing scientist. In short, the channeling and augmenting environments, if successful, lead to a self-transforming environment.

Potential Tested Through Performance

We are persuaded that the ability to plan and complete an experiment, with the arduous and meticulous application required, may well be a valid test of science talent or, if you will, a test of potential in science. It is possible that the artifices of the computer will enable us to prepare similar innovative experiments that will take less time than a full-fledged laboratory experiment and thus be more appealing for most teachers in limited school situations (see later).

We are further persuaded that such a test lends itself to solidly empirical observations and experimentation, that it distinguishes between aimless precocity and the quiet, patient, and persistent qualities undergirding the production of verifiable knowledge, the hallmark of what we know as science.

Thus, doing an investigation or experimentation in science and mathematics might well be a test of the ability to *perform* in science and thus might be considered a test of an ability in *discovery* (synonym: creativity). In a sense, such an approach offers a whole way of looking at the whole individual. Also, in this sense, we are stating a point of view that has pervaded the works of philosophers and students of human development. (On this, see Liam Hudson, 1966.) In any event, until another test is devised that predicts more adequately an individual's talent in science, we propose that doing an investigation that results in an originative work on the high school level will serve as a useful predictor of science talent. (See also "Science Talent: In an Ecology of Achievement" in this volume.)

A Pervasive Problem and a Possible Solution

You will not have failed to note that, although the generation of problems for study was not a problem in itself, the problems generated were nevertheless to be found mainly in the field of biology and ecology—some in astronomy. It is worth probing the major reasons.

Biology *was* the first true laboratory-, discussion-, and seminar-centered course for the great majority of the students in our school—some 97 percent. It is credible to assume that the motivations were of a high order to try individual research early. Very few of the students deferred their research to a later year. Indeed, a questionnaire study and follow-up discussion with the students in the augmenting environment (most of the 354 in the 10-year investigation we conducted) indicated that there was necessary and sufficient reason for their early commitment. Some offered the potent reason that later they would be taken up with a full program of electives as well as with preparation for junior and senior Scholastic Aptitude Tests.

Further, more as a matter of custom than a reality founded in fact, our mentors "felt" that students should have a decent exposure to a year's work in science, say in biology, before experimentation was permitted in a particular sector. This seems hardly convincing, but there it is.

But in recent years another art of investigation has emerged: the computer, with its fascinating promise of rapid engagement of statistic with statistic, fact with fact, and mind with mind. Further, the computer offers speedier recognition of variables, facile recording, and fairly prompt return and recording of data. Of course, the arts of investigation are also developed in such experience in doing the customary laboratory exercises. Certainly, hands-on experience is required to do originative work as well. But, in all cases, hands-on experience should rest in a well-ordered empiricism: first in the statement of a concept and, then, in the actual laboratory experience. In any event, microcomputer-based laboratories promise to expand greatly the types of laboratory experiences we have in the schools, not just in biology, but in physics, chemistry, and geology, as well. Further, "Of equal attraction is the creativity that can and will be released in the development of new lab activities" (Abeles, 1985, p. 39).

For example, many high school chemistry courses include the following investigation demonstrating change of state. In the traditional laboratory, the student uses a substance, such as naphthalene or paradichlorobenzene, which is warmed to its liquid state and then allowed to cool. A thermometer is placed into a test tube containing the substance, and then this tube is placed into a second, larger, chilled tube. The student takes readings from the thermometer as the substance cools and places them, along with the time of each reading, on his or her data table. After the conclusion of the activity, the temperature readings as a function of time are recorded manually upon a graph. The student may now observe the cooling curve of the liquid, the plateau formed as the liquid changes to a solid, and the cooling curve of the solid. The mathematical expression of the physical phenomenon is now available and may be used for prediction of the outcome of further experiment. (Prediction, of course, is a necessary component of well-ordered empiricism.)

This same activity may now be accomplished in the microcomputer-based science laboratory. In this case, a probe containing a thermistor is substituted for the thermometer. The information concerning changes in temperature sensed by the probe is translated into electrical information by the thermistor. This information is sent—even as it occurs—to the computer. A program in the computer draws a graph on the screen of the monitor. The changes in temperature are thus placed upon the graph—again, as they happen. The students can see the graph of the cooling curve form at the same time that they watch the substance change physically in the test tube. The relationship between the observed physical phenomenon and its derived mathematical expression is thus observed in close juxtaposition in time (Guertin, Pease, and Smith, 1987). Because of this relationship, the possibility exists for generating new problems for investigation right then and there. Thus, too, the reduction of disordered empiricism and the induction of methods that promise a better ordering of empiricism.

In the physics laboratory, a rapidly increasing number of activities are particularly amenable to microcomputer-based science laboratory philosophy. Light, sound, motion, force, magnetism, and, of course, voltages are all phenomena that can be detected and translated into information the computer can process. Again, this approach opens new ways of undertaking—in effective instruction—investigations that are more difficult and more time consuming using earlier approaches. As a case in point, consider the simple harmonic motion of a mass-and-spring system. Without computer help, a teacher attempting to show this phenomenon would first connect the mass to the spring, hang both from a support, and set the system in motion. The students would be asked to observe the up-and-down motion of the system. To show the sinusoidal properties of the motion, the teacher might ask that a pen, attached to the mass, be allowed to rest upon a moving chart next to the system to record its motion. Or, perhaps, photographs could be taken to determine the position of the mass as a function of time.

In the microcomputer-based science laboratory, another option is available. A sonic transducer is used (Tinker, 1987). This device produces ultrasound pulses, which are sent from the transducer out toward an object. After a series of these pulses have been produced, the transducer changes its mode of operation and waits to receive the echoes of the pulses bouncing off the object. By timing the period from the transmission of the pulses to their return, the computer can then calculate the distance of the object. Using this concept, a sonic transducer is placed upon the floor underneath an oscillating spring-mass system. The program in the computer draws a graph upon the screen of the monitor. As the information is transferred to the computer from the transducer, the observations—in the form of points—appear upon the graph. When the points are connected, the resulting sine wave now describes the motion of the system. The sine wave can be printed easily to keep a record for analysis.

All this takes but a few moments and leaves time for the student to explore—to investigate—by changing the variables within the system. What happens with a different mass? A different spring? A longer or shorter spring? And so on. Hypotheses can be stated, inferences drawn, and then experiments developed to provide the tests. The student has become a planner and observer of the investigation rather than being captured by its more tedious component—the repetition of observation of routine empirical data. The student can then use the computer's devices to manage a decently ordered empiricism. The program that determines the computer's "observation" and "calculations" comes out of the basic concepts and planned methodology (a well-ordered empiricism, as it were) introduced into its memory.

From a pedagogical point of view, there are several interesting hypotheses that can be developed. First, students can often more easily and more efficiently correlate physical phenomena with their mathematical analogues in a microcomputer-based science laboratory than they can in a traditional laboratory arrangement devised in lengthy empirical settings. Second, students can, because of the shorter time periods involved for each activity, repeat the investigation either under the same or different conditions, thus reinforcing their understanding. Third, because of the reduction in the tedium of manual data gathering, students are free to probe at greater depth the phenomena they have observed and to develop better-ordered investigations and hypotheses.

Certainly, much research into the most effective pedagogical uses of the microcomputer-based science laboratory remains ahead. However, there is little question that this new approach to science instruction in the laboratory will have an impact on present techniques and that—as in the case for all science—the more we learn, then the more there will be to learn. That is, new findings do not put one to rest; they stimulate the finding of new problems for investigation; the newer findings stimulate newer problems, etc., etc., etc.

Entry into New Augmenting Environments and a New Ecology of Achievement

Within the 10 years of the 1955 Forest Hills study, of 655 students who had finally elected the channeling environment, 354 had chosen to undertake the experiences of the augmenting environment: They had completed a piece of research. A good number had submitted themselves to the rigors of the Science Talent Search; in comparison with other schools, a decent number were successful. About 170 of the 354 (circa 48 percent) had placed a career in science high on their list of aspirations. We had planned a thoroughgoing follow-up study, but we found that we could not determine with the validity we desired whether the designed environments in science we had described were primarily responsible for the students' eventual choice of careers in science and/or technology. In fact, considering our emphasis on the significant impact of a variety of *ecologies of achievement* on teaching and learning, as well as on the choice of a life's work, we should have expected what our findings might be. Surprisingly, however, even our experience as teachers in colleges and universities, as well as in the schools, did not prepare us for our observations. (On this, see "Science Talent: In an Ecology of Achievement" in this volume.)

It was clear—almost crystal clear—that, after an initial 23 visits to the universities and colleges different students in our program were attending, the undergraduates had formed strong attachments, even in the freshman years, to new mentors. Those former students who kept in touch with us throughout their undergraduate and, indeed, through their graduate years wrote of their attachment to their mentors in college and university (while firmly acknowledging our shared early relationship). Indeed, we suspect strongly it may also be the other way around: Professors and researchers in the university seek out the most talented (synonym: promising) students and try to interest them in the problems with which they, the mature scientists, are concerned. As it should be.

In brief, *our apprentice scientists had left us for another and different augmenting environment and for a distinctly different ecology of achievement.* Perhaps, as a matter of attaining a durable self-concept, they wished to leave adolescence behind on the welcome path to maturity. Besides, they knew they could always call on us; we had promised them so before they left the school.

It is clear that the university environment itself can also become a channeling and augmenting environment, which can add substantially to the students' capacities and well-being in achieving competence in performance. In these matters, it soon outstrips and supplants what the high school has provided. What is important to consider is that schooling may have opened the door, initiating a significant first step for most. Surely, a good number had bent to science before they came to the high school. Surely, this is also true of those who take tests to enter special science schools. Even so, the university environment is in many ways *not* similar to the school environment. There, in the school, the young can find "the second chance," a guarantee of compassionate stewardship of their present and future, the safety valve guaranteed by our principle of *self-selection* (see page 282). In the school, this second chance may be a guarantee of a "revolving door" (an opportunity to leave a program and return to it at a later time) described by Joseph S. Renzulli, Sally M. Reis, and Linda H. Smith (1981). But the augmenting environment within the ecology of achievement in the university is, to use a safe judgment, much more rigorous and, at the same time, in the expectation that the "apprentice scientist" is to fulfill the requirements of the society of scientists, less forgiving. We may explore the situation somewhat as follows.

One member of the group we had elected to study, a quiet student, had in high school pursued the courses of study, the seminars, and the science fairs and congresses with sustained brilliance; others turned to him for leadership. His writings and his paper in the *Forest Hills High School Journal of Science and Mathematics* demonstrated a much-admired literacy. He turned his attention to an interest in biological development of organisms, and his project earned him the place of finalist in the Science Talent Search. Then, he went on to a fine showing in science and mathematics at a university. There was a period of splendid graduate work, a serial progression from a minor to a major university as instructor, as assistant to associate to full professor, all the while with brilliant contributions in a major field. And now this student is tenured in one of the prestigious chairs of science in one of the major universities in the country. Three other students in this group went on to earn Ph.D.s and then to similar—but not equal—academic positions; four more went to different positions in industry. Indeed, we can point to the successes of a decent number of the young who *selected themselves* to undertake the rigors of the program we have described. May we, should we, take credit for the achievement of these remarkable young men and women?

We are persuaded *we may not.* Teachers in the schools supported by colleagues and community and a moment in history are responsible for the ecology of achievement that enables the young who so wish to take advantage of a channeling and augmenting environment during three to four years of their young lives. That is our responsibility as teachers. The route of the young to becoming scientists began with their birth—if not prior to birth—in the environment prepared for them by their parents, who indeed could be said to have been prepared for parenthood. For parents are also a product of *their* heredity and environment in their own moment of history. Once these young men and women leave environments of schooling for the university, they enter a newer ecology of achievement in which their character-rooted habits and passions play a large part in the road they are to travel. We have found these young to be gracious in their consideration of their high school years and their oft-repeated respect for their experiences then. These acts of gratitude we consider as an earnest of their traits and their exemplary behavior in the human and humane mode. However, do our former students not credit us overmuch for doing the work we have always wanted to do as teachers and now take joy in doing?

As are all organisms, the talented have been forged in the external activity of the dyad—of genes interacting with environment. Their environment at first is that of the culture—of family, of peers, of schooling, of society and its institutions. But, in the end, the talented create their own capsule of another special environment—the freedom to do their work in their idiosyncratic way. In short, shaped by the culture's environment, their talent forges not only a *multitude of contributions but also their own environment,* the one in which these contributions are born.

People who display what we finally may assign to "science talent"—and not to any other kind of talent—have attributes that are formed in the trials, in the labors of originative work. These individuals develop congeries of traits necessary to those who contribute to a field of knowledge called *science,* or a subset within the field. These attributes are found, if you will, in the character of those who are called scientists and are so considered by the community that benefits from their contributions. That armament, discernible in the art of investigation we call scientific inquiry, affirms that mode of inquiry or research clearly demonstrated by the ability to do an investigation whose procedure and results are within the area of well-ordered empiricism: Both procedure and results are repeatable.

And thus science is self-correcting because its theories and conclusions are subject to the intense collaborative scrutiny of a scientist's peers: a community. In such a community, one may live a wholesome life of service to others and thus, in Albert Schweitzer's terms, find an opportunity to be "*finished* with oneself." Finished in the surest sense, that is, in the act of becoming and being completed as a human.

It seems clear that the work of scientists and engineers has reordered certain aspects of the world's economy as well as the political and social relationships of nations. Indeed, current movements in curriculum focus on stressing a new and profound integration of the disciplines of the sciences and social sciences grouped under a significant curricular rubric: *Science,*

Technology, Society. Whether the curricular changes required are, or are not, adopted by the schools is not here the substance of our discussion. However, the reader will not have failed to recognize that onrushing discoveries seem to overwhelm us, that the enormous increases in knowledge in science and technology testing our understanding are of a different kind and degree of magnitude than those of the past. Now these masses of new knowledge affect the entire planet and its peoples—not to mention other bodies in space.

To labor the obvious, does not the future of such knowledge rest in good part on the schooling and education of our young? Thus it is that our young have now become a precious resource in a world that yearns for the wisdom to press the increasing impact of science and technology on society into the mold of human and humane purpose. And that in turn rests on men and women who are as compassionate as they are competent.

In Sum

1. In high school, a student's ability to plan and complete an originative work in science and, thus, to demonstrate the capacity for the well-ordered empiricism such work requires, may well be a valid *test of science talent,* or, if you will, a *test of potential in science.* If, as do a number of psychologists and students in the field of giftedness, one considers "success" in the Science Talent Search an index of such potential, then the thesis stated above may be valid because the Science Talent Search requires the completion of such an originative work.

2. It is demonstrated that students who are enrolled in the general population of a heterogeneous school are mature enough to nominate themselves for entry into such a program, forgoing the need of a prerequisite formal test. Students who select themselves freely (a process we call self-selection) for participation in the curricular and instructional devices explicit in the design of the channeling and augmenting environments described in this paper can indeed complete such an originative work. It is also a matter of record that, over a 10-year period (the length of our investigation), Forest Hills High School, whose heterogeneous population embraced all students in the school district applying for enrollment, placed finalists and honorable mentions in the Science Talent Search in numbers comparable to those of the science high schools that selected their students by requiring tests for entrance. It is demonstrable that the Forest Hills students in our program tended to select careers in science in an appreciable number (Brandwein, 1955/1981). The findings of the Science Talent Search concur.

3. To determine the *spectrum of traits in general giftedness,* one needs to demonstrate a certain literacy and numeracy and an ability to attain the

capacities, aptitudes, and skills essential to the arts of gathering, storing, scanning, and retrieving knowledge. Certain abilities in the use of such knowledge (i.e., critical thinking) may also be demonstrated. The congeries of traits of gifted students are now part of a considerable literature (on this see Abraham J. Tannenbaum, 1983).

4. To determine the *more specific trait of science talent,* we propose a necessity to construct an environment in which it can be demonstrated. In our view (Brandwein 1955/1981) and that of Hudson (1966), *a talent in science is demonstrated in a performing art, the art of investigation.* In a similar sense, the ability to perform as a pianist is rarely discerned in the absence of an environment: a piano, the necessary music, and usually an accomplished pianist, such as a teacher, a mentor. And, of course, years of demanding study and practice. In the end, the pianist is enabled to demonstrate his or her art in a performance. However, unlike the traits we ascribe to musical talent, the traits of the science talented may well be a subset of those delineated in the spectrum of general giftedness; these special traits have been described in this volume by Passow, Tannenbaum, and Brandwein (here and elsewhere [1955/1981]).

5. It seems to us that the *environments that are within the architecture of schooling,* namely those of curriculum, instruction, and administration, comprise, in turn, two elements: one, a channeling environment that appeals to gifted students generally but attends to the process and content of well-ordered empiricism; the other, an augmenting environment that calls upon *potential* in science to be turned into *performance.* Both environments are interactive with genetic components. Thus, science talent (and general giftedness as well) is not solely genetic in nature but is the result of a *dyad: the interaction of genes and environment.* This is to emphasize our belief that, without a nurturing environment such as the one described in this paper, science talent may not be demonstrated in high school. (On this, note the dyad described in "Science Talent: In an Ecology of Achievement" in this volume.)

6. To *become a scientist* one needs to accept the rigors of an environment that differs in manner and degree from that of high school. The university, the graduate school, and the privileged laboratory of the scientist subtend a newer ecology of achievement that includes the mature traits of the scholar: character-rooted independence and the ability to live within the subculture of the scientist. One is required to undergo a considerable period of training (in which options are closed) and education (in which options are opened). To succeed means to earn the respect of one's peers and to enter into a life that is centered in contributions to the culture, specifically, and to the world's store of knowledge, generally.

References

Abeles, Sigmund. (1985, Fall/Winter). A new era for the science lab? *Connecticut Journal of Science Teaching*, 38–40.

Beveridge, W. I. B. (1957). *The art of scientific investigation*. New York: W. W. Norton.

Brandwein, Paul F. (1955). *The gifted student as future scientist: The high school student and his commitment to science*. New York: Harcourt Brace. (1981 reprint, with a new preface [Los Angeles: National/State Leadership Training Institute on the Gifted and Talented])

Brandwein, Paul F. (1981). *Memorandum: On renewing schooling and education*. New York: Harcourt Brace Jovanovich.

Brandwein, Paul F. (1986, May). A portrait of gifted young with science talent. *Roeper Review, 8*(4), 235–243.

Bronowski, Jacob. (1956). *Science and human values*. New York: Harper and Row.

Bronowski, Jacob. (1978). *Magic, science, and civilization*. New York: Columbia University Press.

Conant, James B. (1952). *Modern science and modern man*. Garden City, NY: Doubleday Anchor.

Guertin, Arthur, Pease, Robert, and Smith, William. (1987). *A guide to computer use by the science teacher*. Middletown, CT: Wesleyan University Press. (Project to Increase Mastery of Mathematics and Science)

Hudson, Liam. (1966). *Contrary imaginations: A psychological study of the young student*. New York: Schocken Books.

Inhelder, Bärbel, and Piaget, Jean. (1958). *De la logique de l'enfant à la logique de l'adolescent*. Paris: Presses Universitaires de France.

Miller, Richard W. (1987). *Fact and method: Explanation, confirmation, and reality in the natural and the social sciences*. Princeton: Princeton University Press.

Pressey, Sidney L. (1955). Concerning the nature and nurture of genius. *Scientific Monthly, 81*, 123–129.

Renzulli, Joseph S., Reis, Sally M., and Smith, Linda H. (1981). *The revolving door identification model*. Mansfield Center, CT: Creative Learning Press.

Rogers, Carl. (1961). *On becoming a person*. Boston: Houghton Mifflin.

Stanley, Julian C., and Benbow, Camilla Persson. (1986). Youths who reason exceptionally well mathematically. In Robert J. Sternberg and Janet E. Davidson (Eds.), *Conceptions of giftedness.* (pp. 361–387). New York: Cambridge University Press.

Tannenbaum, Abraham J. (1983). *Gifted children: Psychological and educational perspectives*. New York: Macmillan.

Tinker, Robert. (1987, Winter). Building a motion detector—Ultrasonic ranging. *Hands-on, 10* (1), 15. Cambridge, MA: Technical Education Research Centers.

The Role of Content and Process in the Education of Science Teachers

Joseph D. Novak

For decades there has been disagreement regarding the importance of content versus process in science teaching. That is, should learning emphasize the accumulated knowledge in science, or should the emphasis be placed on the methods used by scientists to produce new knowledge? Most science instruction for the past 100 years or more has emphasized learning "facts," which Joseph J. Schwab (1962) characterized pejoratively as memorizing a "rhetoric of conclusions" (p. 24).

Emerging Concepts About Science Education

This kind of instruction by memorization has failed to achieve the goals of general education for two reasons: new knowledge has been accumulating at an accelerating rate, often replacing or contradicting earlier "facts," and information learned by rote is usually forgotten in two to four weeks.

Another criticism has been that emphasis on memorizing "facts" not only fails to give learners lasting and usable knowledge but also, at best, gives no insight into how scientists work and, at worst, leads to a distorted view of the nature of scientific inquiry. Schwab and Paul F. Brandwein (1962) have argued that science teaching should place emphasis on the nature of scientific inquiry and not on simply gathering data. And yet every study of the teaching of science done in the past decade has shown that most science lessons focus primarily on the acquisition of the facts of science. Courses for gifted students often do little more than increase the number of facts to be memorized. Twenty years of federally supported projects in curriculum development and programs for educating teachers have not resulted in a discernable improvement in science instruction in our schools (U.S. General Accounting Office, 1984).

Why have we failed? Is there a basis for improvement? I believe the answers lie in newly emerging knowledge about learners and learning, better understanding of epistemology—the nature of knowledge—and new educational strategies to help teachers help students "learn how to learn." Further, a body of evidence from research suggests that science education can be improved. Arguments over emphasis on content versus process will collapse with the recognition that how scientists construct knowledge is a process complementary to how students learn science.

There is today what Marcia C. Linn (1987) describes as "an emerging consensus" among psychologists, philosophers, and science educators. This consensus includes a movement away from behavioral psychology and "laws of learning," which had relevance to animals' learning in mazes or in Skinner boxes and some application to students' memorizing isolated bits of information by rote. Behavioral psychology not only ignored the important role that feelings play in how and when students learn but also deemphasized the positive and negative influences of students' prior experience upon their acquisition of new knowledge (see Ausubel, 1968, and Ausubel, Novak, and Hanesian, 1986). The new consensus in psychology, on the contrary, recognizes how invalid ideas or misconceptions held by students can negatively affect their understanding. Teachers and researchers who subscribe to this consensus now know that learners must be helped to take charge of the reorganization and elaboration of their own conceptual frames of reference. When students are helped to recognize that they have the power to learn new ideas and to use these ideas in problem solving, positive feelings and attitudes toward learning—and toward themselves—can result. Beginning with the pioneering work of developmental psychologist Jean Piaget in the 1920s and continuing in most psychological research today, the consensus is that, while learners must construct their own meanings for ideas about how the world works, they can profit from assistance both as to how their learning takes place and how it can be enhanced.

A New Consensus

A new consensus is also emerging in the domain of philosophy. For three centuries, the dominant view was that the objective of scientists was to "discover" the laws of nature that explain "reality" by "objective" observation and experimentation. The emerging consensus in epistemology holds that, just as the individual constructs his or her own view of science, the scientist works with existing concepts and constructs and elaborates new ideas to account for what is observed. But epistemologists now recognize that the current beliefs held by scientists empower or constrain them in selecting forms of inquiry, in choosing the phenomena to observe, in designing ways to record events or objects, in deciding which questions to ask, and in selecting the ways to transform records—to *construct claims*—about how the world works. Moreover, the contemporary "constructivist" epistemologist recognizes that emotions play a key role in every aspect of the modes of inquiry scientists choose to employ. The idea that scientists study nature without interference from their feelings or prior conceptualizations is rejected. Instead, the "biases" of the genius help guide him or her to ask the "right" questions, choose the "best" way to make records, and construct the "best" claims for knowledge. (See, for example, Watson, 1968; Keller and Freeman, 1983; and Gunther S. Stent, in this volume.)

Years ago Michael Polyani (1956) wrote about the importance of "passion" in science; though his ideas were then widely out of step with prevailing positivistic epistemology, they were prophetic of current constructivist views. Unfortunately, in many textbooks and most classrooms, the message students frequently receive is rooted in the positivist dogma that there is but one "correct" answer and their task is to memorize it. This pervasive positivism justifies the widespread practice of fact giving and student memorizing that characterizes much science instruction in schools.

Because our views of the world keep changing, it is important to recognize that constructivism does not imply that nothing is worth learning. For learners and for scientists—constraints at hand—working in given disciplines, current conceptual ideas are all that we have to make sense of the world. These concepts permit us to find answers to problems and to create new knowledge. Constructivism holds that this process is never ending for the individual, whether scientist or not, who seeks to understand his or her world. Much of the excitement and reward in science and other forms of scholarship come from the recognition that there is always a new challenge ahead, even when present questions appear to be answered.

The theory of instruction that emerges from the new consensus is strikingly different from that commonly observed in schools. Table 1 shows the contrast between "traditional" and "constructivist" views of the classroom. Basically, the new view recognizes not only individual differences among learners but also those of gender, race, culture—factors embracing heredity and personality. The new perspective is optimistic with regard to human potential, viewing individual limitations as induced by culture (family, school, or society) rather than as inherent or innate. John B. Carroll's (1963) and Benjamin Bloom's (1976) ideas that 90 percent of students can "master" most school learning tasks are expanded to include new ideas about the

Table 1

Traditional Patterns
in Education of Teachers:
In Relation to Five Elements [a]

Learner	Teacher	Curriculum	Context	Evaluation
Task is to acquire information (usually by rote learning).	Management and class control emphasized.	Fixed, textbook centered.	Schooling is good. Minor improvements may be needed.	"Objective" tests are the key to evaluation, with grades assigned "on a curve."
Emphasis on lesson planning focused on discipline, not learner's prior knowledge.	View that teachers cause learning.	Emphasis on coverage techniques.	Children should do as they are told.	Frequent testing helps students meet course objectives.
Failure regarded as lack of aptitude or motivation.	Motivation strategies emphasize clear statement of rewards and punishments.	View that knowledge is truth to be learned (i.e., memorized).	School curriculum is generally okay, but more emphasis on "basics" is needed.	Scores on standardized state or publishers' tests are good criteria of success.
Use of "objective" tests validates view of learner as "empty vessel" to be filled with information.	Teacher charisma is a desired goal.	Little planning or regard for student's feelings.	Teachers should be rewarded according to standardized test scores received by their pupils.	Time-consuming evaluation methods are not worth the effort (e.g., essay exams, group project reports).
Group instruction validates view that failure is due to lack of aptitude.	Audiovisual aids, computers seen as information givers rather than as tools to help in meaning making.	Subject matter taught and testing should show close to one-to-one correspondence.	Years of service and college credits/degrees earned are primary basis for salary levels.	"Test item banks"—collections of test questions "covering" various subject matters—are a primary resource for teacher-made tests, together with tests prepared by book publishers.
Rewards and punishments are principal motivators for learning.	Lecturing, test writing skills emphasized.	School, state, or university exams set the criteria for what is covered.	Educational theory and research is of little relevance and value to teachers or program planners.	
	Little concern for curriculum development by teachers.	Publishers are responsible curriculum developers.	Administration should run the schools.	Facts must be learned before understanding can develop; hence, tests should stress knowledge of facts.

[a]These five elements are my modification of Schwab's (1973) "commonplaces": (1) learner, (2) teacher, (3) subject matter, and (4) social milieu. I have added (5) evaluation, because it plays a dominant, indeed, often a controlling role in schooling.

nature of learning and the construction of knowledge. Finally, new methods of teaching employ metacognitive tools to help students learn how to learn.

Metacognitive Tools

D. Bob Gowin and I have devised two tools that help teachers and students better understand how new knowledge is learned and created. These are "meta" tools because they transcend subject matter and assist all kinds of learners. *Concept mapping* emerged from our need to make better records of

Table 1 (continued)

Constructivist Patterns
in Education of Teachers:
In Relation to Five Elements

Learner	Teacher	Curriculum	Context	Evaluation
Learner must make new meanings based on his/her prior knowledge.	Emphasis on finding out what the learner already knows.	Emphasis on major conceptual ideas and skills.	Schooling emphasizing rote learning is "domesticating."	Progress of students should be monitored with files containing a broad range of performance indicators.
Meaningful learning is primary basis for positive motivation and sense of empowerment.	Research and theory guide practice. Clear distinction	Recognition of diversity of learners and need for variety in learning resources.	Schooling emphasizing meaningful learning and creativity is empowering.	A broad range of evaluation measures is needed.
Teacher skills needed for appraising student's prior knowledge (e.g., pretests, concept maps, occasional interviews).	between topical or "logical" organization of subject matter and "psychological" organization. Use of concept maps to help with latter.	Efforts in student involvement in planning and executing instructional program. Emphasis on evolving nature of knowledge.	Much of the school curriculum is anachronistic, and major revisions in curricula are needed. Teacher preparation should be viewed as lifelong	Objective tests measure only a small percentage (about 10 percent) of aptitudes and achievement relevant to real-life application.
Learners need help to learn how to learn.	Techniques needed for helping students learn how to learn.	Wide variety of learning approaches, with flexible evaluation.	with continuing efforts for appraisal and "renewal."	Evaluation measures should help students and teachers identify conceptual prob-
Human potential is much greater than usually manifest.	Optimistic view of human potential. Lack of motivation seen as de-	Confidence in meaningful learning as preparation for standardized exams.	"Career ladders" are needed to keep the most talented teachers in classrooms and	lems and work toward their resolution (e.g., concept maps).
Feelings are important.	rived in large part from lack of meaning/	Emphasis on empowering learners	help them to help their peers.	Evaluation should help students take responsibility for
Learning is the responsibility of the learner.	understanding. Teacher is responsible for sharing meanings with/between learners.	rather than "coverage" of material.	Teaching practice should be theory and research based and evaluated.	their own learning (e.g., use of journals, self-report measures, concept maps, etc.).
	Gaining skills is lifelong process.		Major decisions should involve teachers, parents, and administration.	Teachers should conduct occasional in-depth interviews with students.

what a learner knew before and after instruction. Figure 1 shows two concept maps constructed from interviews; these maps represent a student's— Phil's—knowledge of the particulate nature of matter in grades two and twelve.

Primary grade children can learn, in a matter of a few minutes up to a few hours, how to make their own concept maps from concept (word) lists or from texts. After students have had years of schooling based predominantly on rote learning, however, we find that—at the upper elementary school to college level—they may require hours of instruction and weeks of practice

Figure 1

Two Concept Maps

Phil—Grade 2

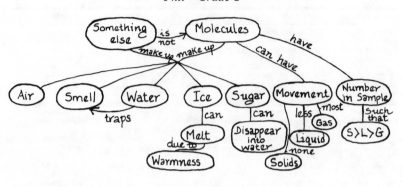

Phil—Grade 12

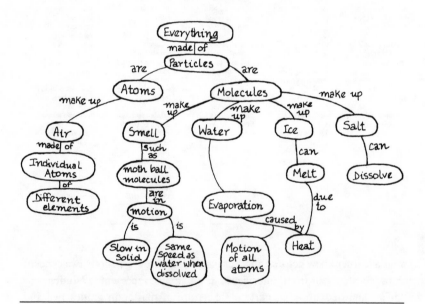

Two concept maps drawn from interviews with a student (Phil) in grade 2 (top) and grade 12 (bottom). Note that even after junior high school science and high school biology, physics, and chemistry, Phil has not integrated concepts of atoms and molecules with states of matter nor corrected his misconception that sugar or smell molecules are "in" water molecules.

before being able proficiently to construct their own concept maps. Our experience has been replicated many times by teachers at all grade levels and in all subject matters—including sports, literature, and mathematics—not only science. Concept maps can also be enormously helpful in designing new instructional programs and planning new textbooks, films, or computer programs. (See Novak and Gowin, 1984, chapter 4.)

I have used concept maps as the principal evaluation tool in my under-graduate and graduate classes for the past decade. Unlike true-false or multiple-choice exams, concept maps place the responsibility on the learner to recognize and illustrate important ideas, principles, structures of knowl-edge, and statements of values. It would take thousands of multiple-choice questions to test for the ideas and relationships shown in a map with 30 or 40 concepts; yet those questions would still not reveal the unique and creative relationships recognized by better learners. Gifted students work particu-larly well with concept maps, which provide ample opportunities for demon-strating breadth and depth of knowledge as well as creative insights.

A second metacognitive tool invented by my colleague, Gowin, is the Vee heuristic. Figure 2 shows the general form of the Vee, and figure 3 shows a Vee constructed by students in a junior high school science class. Both the students' versions and Gowin's Vee's were constructed with the help of a computer program, which my colleagues and I developed.*

The Vee heuristic, based on constructivist epistemology, helps to illustrate the many conceptual and procedural elements involved in constructing knowledge. Each element interacts with and modifies every other element in the process of probing the various aspects making up a single area. If, in response to this heuristic, we change our question, the event we observe, or the theory we apply, different claims are to be made for the knowledge and/ or values that emerge. A learner's or researcher's cognitive structure can be shown as a complex concept map on the left side of the Vee; this device helps to illustrate why two learners or two researchers may draw different conclu-sions from the same observations. If the learners choose to make different records as well, they can make widely different claims for their conclusions, even in response to the same question about the same event. For example, early chemists saw the change in burning wood as the loss of phlogiston, but today, in response to the same question—"What happens when wood burns?"—people think primarily of the oxidation of carbon and hydrogen.

In the pursuit of knowledge, it is often necessary to devise new ways of recording or approaching experiments or results. The Leyden jar, the oscillo-scope, and the laser are a few examples of these efforts. What the history of

---

*Our computer program to help students construct concept maps is available through Exceller Software Corporation, Cornell Research Park, Ithaca, NY 14850.

Figure 2

Focus Questions

There are usually only one or a few major questions we seek to answer regarding an event or set of objects. These questions serve to "focus" our attention on the phenomenon we wish to study and point to *possible* knowledge claims (usually called hypotheses) and value claims that may arise from the inquiry. The "focus questions" help us to define the phenomenon we seek to study.

A

Philosophy: This is our *broad guiding view* of how we believe the universe works and how the phenomena we study relate to the range of questions that could be addressed.

Example: Constructivist philosophy holds that our concepts, principles, and theories continue to evolve over time.

B

Theory: Theories *explain why* phenomena appear as they do or operate as they do. Theories tend to change only slowly over time, but no theory is perfect in explaining any phenomenon. New theories are constructed by highly creative persons, and these compete with old theories for years or centuries.

Example: Matter is a mode of particles whose mass is conserved in any transformation.

C

Principles: Principles *describe how* objects appear or events work. Each involves two or more concepts and some specified relationships between the events or objects observed. Principles sometimes are expressed as mathematical formulas, but in the real world, all events or objects behave in ways beyond that specified by formulas, at least in extreme cases.

Example: Gas volume is proportional to pressure, under standard conditions.

D

Concepts: These are *perceived regularities in events or objects,* or records of events or objects, designated by a label. What we *perceive* depends upon the concepts, principles, and theories we hold for given events or objects, as well as upon the validity of our records.

Example: Concepts of solids, liquids, and gases are based upon our principles of atomic/molecular structure and relationship between temperature and states of matter.

E

Value Claims: These designate the value or worth of new knowledge. Worth can derive from improvement of human lives or power for advancing new inquiries and new learning.

Example: Understanding the relationship between force, mass, weight, and quantity has many useful applications.

F

Knowledge Claims: These describe the regularities and/or relationships we see in records or transformed records of events or objects. They are also *hypotheses* before the inquiry which may be supported or negated with the data obtained. Knowledge claims are the basis for proposing new or modified concepts, principles, or on rare occasions, theories.

Example: In spite of change in form of matter, mass is conserved.

G

Transformations: These are the ways we organize or process our records to assist us in making knowledge and/or value claims about phenomena studied. Concepts, principles, and theories guide our decisions as to the most productive way to organize or transform our records. Transformed records are also called *data.*

Example: Computation of ratios of mass are recorded before and after physical or chemical changes in substances.

H

Records: These are the observations we make of events or objects, either directly with our senses or by using instruments. Sometimes records are called *raw* data. Good records are *facts* or *valid* indicators of what we are observing. Faulty records can lead to faulty knowledge or value claims. Understanding our record-making tools can involve concepts, principles, and theories relating to our instrument. Measures can be faulty, if the latter are faulty.

Example: A poor balance can give faulty measurements or weights before and after a change in materials is observed.

I

Events or Objects: the objects or events we *choose* to study. Our purpose is to understand better how and why they appear or behave as they do.

Example: Ice, water, and salt are weighed in various conditions.

Ten elements involved in the construction of knowledge. Each element interacts with every other element in the process of knowledge construction. An inquiry could begin with any element on the Vee, although most school science tends to begin with record making and questions. Specific recognition of concepts, principles, theories, and philosophy guiding the inquiry is usually absent.

Figure 3

Focus Questions

Does the process of dissolving, melting, mixing, or generating of gases change the mass of the materials being dealt with?

S. Szabadi

A
Philosophy

The society that we live in is scientifically and technologically oriented. We depend on science to give us answers to problems that face our society.

In search for a cure for cancer the scientists are expected to find a "cure." Many societies do not feel this way. They may feel that people die or get sick because of the fate of the gods.

D. Larison

B
Theories

Particle Theory
Kinetic Theory

M. Hylwa

C
Principles

1) 95 percent of gases are empty space.
2) Gases, liquids, and solids are made of particles.
3) Heat is particle motion.
4) Gases, liquids, and solids expand when heated.
5) Particles move in random directions.
6) Liquids have less empty space than gases.
7) The triple beam balance is accurate to $+/-.3$ g.

K. Kirby

D
Concept

gas	balance	ice
melting	water	95 percent
salt	yellow	white
clear	tablet	container
shaking	warm	motion
solids	liquids	expansion
particles	empty space	mix
pour	collisions	air
escape	dissolve	solutions A&B
bubbles	volume	mass

S. Szabadi

E
Value Claims

The concept of mass was invented because gravity changes the weight of an object; gravity cannot change the mass of the object. The concept of mass depends on the particles in the object. The more particles, or the bigger the particles, will have an outcome on the mass.

K. Kirby
S. Szabadi

F
Knowledge Claims

The mass does not change if the stuff is dissolved, mixed, melted, or gas is generated.

J. Smith

G
Transformations

LAB	Kirby	Matt	June	Oaks
Ice	-1.2g	+.5g	+5.3g	-.1g
Salt	-/+0g	+41.2g	-3.8g	-.3g
Solutions A&B	-/+0g	-3.4g	-3.4g	+.2g
Alka Seltzer™	-/+0g	-.2g	+0g	+0g

K. Kirby

H
Records

LAB	Kirby	Matt	June	Oaks
Ice				
before	40.4g	23.3g	10.6g	23.7g
after	39.2g	23.8g	15.9g	23.6g
Salt				
before	81.5g	25.1g	54.1g	40.3g
after	81.5g	66.3g	50.3g	40.0g
Solutions A&B				
before	31.2g	29.5g	31.3g	29.1g
after	31.2g	29.3g	27.9g	29.3g
Alka Seltzer™				
before	39.7g	44.3g	43.7g	44.2g
after	39.7g	44.1g	43.7g	44.2g

J. Smith

I
Events and Objects

Objects:

Ice cubes, containers, salt, A&B solutions, Alka Seltzer™, balance, and water.

Events:

Lab 1) Weigh the salt and water with containers. Mix salt and water and reweigh.

Lab 2) Weigh ice and container. Melt ice with hands and reweigh.

Lab 3) Weigh solutions A&B together, but in separate containers. Mix the two and reweigh.

Lab 4) Weigh water in container and tablet in cap. Put tablet in container, shake, and reweigh.

M. Hylwa

A Vee diagram constructed by students in an eighth grade general science laboratory dealing with states of matter.

science and technology has shown is that there are intimate relationships between efforts in the discovery of new knowledge and the creation of new technology. One of the most dramatic efforts in progress now is the work to understand superconductivity, the hoped-for product being materials, effective at or near room temperature, which can transmit long-distance power and huge magnetic forces with minimal electrical energy. And, in biology, our efforts to understand the nature of genetic material and control of biological processes is leading to the rapidly growing field of biotechnology. In the effort to understand the creation of this technology and in the decisions that derive from its application, the Vee heuristic can be a valuable tool.

Gowin and I have found resistance to the use of the Vee on the part of both students and teachers, especially older students. The fragile, tentative, and fluid nature of the claims made for knowledge derived through the Vee is unsettling to both teachers and students who have spent years seeking the "right" answers. We have found a negative relationship between the extent to which teachers or students are committed to rote learning and their acceptance of the Vee heuristic; the reverse, fortunately, is also true.

There are numerous courses in "study skills" or "thinking skills" promulgated in schools today, but many of these are predominantly techniques to accelerate rote learning. Unfortunately, memorized material, unless repeatedly rehearsed, is soon forgotten, with or without the use of study skills. Test scores may improve briefly, but the individual's capacity to understand and control the subject matter does not occur through the aid of study skills alone. That is, problem-solving abilities are not advanced through rote learning; in fact, the very nature of problem solving demands a flexible approach to critical thinking. In today's climate of "accountability"—with student "progress" mostly appraised by multiple-choice tests of dubious validity—training in study skills may show striking short-term benefits. This can mean that work in study skills can divert efforts away from more powerful learning strategies. Some schools spend so much time teaching "thinking skills" that they curtail instruction in the basic concepts and the content necessary to form them. Concept mapping and Vee diagraming are useful only *in conjunction* with instruction in subject matter and thus complement rather than compete with school instructional time in teaching science or any other subject matter.

New Methods in the Education of Teachers
It should be evident from the above that there should no longer be a debate about the extent of emphasis on teaching the content of science vs. teaching the processes of science. If a constructivist perspective guides our work, and

especially if we use compatible metacognitive tools, there is no reasonable way to teach the processes of science without simultaneously teaching its concepts and principles, which are part and parcel of the strategies of instruction described here. Conversely, any attempt to teach the content of science that does not consider its complex "conceptual web," its evolving nature, is destined to failure, provided that our objective is *meaningful,* rather than verbatim, rote learning. Constructivist teaching, rooted in psychology, is mutually supportive of and necessary for constructivist teaching, rooted in epistemology. In this case, one hand washes the other (Novak, 1987, July).

The danger lies in instruction or education of teachers that ignores or contradicts this symbiosis. And this is where we science educators face our greatest challenge in most college and university science teaching as it exists today. Most postsecondary science courses encourage or require rote learning by the ways they test students' memorization of myriads of details with little or no emphasis on the major concepts and evolving character of knowledge in the field. How, then, do we break the cycle? How do teachers avoid the trap of teaching as they were taught?

There are encouraging developments in both preservice and inservice teacher education. In England, at the University of Leeds, Rosalind Driver and her colleagues are teaching teachers modes of interviewing students; teachers are learning particularly to identify their students' valid and invalid conceptions (or "alternative conceptual frameworks"). Similar work is being done in Israel, Canada, and the U.S. At the University of British Columbia, the University of Kansas, and other colleges and universities in the U.S., prospective science teachers are also learning to use metacognitive tools. There is much ferment and excitement in many countries among those who educate science teachers: A worldwide revolution of a sort is in progress.

In 1987 at Cornell University, we began again the education of science and mathematics teachers. New colleagues committed to constructivist ideas and involved in classroom-based research are at the heart of our program. We are making a major effort to recruit outstanding students into a five-year Master of Arts in Teaching program, beginning in the junior year. Ideally, we should like to begin with freshmen, but few students come to Cornell to become science teachers. There is a pool of second-semester sophomores, however, who want to do something useful with their lives and who are talented, committed, and, often, frustrated with the memorization requirements that characterize many courses populated by premedical students. These are the candidates we are seeking. They will be offered a six-semester seminar/practicum program, learning early on to "reconceptualize" the subject matter of science courses they are taking; they will be encouraged to learn through the modes I have described, rather than by rote, even if their work leads to a grade of B or B+ rather than A. (Too many tests require *verbatim* answers, and students who synthesize their own answers are often penalized.) These students will have the opportunity to interview fellow university students, high school pupils, professors, and schoolteachers in order to understand the rich and varied conceptual frameworks that each brings to new learning events. These science teachers of tomorrow will study and reflect upon the complex social milieu (Schwab's fourth "commonplace" of education [1973]) and will try to understand the interplay between student, school, home, and community that shapes the conceptual frameworks, the attitudes, even the nature of students' ways of learning. These science education students will have early and repeated opportunities to intervene in the education of children, first as tutors or members of a team and later as "teachers" in charge of classes at collaborating schools.

Will this kind of preparation produce better science teachers? Will it make a difference in the students' achievement, their self-esteem, and their understanding of science—differences that will be felt when they go out into the schools to teach? Will it help our teachers unleash the enormous potentials of gifted students? We hope so. We plan to conduct an intensive ongoing program of evaluation to assess both the strengths and weaknesses of this new program—and to build upon the former and ameliorate the latter. Ours is but one effort to achieve the goals presented in this book.

We are optimistic.

References

Ausubel, David. (1968). *Educational psychology: A cognitive view*. New York: Holt, Rinehart, and Winston.

Ausubel, David, Novak, Joseph D., and Hanesian, Helen. (1986). *Educational psychology: A cognitive view* (2nd ed.). New York: Warbel and Peck.

Bloom, Benjamin. (1976). *Human characteristics and school learning*. New York: McGraw-Hill.

Carroll, John B. (1963). A model of school learning. *Teachers College Record, 64*, 723–733.

Keller, Evelyn Fox. (1983). *A feeling for the organism: The life and work of Barbara McClintock*. New York: W. H. Freeman.

Linn, Marcia C. (1987). Establishing a research base for science education: Challenges, trends, and recommendations. *Journal of Research in Science Teaching, 24*(3), 191–216.

Novak, Joseph D. (1987, July). *Human constructivism: Toward a unity of psychological and epistemological meaning making*. Paper presented at the Second International Seminar on Misconceptions and Educational Strategies in Science and Mathematics Education, Department of Education, Cornell University, Ithaca, NY.

Novak, Joseph D., and Gowin, D. Bob. (1984). *Learning how to learn*. New York: Cambridge University Press.

Polyani, Michael. (1956). Passion and controversy in science. *The Lancet, 270*, 921–925.

Schwab, Joseph J., and Brandwein, Paul F. (1962). *The teaching of science* (The Inglis and Burton Lectures.) Cambridge, MA: Harvard University Press.

Schwab, Joseph J. (1973). Practical 3: Translation into curriculum. *School Review, 81*(4), 501–522.

U.S. General Accounting Office. (1984, March 6). *New directions for federal programs to aid mathematics and science teaching* (GAO Report No. PEMD, 84–85). Washington, DC: Author.

Watson, James D. (1968). *The double helix*. New York: Signet Books, New American Library.

Cooperative Learning and the Gifted Science Student

Roger T. Johnson
David W. Johnson

How should educators best serve gifted science students? We want to accelerate their learning, to enrich their experiences, and to broaden their thinking and skills. We want to push them to be the best they can academically and to integrate them into their peer group so that they are accepted and appreciated. Understanding how gifted students in science interact with each other and with the broader student population can lead to improved learning environments for all students.

Students interact with each other in three fundamental patterns: They compete to see who is best; they work alone toward set criteria without competing with each other; or they work together cooperatively, concerned about each others' learning as well as their own. In American classrooms, most students, but especially gifted ones, tend to compete with each other, to try to do better than the other students in the class. Traditional practices of grading on the curve, eliminating student-to-student talk during class time, requiring individual work, and seating students in rows that focus on the teacher have given students the message that they should not talk among themselves, and, if they want to do well, they would be well advised to surpass their classmates to reach the "top" of their class.

Given this situation, it is fascinating—and depressing—to realize that research on how people best learn indicates that isolation and competition create much less effective learning environments than cooperation, where students work together to help each other understand and learn. The practice of cooperating while learning, although not encouraged over the last 40 years in most schools, is an old idea and probably dates back to learning in the family settings of our ancestors. In more recent times, the helping environments of our "one-room schoolhouses" popularized by Francis Parker in the early 1800s were approved and affirmed by the philosophy of John Dewey. The importance of a supportive social environment for learning has also been recognized by many who have studied learning. Bärbel Inhelder and Jean Piaget, for example, listed social transmission as a requirement for cognitive growth and believed in social cooperation as an active part of learning. Extensive research on student-to-student interaction dating back into the 1800s validates cooperative learning and suggests that, for optimum learning, students need to care about each others' progress, talk with each other about the material, and learn together for much of the time.

Cooperative learning is also a much more realistic way of learning science if one looks at how scientists tend to work. It is somewhat unusual to find single-authored research reports in science because so many scientists find it productive to work in research teams, which in some cases are made up of individuals not only coming from several different countries but also contributing many diverse skills. In addition, we have been told several times by members of the business community that of major concern in hiring engineers is not only technical competence but also the candidate's ability to work effectively as part of a project team. Almost everything a business does has to be interdependent to be successful, and a large share of an engineer's contribution is in the context of a cooperative effort.

What We Already Know

Much research on student-to-student interaction has a direct bearing on science instruction. Several fairly recent reviews (Deutsch, 1962; Johnson and Johnson, 1982, 1983; Sharan, 1980; Slavin, 1977) and two meta-analyses (Johnson, Johnson, and Maruyama, 1983; Johnson, Maruyama, Johnson, Nelson, and Skon, 1981) summarize this research. These studies suggest the following four advantages of such learning methods.

1. Cooperative learning experiences promote more learning and more retention than competitive or individualistic learning experiences. When students talk through what they are learning with each other, they not only learn more and remember it better than they do through other approaches, but they are also more likely to develop a strategy or procedure for learning. While the greatest gain in achievement is in the lower third of the class, achievement goes up for middle- and high-ability students as well. Interaction within a heterogeneous group provides an area for the gifted students to explain why they think the way they do and how they arrived at the answer.

This not only deepens understanding of the material but also encourages the awareness and development of strategies for doing this kind of assignment. Several studies suggest that gifted science students were able to arrive at reasonable answers working alone but were unable to explain how they did so. Under the cooperative experimental conditions, most of the students in the group knew the answer and had a conscious strategy for finding it. Students learn material better when they have to explain it to others and often find their methods becoming explicit when it is necessary to communicate to someone else how to do the problem.

2. The more difficult the material, the wider the gap in achievement favoring cooperative over competitive or individualistic learning. Cooperative learning groups are at their best when the task requires problem solving, decision making, conceptual learning, or divergent thinking. Thus, the kinds of learning we want gifted science students to master are appropriate in heterogeneous, cooperative learning groups.

3. Cooperative learning experiences tend to create higher levels of self-esteem and healthier processes for achieving a sense of self-worth than do competitive or individualistic experiences. In addition, students are positive about learning in cooperative groups and enjoy the peer support and the opportunity to celebrate successes as a group. It is important that gifted science students succeed and thus are encouraged to extend their work in science.

4. Acceptance of differences, which comes with successful cooperative experiences, is important to gifted science students. It is ironic that, at the same time we are working to integrate students with diverse ethnic, racial, and economic backgrounds and trying to destroy barriers against female science students or children with handicaps, some are suggesting isolation of the gifted. While there are without a doubt some excellent pull-out experiences for gifted science students, we must strive to integrate gifted science students with their broader peer group. Like all students, the gifted want to be popular and to be accepted. This is not accomplished simply by placing high, middle, and struggling students together in the same science classroom. There needs to be heterogeneous, cooperative interaction.

In predominantly competitive classrooms, wise students learn to avoid two situations if they can: always losing and always winning. Both high- and low-ability students can be stereotyped and isolated by competition. Think of the difference for gifted science students between competitive classrooms, where they might be resented because they make other students feel inferior, and cooperative classrooms, where gifted students can benefit everyone. Gifted students do not become isolated by choice; almost everyone would rather be part of a supportive, productive group than be a loner. In science in particular, gifted isolates would have extreme difficulty working on today's project-oriented research and technology teams.

The important finding that stands out in reviewing the extensive research about student-to-student interaction is its consistency in supporting cooperation. More than 500 studies spanning eight decades examine a wide range of academic and nonacademic tasks with positive outcomes in terms of achievement, self-esteem, attitudes toward subject matter, motivation, mental health, and a host of other factors. The research also covers a wide range of age levels. We recently pulled out 133 studies that focused on adult (posthigh school) learning and found consistent support for cooperative interaction. We use the same predominantly cooperative approach in the honors class we teach at the University of Minnesota as do the elementary and secondary schoolteachers with whom we work in classrooms across the country. Increasingly, we are encouraging professors in higher education to provide a more active, cooperative learning environment for adults.

What Does It Mean?

For these and many other reasons, the research reports on student-to-student interaction suggest directions for working with gifted science students. Such individuals working in heterogeneous, cooperative groups achieve as well as or better than they do working individualistically or competitively. Studies of students grouped homogeneously in cooperative groups seem to indicate that loss of heterogeneity takes something important away from gifted students as well as from less able ones. A richness results when students come from different perspectives and backgrounds. Often groups of gifted students working together come to agreement too quickly and don't dig as deeply into the material as they do in more diverse groups.

However, in spite of their strengths, cooperative learning groups are not appropriate all of the time. The teacher still teaches, and students still listen, read, and think independently. The cooperative interaction is at its best when new information needs to be processed cognitively and examined critically. Certainly, some time to work individually and maybe even some carefully structured, low-key competition is also part of a flexible learning environment. One of the best integrations of individualistic and cooperative work happens when students are assigned different parts of a task and required to bring their contributions together and synthesize them so that each group member sees the whole picture. It is also important to keep in mind that the cooperative learning group, made up of individuals, has as its goal that each member understand the material and contribute. Each student, bringing individual talents and perspectives, is accountable in the cooperative learning experience and therefore leaves the group stronger. That's why learning groups are formed—to strengthen individuals' qualities such as achievement, self-esteem, leadership skills, and positive attitudes.

The classroom where students are learning to work in all three interaction patterns would be ideal. The cooperative umbrella is important in that it creates the environment where gifted science students, in fact all students, are at their best, learning as much as possible and acquiring the skills needed to work in the real-world settings of research and project teams.

From the research on student-to-student interaction, several suggestions to science educators seem clear: First, teachers should give gifted science students structured, cooperative learning experiences in heterogeneous settings, so other students perceive them as real people—assets to their groups and the class—not as stereotypical superachievers. (This is not to derogate the real-world setting where some scientists prefer to work as isolates, knowing full well that their work eventually demands communication in writing or reporting to their peers.) Second, teachers should recognize that cooperative experiences in mixed groups not only tend to raise the cognitive level upon which talented students and their classmates understand the material but also will help them develop collaborative skills in communication.

Finally, teachers should provide structured experiences with appropriate competitive and individualistic learning and research opportunities so that both gifted science students and others develop the skills to work in all three interaction patterns successfully. (Both competitive and individualistic experiences tend to be more effective when they occur under an umbrella of cooperation.)

What Are Cooperative Learning Groups?

There is a considerable difference between just putting students together to do a lab and structuring cooperative work. In cooperative relationships, the group members share a clear group goal that binds them together. An individual student in a cooperative learning group can achieve the goal only if the other members also achieve it. This sink-or-swim together motivation encourages members to seek outcomes beneficial to all. But four students working individually at the same lab table, sharing equipment, and talking to each other do not necessarily form a cooperative group. A clear, positive interdependence needs to be present where students considering the problem being studied understand that they all have the opportunity to reach an appropriate conclusion, if there is one, and that each must be able to explain the finding to each other and to the class.

Specific strategies for structuring appropriate cooperative interaction and teaching cooperative skills are described in *Circles of Learning* (Johnson, Johnson, and Holubec, 1984) and in *Learning Together and Alone: Cooperative, Competitive, and Individualistic Learning* (Johnson and Johnson,

1987). Five basic aspects of cooperative learning distinguish it from simply placing students in groups:

1. Students are structured into small, heterogeneous groups of two to four to work through an assignment together, sharing ideas and working as a group for the best possible solutions. The students gain further understanding of the problem as they work and stay on task. It is important that classroom tasks start in small groups not only to give each student more opportunity to talk (and, therefore, to learn the material better), but also because it is easier for most students to share their ideas in a 3-some than a 30-some. The ensuing class discussion is widespread and inclined to be more thoughtful, if it is first "practiced" in a small cooperative group.

2. Students should have a strong sense of interdependence, share and work toward a mutual goal, and receive rewards only if the group is successful. For example, one physics teacher structures all of his quizzes cooperatively. Groups of three study together for a quiz that is taken individually. Each student gets a base score on the quiz, but if everyone in the group meets the teacher's mastery criteria (above 80 percent correct), then each group member gets five bonus points. While bonus points usually do not seriously affect the base score, they give students a chance to celebrate the group's success and reinforce the goal that everyone master the material. This same teacher emphasizes cooperation even further by awarding 10 bonus points to each student if everyone in the class achieves the mastery criteria on any quiz. One benefit of the bonus-point group reward is that, without penalizing the high-scoring student who has already done well on the individual quiz, it provides an incentive for all students to help each other.

3. Each group member feels a strong sense of individual accountability, a conviction that each needs to know the material and be able to explain it. One method for checking individual accountability is to have the teacher move from group to group, occasionally selecting one student to explain his or her work without help from the others. If the student has difficulty explaining, the teacher lets all group members know they are not finished until everyone understands and can explain the problem and the solution at hand. Through these individual oral exams, students realize that it is dangerous to try to "hitchhike," and that it is important that students recognize the contributions of each group member.

4. The instructor monitors the work of the groups, sometimes using a systematic observation sheet to record specific behaviors, sometimes actively listening and noting both how the group is functioning interpersonally and how the students are progressing on the task academically.

5. Groups evaluate how well they are working together in completing the instructional tasks. This group processing needs to be done in a way that goes beyond a quick, superficial evaluation of the group's work. One specific way to structure the processing so that students go into some depth is to have

students rate themselves on a scale of 1 to 10 on how actively they participated, how well they listened to each other, or how well they supported each others' ideas. After they choose a number for themselves (1 meaning not at all and 10 meaning extremely well), they share their numbers and explain their ratings.

Some Examples of Cooperative Learning Groups

Heterogeneous cooperative learning groups would be workable problem-solving situations in the classroom or laboratory. Teachers could follow some of these procedures:

• Form small, heterogeneous groups of two or three by counting off randomly or by structuring the groups to include the desired differences in gender, ability, ethnic background, and so forth.

• Give each group a single lab sheet and set of equipment and instruct the groups to work toward a process, concept, or finding that each member could defend. Endorsements mean that each student could represent the group in explaining why a given finding is appropriate.

• Let the students know that all group members are obliged to share their ideas, actively listen to each other, use quiet voices, stay with their group, and work to make their joint report as accurate as possible. Monitor each group's work and occasionally select one member to represent the group by explaining the last step of the lab or the finding.

• Select one or two behaviors to monitor with care and at intervals systematically collect data from among the groups when you see or hear that behavior. Focus, perhaps, on the gifted students only during one of these sessions to see what roles they take in heterogeneous groups and what interactive skills they exhibit—or need help with.

• Give each group five minutes or so at the end of class to determine how well the members worked cooperatively. Make sure they have enough structure to make this processing real, perhaps asking for a written summary on the lab sheet signed by all group members. Give your own feedback on the targeted behaviors at this time.

Another way of getting started with cooperative learning groups is to structure cooperation around a key lecture, film, or speaker in one or all of the following ways:

• Form small, heterogeneous groups of three before the lecture or film and ask the students to formulate a question that they think will be central to the process, concept, or finding being presented. Give the groups three or four minutes to speculate and focus on the material.

• Stop periodically during the lecture or film (every 12 to 15 minutes) and ask the students to check the understanding of each group member on specific concepts or processes. These short discussion breaks not only involve students actively but also renew their interest.

• At the end of the lecture or film, give to the group members to do together a specific task that extends or applies the information they received. For example, ask them to list the key messages, choose the most important three, and put them in order of rank. When all members have agreed on the sequence, have them sign their report.

In Brief

Teachers and parents of all science students, including gifted ones, share three major concerns: Are the children learning as well as they can? Are they integrated into their peer group, accepted and appreciated by other students? Will what they are learning contribute to their success after school is finished? One of the key messages from the student-to-student interaction research is that gifted students in science classes learn more, are happier, and are more accepted by peers, if a good part of their learning occurs in heterogeneous, cooperative settings. When students challenge each other, require verbal explanation from one another, and give each other peer support, the learning experience deepens.

From the research also comes the message that it seems unnecessary to isolate gifted science students from their broader peer group, even though some key resources in pull-out programs may be especially effective for the gifted. However, it is vital for gifted students to maintain positive contact with the other students in science, so that the former can work from a broad base of support, be accepted and appreciated by peers, and develop the key collaborative skills that come from working in a heterogeneous group setting.

After school days are over, the leadership skills gained in working in the cooperative groups may be as important as the science learned. In addition, all children profit from working with gifted students and vice versa. Both groups deserve a chance to interact cooperatively.

References

Deutsch, Morton. (1962). Cooperation and trust: Some theoretical notes. In M. R. Jones (Ed.), *Nebraska Symposium on Motivation* (pp. 275–319). Lincoln: University of Nebraska Press.

Inhelder, Bärbel, and Piaget, Jean. (1958). *De la logique de l'enfant à la logique de l'adolescent.* Paris: Presses Universitaires de France.

Johnson, David W., and Johnson, Roger T. (1983). The socialization and achievement crisis: Are cooperative learning experiences the solution? In L. Bickman (Ed.), *Applied social psychology annual, 4* (pp. 119–164). Beverly Hills, CA: Sage.

Johnson, David W., and Johnson, Roger T. (1987). *Learning together and alone: Cooperative, competitive, and individualistic learning* (2nd ed.). Englewood Cliffs, NJ: Prentice-Hall.

Johnson, David W., Johnson, Roger T., and Holubec, Edythe. (1984). *Circles of learning: Cooperation in the classroom* (rev. ed.). Edina, MN: Interaction Book Company.

Johnson, David W., Johnson, Roger T., and Maruyama, Geoffrey. (1983, Spring). Interdependence and interpersonal attraction among heterogeneous and homogeneous individuals: A theoretical formulation and a meta-analysis of the research. *Review of Educational Research, 53*(1), 5–54.

Johnson, David W., Maruyama, Geoffrey, Johnson, Roger T., Nelson, Deborah, and Skon, Linda. (1981). Effects of cooperative, competitive, and individualistic goal structures on achievement: A meta-analysis. *Psychological Bulletin, 89*(1), 47–62.

Johnson, Roger T., and Johnson, David W. (1982). What research says about student/student interaction in science classrooms. In Mary Budd Rowe (Ed.), *Education in the 80s: Science* (pp. 25–37). Washington, DC: National Education Association.

Sharan, Shlomo. (1980, Summer). Cooperative learning in small groups: Recent methods and effects on achievement, attitudes, and ethnic relations. *Review of Educational Research, 50*(2), 241–271.

Slavin, Robert. (1977, Fall). Classroom reward structure: An analytic and practical review. *Review of Educational Research, 47*(4), 633–650.

Necessary Writing in the Sciences

Robert A. Day

It has been said many times: A scientific experiment is not complete until the results have been published. Thus, to "do" science, one must also "write" science. And good science students, those who are likely to mature into good scientists, learn that the writing is as important as the doing. They learn to weigh their words as carefully as they weigh their reagents.

Thus, writing by scientists is "necessary." True, scholars in all fields must "publish or perish" and therefore must engage in necessary writing. However, science has developed a unique style of communicating research results. New research results are presented in a rigid but highly logical format. In fact, this style of writing has become so successful in the past 50 years or so that it is no longer unique to the sciences; increasingly, this IMRAD format (Introduction, Methods, Results, and Discussion) is being used by professionals in many fields.

"Necessary" Definition

At the outset, it is important to distinguish between "scientific writing" and "science writing." These terms are often (and confusingly) used synonymously. In my opinion, scientific writing is the term for the "necessary" writing of the research scientists, that is, the description of the results of experiments. Scientific writing can be defined as the communication of research results from research scientists within a particular discipline to other scientists within that same discipline. The writer of a scientific paper is a scientist. The reader of a scientific paper is a scientist. The style and organization of scientific papers have been prescribed in a number of manuals (Council of Biology Editors [CBE] Style Manual Committee, 1983) and books (e.g., Day, 1988). In fact, a *standard* has been published prescribing the format of scientific papers (American National Standards Institute, Inc., 1979).

Science writing, on the other hand, is writing *about* science. The author *may* be a scientist, but in this instance the author is writing for a *general* audience. By "general," I mean an audience other than the author's professional peers. Perhaps this distinction can be best understood in terms of audience analysis. The audience for scientific writing is very small and is essentially limited to those scientists working in the same subdiscipline as the author. The audience for science writing is much broader and includes scientists from many disciplines, students of the sciences, and the public interested in the sciences.

Science Writing

This paper concerns scientific ("necessary") writing, rather than science writing. However, I offer an immediate caveat: Science writing is by no means "unnecessary" and is in no way inferior to scientific writing. It simply has a different form (usually books rather than journal articles), purpose, and audience.

Certainly, serious teachers should read and encourage their students to read at least some of the truly excellent books about science. One thinks immediately of the wide-ranging essays of Lewis Thomas (1975, 1979) and Carl Sagan (1977, 1979). The books and reviews by Stephen Jay Gould (1983, 1987) recommend themselves to anyone with an inquiring mind. Many students would be turned on (and some would be turned off) by the compulsive enthusiasm expressed by James Watson in *The Double Helix* (1968). And what better way is there to learn about the fascinating history of science than by reading the biographies and autobiographies of famous scientists? Among my favorites are René Dubos' *Life of Pasteur* (1958), Hans Zinsser's *As I Remember Him* (1940), and S. E. Luria's *A Slot Machine, A Broken Test Tube* (1984).

All serious students in the sciences can enjoy the rich literature about science. Some students may want to go on to establish a career in science writing. Fortunately, our increasingly technological society has an increasing need for science communicators. In journalism, in business, in government, and in academe, people skilled in communicating scientific materials are needed. A useful book for anyone interested in this burgeoning field is Barbara Gastel's *Presenting Science to the Public* (1983).

Scientific Writing

Scientific writing, that is, "the first publication of original research results" (Day, 1988), is now an extremely rigid style of communication. Some scientists complain about this rigidity and the resulting "dullness" of scientific papers. Most scientists, however, especially after extensive experience, agree that this rigid yet logical system of writing makes it relatively easy for authors to write their papers, for referees and editors to review them, and for readers to read them.

This rigid style of "necessary" writing is still relatively new. Perhaps that explains why the principles of "necessary" writing are not yet taught in our schools. Few colleges and universities, and even fewer secondary schools, teach courses that explain and define modern scientific writing.

But the principles of "necessary" writing can be taught, and they should be taught, even if they were used only in the sciences. However, this new method of communicating information has already spread to engineering and several of the social sciences and now has its adherents even in the arts and humanities.

To start, and this could be done in the secondary schools, laboratory science courses could require that laboratory reports be written in the format of scientific papers. Students could be taught the logic of separating problems, methods, facts, and conclusions. If such students go on to become scientists, they will be off to a good start. If they go on into any other professional field, they will have learned a system of logical thinking and writing that will often be of use.

Signals and Communication

I suppose we have all heard this question: "If a tree falls in a forest, and there is no one there to hear it, does it make a sound?" The answer is no. Dictionaries confirm this answer. For example, *Webster's Ninth New Collegiate Dictionary* gives as the first two definitions of "sound": "1 *a:* the sensation perceived by the sense of hearing *b:* a particular auditory impression." Sound is more than "pressure waves," and indeed there can be no sound without a hearer.

Thus, "communication" is a two-way process. A signal of any kind is useless unless it is both received *and* understood by its intended audience. If we keep this definition in mind, it will guide us to effective ways to write, to edit, and to publish the literature of science.

Historical Background

Because the means of effective written communication were not yet available, early scholars tended to communicate orally. The early Greeks, especially Socrates and his contemporaries, became proficient in speaking effectively to both students and peers. That early oral tradition is still strong among scientists and other scholars today. Conferences, seminars, workshops, national and international congresses, and meetings of all kinds are still the most-used form of communication among scientists. Modern science, however, demands a way of recording, preserving, and retrieving massive amounts of information. Thus, science has come to depend upon its written records to record its progress.

The world's first scientific journals were published in 1665, when, coincidentally, two journals first appeared—the *Journal des Scavans* in France and the *Philosophical Transactions of the Royal Society of London* in England. These journals were both successful. (Both are still being published today.) The "journal" idea spread to Germany and to many other countries. By 1750, there were about 10 scientific journals in the world, a figure that would increase by a factor of 10 during every succeeding half-century (de Solla Price, 1961/1975). In the United States, Ben Franklin was instrumental in establishing the American Philosophical Society in 1743, and the Society began publishing its *Transactions* in 1771. Today, something like 70,000 scientific and technical journals are being published around the world (King, McDonald, and Roderer, 1981).

The early scientific journals were "letters" journals. This is not surprising, inasmuch as scientists, prior to 1665, were forced to communicate by letter (and of course orally). Fortunately, the letter format served early science very well. Most letters were written in a simple chronological order—First I did this, then I did that, etc.—and this style of writing was appropriate for the relatively simple observations being made and recorded. In fact, even today, this kind of descriptive reporting is still much used in science; most clinical case reports and most geological field studies, for example, are written in this clear, straightforward, chronological way.

Writing Gets Organized

The modern scientific paper began to take shape during the last half of the last century. Perhaps much of the credit should go to those great experimentalists, Robert Koch and Louis Pasteur. "Koch's Postulates" were important not only in demonstrating the germ theory of disease but also in establishing reproducibility as an essential element of scientific research. Pasteur also gets credit for laying much of the groundwork for modern microbiology and biochemistry, while at the same time pointing to a new way of writing science. In order to defend his work against its many detractors, Pasteur described his experiments in exquisite detail. Thus was born the "Methods" section of the modern research paper. To contend with the long-held and passionately defended theory of spontaneous generation, Pasteur not only developed pure-culture techniques, but also described them so carefully that any reasonably careful peer could reproduce his experiments. Thus, Pasteur showed the practical need for a segregated methods section in a scientific paper, and he confirmed that *reproducibility of experiments* is essential to the conduct of science.

Pasteur's work did indeed administer the coup de grace to the theory of spontaneous generation. And his work, combined with that of his contemporary, Charles Darwin, made it possible to define life in scientific terms. As succeeding scientists accepted these new principles, the end of serious belief in magic and religious superstition seemed to be at hand. In our day, however, millions of people continue to believe in such unscientific notions as astrology and "scientific creationism."

In the first half of this century, modern science and the recording of modern science became increasingly sophisticated. In medicine, especially, a remarkable revolution occurred. Our understanding of infectious disease, now that the germ theory of disease was in place thanks to Koch and Pasteur, grew dramatically. The work of Paul Ehrlich in the early 1900s (Salvarsan) was followed by that of Gerhard Domagk in the 1930s (sulfa drugs), introducing what came to be known as the miracle drugs. Then, during World War II, came penicillin (originally discovered in 1929 by Alexander Fleming) and, a few years later, streptomycin, the tetracyclines, and others. The great killers of the past—tuberculosis, septicemia, diphtheria, typhoid, and the plagues—could now be controlled.

The fantastically good news that was coming out of our medical research laboratories during and after World War II made it attractive to increase our investment in research and development. This positive spur to research was soon joined by a negative one when the Soviets flew Sputnik around the earth in 1957. U.S. politicians, obsessed with fear that the Soviets would get ahead of us, began to throw money at the science establishment as never before.

Today's Scientific Writing

Money paid for research. Research produced papers. Mountains of them. The result was powerful pressure on the journals. Journal editors, in self-defense if for no other reason, began to demand that manuscripts be reasonably written and organized. The system of organization that had been slowly progressing since the latter part of the 19th century now came to its simplest expression. This system of writing scientific papers became known as "IMRAD," an acronym standing for Introduction, Methods, Results, and Discussion.

This new system, now more or less rigidly required by most scientific research journals, has a simple logic to it that most scientists (and especially editors) have come to appreciate. IMRAD can be defined in question form as follows: What was the question (problem) that was studied? The answer is the Introduction. How was the question studied? The answer is given in the Methods section of the paper. What were the findings? The answer is the Results. What do these findings mean? The answer is the Discussion. The vast majority of scientific papers being written and published today are organized in this simple four-part structure. Furthermore, there is general agreement regarding the appropriate components of each of the four main sections.

The purpose of the *Introduction* is to state, with all possible clarity, the problem studied. To accomplish this, the problem or question studied must be placed within its contextual background. This means that a brief summary of what is already known (i.e., a survey of the pertinent literature) is usually required.

The purpose of the *Methods* section is clear. The author should carefully describe the methods and materials used in the experiments, resisting any temptation to prematurely mention data (to be reserved for Results) or conclusions (to be held for the Discussion). What is mandatory is that the Methods be written with such precision and completeness that a competent colleague could repeat the experiments and obtain the same or similar results.

The *Results* section is reserved for the author's own results. This section is often short, even though it provides the *new* information that is the focal point of the whole paper. The Results should not give explanatory preludes (properly a part of the Introduction) and should not give conclusions (the purpose of the Discussion).

The *Discussion* section is the section of a modern research paper that tends to be difficult to write. A number of components are likely to be necessary, including the following: discussion of any principles, relationships, and generalizations shown by the results; exceptions and unsettled points; agreement (or lack thereof) with previously published work; theoretical and practical applications; conclusions; and evidence for *each* conclusion.

The simple logic of IMRAD makes it relatively easy for the writer of a paper to organize the manuscript. This organization makes it easy for editors and referees to evaluate the manuscript. And this rigid system gives the reader a road map to follow in digesting the contents of a paper.

Should the writer have the freedom to add variation and ornamentation to the paper, in an attempt to add interest and to avoid dullness? The answer is no. "Necessary" writing should not be a literary product. This kind of writing is "necessary" because it carries a *message*. The message must be both received and understood, or the tree will fall silently in the forest.

Necessary writing is the transmission of a clear signal to a recipient. The words of the message should be as clear and simple and well-ordered as possible. In necessary writing, there is no room for and no need for the confusion that can be introduced by ornamentation. Literary devices such as metaphors, similes, and idiomatic expressions should be used rarely if at all.

Science is simply too important to be communicated in anything other than words with the most precise meaning possible. And that clear, certain meaning should be designed not just for the author's peers, but also for students just embarking upon their careers, for scientists reading outside their own narrow discipline (and interdisciplinary research is becoming ever-more necessary), and *especially* for those readers (the majority) whose native language is other than English. Now that English has clearly become the international language of science, we should all do our best to use the language effectively and clearly for our common good.

In conclusion, necessary writing is not designed for entertainment. However, necessary writing, designed to transmit information, need not be ugly. Necessary writing must be as clear and simple as possible. Plato said it well: "Beauty of style and harmony and grace and good rhythm depend on simplicity."

References

American National Standards Institute. (1979). *American national standard for the preparation of scientific papers for written or oral presentation* (ANSI Z39.16–1979). New York: Author.

Council of Biology Editors [CBE] Style Manual Committee. (1983). *CBE style manual: A guide for authors, editors, and publishers in the biological sciences* (5th ed.). Bethesda, MD: Author.

Day, Robert A. (1988). *How to write and publish a scientific paper* (3rd ed.). Phoenix, AZ: Oryx Press.

De Solla Price, Derek. (1961/1975). *Science since Babylon* (enlarged ed.). New Haven: Yale University Press.

Dubos, René. (1950). *Louis Pasteur, free lance of science*. Boston: Little, Brown.

Gastel, Barbara. (1983). *Presenting science to the public*. Philadelphia: ISI Press.

Gould, Stephen Jay. (1983). *Hen's teeth and horse's toes*. New York: W. W. Norton.

Gould, Stephen Jay. (1987). *An urchin in the storm: Essays about books and ideas*. New York: W. W. Norton.

King, Donald W., McDonald, Dennis D., and Roderer, Nancy K. (1981). *Scientific journals in the United States*. Stroudsburg, PA: Hutchinson Ross.

Luria, S. E. (1984). *A slot machine, a broken test tube: An autobiography*. New York: Harper and Row.

Sagan, Carl. (1977). *The dragons of Eden: Speculations on the evolution of human intelligence*. New York: Random House.

Sagan, Carl. (1979). *Broca's brain: Reflections on the romance of science*. New York: Random House.

Thomas, Lewis. (1975). *The lives of a cell*. New York: Bantam Books.

Thomas, Lewis. (1979). *The Medusa and the snail: More notes of a biology watcher*. New York: Viking.

Watson, James D. (1968). *The double helix: A personal account of the discovery of the structure of DNA*. New York: Atheneum.

Zinsser, Hans. (1940). *As I remember him: The biography of R. S.* Boston: Little, Brown.

Part IV

Personal Reflections: From Gifts to Talents

The editors invited a number of scientists and science teachers to describe briefly—in any literary form that appealed—the event or events that led them to their work. While our letter of invitation (reproduced below in the section containing NSTA's exchanges with Isaac Asimov) asked about the kinds of environments that inspired choices, most respondents chose not to tell us *why* they became scientists or teachers but *how*. Of course, one moment does not shape a life. Surely, many, many experiences conspire to result in individual choice. The chosen moments, however, are instructive. One common thread is gratitude to those whose previous work made later discovery possible: to mentors, to colleagues, to students, to parents, to friends. All directions are unique. All offer promise.

Gerald Skoog
Contributing Editor, Part IV

Dear Mr. Asimov . . . .
Dear NSTA

Isaac Asimov

Editor's note: Dr. Asimov chose an epistolary response to our request. The exchange appears below.

The Invitation

May 15, 1987: NSTA Association Editor Deborah C. Fort solicited Dr. Asimov's contribution.

NATIONAL SCIENCE TEACHERS ASSOCIATION

1742 Connecticut Avenue, NW, Washington, DC 20009 (202) 328-5800

May 15, 1987

Dear Mr Asimov:

The National Science Teachers Association is gathering a last series of papers for the third section of its upcoming book tentatively titled Gifted Young in Science--Potential to Performance. The permeating belief of all four parts of the volume is our conviction that teachers of science, like all teachers, can and must create environments that nurture and evoke the potential of all their students. However, its particular focus is on settings that encourage the young who as scientists, engineers, technologists, and teachers will conserve, transmit, correct, and expand the variety of cultures rooted in the sciences.

As architect or as beneficiary, you have participated in such an environment. With your help, science teachers could learn from your experience. Would you be willing to think back and describe the time(s) and place(s) where your interest and achievement in performing or teaching science crystallized, deepened, or was sustained? What science environment either enriched your life or that of your students or led you as a student to incline to science?

Knowing how much you do, we suggest only some 500 to 1,000 words. You may approach the task historically, pinpointing significant factors within a specific environment. Or you may decide to write a letter--to teachers, to children, to the future, to whomever. Choose any device that pleases you.

Regardless of its form, however, we are asking from each of the contributors to this section, which we think we will title "From Gifts to Talents," a personal message about how your particular environment led to your vocation.

If this idea appeals, won't you inform us soon by mailing the enclosed postcard to Deborah Fort? We are providing a table of contents.

We will look for your essay in early October, 1987.

Sincerely yours,

Deborah C. Fort
Deborah C. Fort
Association
Editor

Gerald Skoog
Contributing
Editor,
Part IV

Paul F. Brandwein
A. Harry Passow
Editors

Bill G. Aldridge, Executive Director

OFFICERS AND
BOARD OF DIRECTORS
Mary Budd Rowe, President
University of Florida, Gainesville
LaVerne L. Motz, President-elect
Oakland, MI Public Schools, Pontiac
LeRoy R. Lee, Retiring President
Wisconsin Academy of Sciences,
Arts, and Letters, Madison, WI

DIVISION DIRECTORS
Bonita R. Talbert-Wylie
Preschool, Elementary School
Evanston, MN Elementary School
Sam S. Chaston
Middle Junior High School
Scottsdale, IN Junior High School
George K. Hague, Jr
High School
St. Mark's School of Texas, Dallas
John J. Rusch
College
University of Wisconsin-Superior
William C. Kyle, Jr
Research
Purdue University, West Lafayette, IN
Gary A. Nakagiri
Supervision
San Mateo County, CA Public Schools
Redwood City
Michael I. Padilla
Teacher Education
University of Georgia, Athens

DISTRICT DIRECTORS
David Lopath, District I
Morgan High School, Clinton, CT
John E. Evans, Jr., District II
Fairmount Junior High School
Philadelphia, PA
Harriet B. Donofrio, District III
Cape Henlopen High School, Lewes, DE
Anne F. Barefoot, District IV
Whiteville High School, Whiteville, NC
Donna Bentley, District V
Alabama State Department of Education
Kay Wagner, District VI
Ohio State Department of Education
Jon L. Harkness, District VII
Wausau West High School, Wausau, WI
John E. Penick, District VIII
University of Iowa, Iowa City
Barbara K. Foots, District IX
Houston, TX Independent School District
Elizabeth A. Horsch, District X
Kelly Walsh High School, Casper, WY
Glyn A. Bryce, District XI
Whitford Intermediate, Beaverton, OR
David Harbster, District XII
Chandler, AZ

DIVISION AFFILIATE PRESIDENTS
I. Joyce Swartney, AETS
State University College, Buffalo, NY
Linda Froschauer, CESI
Weston Middle School, Weston, CT
David A. Kennedy, CSSS
Washington State Dept. of Instruction
Linda DeTure, NARST
Rollins College, Winter Park, FL
Kenneth Russell Roy, NSSA
Glastonbury, CT Public Schools
William Frase, SCST
University of Cincinnati, OH

An Affiliate of the American
Association for the
Advancement of Science

36th NSTA National Convention, St. Louis, Missouri, April 7-10, 1988

Ambivalence
May 18, 1987: Dr. Asimov returned his enclosed postcard with this message.

> 18 May 1987
>
> Dear Ms. Fort,
>
> I suspect that what you have sent me is a form letter sent out in the same mail to many others. There seems a curious lack of personalness about it. This means that it would probably make little difference to you, if one or more of is didn't come through.
>
> In my own case, thing are so atypical that I honestly don't believe that what I say can be of any use to others. Therefore I would rather not unless there is some particular reason why you want me. If so, do write me a personal letter and not a form.
>
> *Isaac Asimov*

Reinvitation
May 28, 1987: Dr. Fort replied as follows.

NATIONAL SCIENCE TEACHERS ASSOCIATION
1742 Connecticut Avenue, NW, Washington, DC 20009 (202) 328-58⬚⬚

May 28, 1987

Bill C. Aldridge, Executive Director

OFFICERS AND
BOARD OF DIRECTORS
Mary Budd Rowe, President
University of Florida, Gainesville
LaMoine L. Motz, President-elect
Oakland, MI Public Schools, Pontiac
LeRoy R. Lee, Retiring President
Wisconsin Academy of Sciences,
Arts, and Letters, Madison, WI

DIVISION DIRECTORS
Bonita B. Talbot-Wylie
Preschool/Elementary School
Excelsior, MN Elementary School
Sam S. Chaton
Middle/Junior High School
Scottsburg, IN Junior High School
George R. Hague, Jr
High School
St. Mark's School of Texas, Dallas
John J. Rusch
College
University of Wisconsin-Superior
William C. Kyle, Jr
Research
Purdue University, West Lafayette, IN
Gary A. Nakagiri
Supervision
San Mateo County, CA Public Schools
Redwood City
Michael J. Padilla
Teacher Education
University of Georgia, Athens

DISTRICT DIRECTORS
David Lopath, District I
Morgan High School, Clinton, CT
John E. Evans, Jr., District II
Fitzsimons Junior High School
Philadelphia, PA
Harriet B. Donadino, District III
Cape Henlopen High School, Lewes, DE
Anne F. Barefoot, District IV
Whiteville High School, Whiteville, NC
Donna Bentley, District V
Alabama State Department of Education
Kay Wagner, District VI
Ohio State Department of Education
John L. Harkness, District VII
Wausau West High School, Wausau, WI
John L. Penick, District VIII
University of Iowa, Iowa City
Barbara K. Foots, District IX
Houston, TX Independent School District
Elizabeth A. Horsch, District X
Kelly Walsh High School, Casper, WY
Glyn A. Bruce, District XI
Whitford Intermediate, Beaverton, OR
David Harbster, District XII
Chandler, AZ

DIVISION AFFILIATE PRESIDENTS
J. Joyce Swartney, AETS
State University College, Buffalo, NY
Linda Froschauer, CESI
Weston Middle School, Weston, CT
David A. Kennedy, CSSS
Washington State Dept. of Instruction
Linda DeTure, NARST
Rollins College, Winter Park, FL
Kenneth Russell Roy, NSSA
Glastonbury, CT Public Schools
William Frase, SCST
University of Cincinnati, OH

An Affiliate of the American
Association for the
Advancement of Science

Dear Mr. Asimov:

Thank you for your postcard. I plead guilty to your charge of sending out boilerplate but not to "many others" and promise immediate reform. We sent an initial general letter to the seven scientists and teachers invited because we wanted them to start with the same information. Where each goes from our book's environmental thesis will be different, to say the least.

All we'd like from you is a two to four page view of why one particular Russian immigrant grew up to write hundreds of books on dozens of subjects. If you can enlighten millions of science laymen (lay people?) through fact and fiction about mysteries as dark as black holes, while still remaining "as youthful, as lively, and as lovable as ever" and growing "more handsome with each year," the paper we solicit should be a piece of cake. I'm sure that, as you predict, your experience will be "atypical," but (I disagree) hardly useless. If you can teach us what or who in your environment combined with your raw material to fashion what George Gaylord Simpson calls a "national treasure," think of the implications.

We will take our chances on your being an irreproducible result.

I look forward to hearing from you.

Personally yours,

Deborah Fort
Deborah Fort

cc: Harry Passow
    Paul Brandwein
    Gerald Skoog

36th NSTA National Convention, St. Louis, Missouri, April 7-10, 1988

Second Attempt

June 1, 1987: Dr. Asimov "tried again to explain."

ISAAC ASIMOV

---

1 June 1987

Deborah C. Fort
National Science Teachers Assoc.
1742 Connecticut Avenue, N. W.
Washington, D.C., 20009

Dear Ms. Fort,

Let me try to explain again, though it will be terribly embarrassing. I <u>never</u> organized my life to achieve some goal. I went along in carefree manner doing exactly what I wanted to do, and here I am. How I got here is a mystery to me.

I never learned how to write -- I knew how from the start. As Alexander Pope said:

As yet a child, nor yet a fool to fame,
I lisp'd in numbers, for the numbers came.

("numbers" is metrical verse.)

How am I going to explain that? How am I going to spend 500 to 1000 words saying "I wasn't pushed, I wasn't guided, I just wrote."? How am I going to say, "You can do it, too; all you have to do is be me."? Of what use will that be to science students, science writers, or science teachers? It will just induce more bitter comments about my "conceit" and "vanity."

Isaac Asimov

A New Appeal

June 15, 1987: NSTA Executive Director Bill G. Aldridge joined the correspondence.

NATIONAL SCIENCE TEACHERS ASSOCIATION
1742 Connecticut Avenue, NW, Washington, DC 20009 (202) 328-5800

Bill G. Aldridge, *Executive Director*

OFFICERS AND
BOARD OF DIRECTORS
Mars Budd Rowe, *President*
University of Florida, Gainesville
LaMoine L. Motz, *President-elect*
Oakland Intermediate Public Schools, Pontiac
LeRoi R. Lee, *Retiring President*
Wisconsin Academy of Sciences,
Arts, and Letters, Madison, WI

DIVISION DIRECTORS
Bonita R. Talbot-Wylie
Preschool-Elementary School
Excelsior, MN Elementary School
Sam S. Chattin
Middle Junior High School
Scottsburg, IN Junior High School
George R. Hague, Jr.
High School
St. Mark's School of Texas, Dallas
John J. Rusch
College
University of Wisconsin-Superior
William C. Kyle, Jr.
Research
Purdue University, West Lafayette, IN
Gary A. Nakagiri
Supervision
San Mateo County, CA Public Schools
Redwood City
Michael J. Padilla
Teacher Education
University of Georgia, Athens

DISTRICT DIRECTORS
David Lopach, *District I*
Morgan High School, Clinton, CT
John E. Evans, Jr. *District II*
Fitzsimons Junior High School
Philadelphia, PA
Harriet B. Donofrio, *District III*
Cape Henlopen High School, Lewes, DE
Anne F. Barefoot, *District IV*
Whiteville High School, Whiteville, NC
Donna Bentley, *District V*
Alabama State Department of Education
Kay Wagner, *District VI*
Ohio State Department of Education
Jon L. Harkness, *District VII*
Wausau West High School, Wausau, WI
John E. Penick, *District VIII*
University of Iowa, Iowa City
Barbara K. Foots, *District IX*
Houston, TX Independent School District
Elizabeth A. Horsch, *District X*
Kelly Walsh High School, Casper, WY
Glen A. Brice, *District XI*
Whitford Intermediate, Beaverton, OR
David Harbster, *District XII*
Chandler, AZ

DIVISION AFFILIATE PRESIDENTS
J. Joyce Swartney, AETS
State University College, Buffalo, NY
Linda Froschauer, CESI
Weston Middle School, Weston, CT
David A. Kennedy, CSSS
Washington State Dept. of Instruction
Linda DeFure, NARST
Rollins College, Winter Park, FL
Kenneth Russell Roy, NSSA
Glastonbury, CT Public Schools
William Frase, SCST
University of Cincinnati, OH

An Affiliate of the American
Association for the
Advancement of Science

June 15, 1987

Dear Isaac:

I have just reviewed with some amusement the correspondence between Deborah Fort and you in regard to a small piece we requested for our forthcoming book on science education for gifted students.

So far, within these letters alone, we have almost 500 excellent words from you. I can readily believe that you were a writer from the very start. But come on, Isaac, isn't there anything in your environment that excited your early interst in science? No parent, teacher, family friend, library, book, observation, or anything? Would you be the same if you had been reared in a dark African jungle, without books, schools, or the intellectual and cultural richness of your own background?

We really are trying to offer something helpful, something neither trivial or prescriptive. You are a phenomenon; we are looking for some hint as to your origins, to the cause or causes of your genius, even to what has influenced your admittedly unique path. I am quite serious that within our exchange of letters an answer to our original questions is forming. May we continue to correspond until the image gets clearer? We will of course send you a "manuscript" before the book is published.

It has been too long since you spoke to our convention in Boston. We must schedule another nearby, and you know full well I will then pester you until you agree to address the science teachers again.

In the meantime, take good care of yourself.

Sincerely,

Bill G. Aldridge
Executive Director

cc: Deborah Fort

36th NSTA National Convention, St. Louis, Missouri, April 7-10, 1988

Third Try

June 15, 1987: Dr. Asimov elaborated further.

ISAAC ASIMOV

---

18 June 1987

Bill G. Aldridge
National Science Teachers Association
1742 Connecticut Avenue, N.W.
Washington, D.C., 20009

Dear Bill,

I feel that you are becoming annoyed with me.

All right, then. When I was a child, we were poor and there were no books in the house. Comic books didn't exist, television didn't exist, I wasn't allowed to touch the radio and I wasn't allowed to read the magazines in my father's candy store. To keep me busy my parents got me a library card when I was six. My parents could not at that time read or speak English except in a very primitive way, so they could give me no guidance. I therefore took out books indiscriminately (choosing long books over short ones, because they would last longer) and read everything. In this way I grew interested in everything. When I was 9 years old, I finally persuaded my father to let me read science fiction magazines. That reinforced my interest in science. And when I was 11, I began to write. (Incidentally the library card at 6 is not a typo. I had taught myself to read when I was 5.)

Now what on Earth good is that going to do anybody?

Isaac

Gratitude

July 6, 1987: Mr. Aldridge began to sum up.

Bill Aldridge
Executive Director

NATIONAL SCIENCE TEACHERS ASSOCIATION
1742 Connecticut Avenue, N.W.
Washington, D.C. 20009

July 6, 1987

Dear Isaac:

Thanks for your note of June 18. Believe it or not, what you have said is going to do some good! Some of us got a normal distribution of genes, and we were able to achieve some measure of success, intellectually, as a result of our environment. You clearly had a superb set of genes, and your achievements were almost a natural consequence. Yet we need to identify the rare children like you were and do what we can to stimulate them to do science. We must do so for very selfish reasons: Solving the problems that plague us will require great minds. Cancer, AIDS, diabetes etc. have causes & cures that can be found by the original, unusual mind.

Best wishes

Bill

Finis

December 16, 1988: But Dr. Asimov had the last word.

ISAAC ASIMOV

16 December 1988

Deborah C. Fort
National Science Teachers Association
1742 Connecticut Avenue, N. W.
Washington, D.C., 20009

Dear Ms. Fort,

I have your letter of 14 December and the enclosures, including the correspondence between myself on one side and Bill and you on the other.

I have no objection to its being published as you have it. No doubt if will make fine comic relief for an otherwise serious book.

I cannot help but feel that, considering my skill at writing fiction, I would have been able to have invented a story about how I came to be me that wouldn't have left a dry eye in the house, I could have outmatched the Little Match Girl, However, I have this hang-dog attachment to the truth. It spoils everything.

[signature: Isaac Asimov]

Still in My Dinosaur Phase

Stephen Jay Gould

Many scholars can trace their personal roots to a lonely childhood spent reading. I was too busy on the local stickball court for such a life. My minimal taste in reading ran to the cardboard and inspirational—such fine works as *The Little Engine That Could* (I was also small), and Joe DiMaggio's *Lucky to Be a Yankee* (he was my hero, and I planned to replace him in centerfield, until a guy named Mantle came along). My nadir was the time I presented an oral book report on *Les Misérables,* having read only the Classic Comics version (I loved the pictures of Jean Valjean in the Paris sewers). I was caught because I had no idea of the book's real length—and the teacher knew I could not have read it in the time between assignment and presentation.

I don't mean to exaggerate. I was no wiseacre, or particularly streetwise Philistine. I was timid and insecure, but I knew exactly what I wanted to do as an adult. I would study dinosaurs, a firm conviction inspired by one supreme moment of childhood terror dissipated by fascination—my first look at Tyrannosaurus rex in the American Museum of Natural History.

There is nothing uncommon about childhood passion for dinosaurs; paleontology is but one stage in the standard sequence of pirate, policeman, fireman—something to pass through, and then move onward. The rarity is persistence into adulthood of a permanent childlike pleasure.

I don't know what diverts dinosaur fanatics. For most, I suppose, the world is simply too full of wonderful things to permit a long dalliance with any particular obsession. But for me and others more awestruck, intent, and committed, the main barrier to persistence must be absence of a larger context to absorb a purely visceral fascination. Dinosaurs may be "big, fierce, and extinct"—the reasons cited by my colleague, the psychologist Sheldon White, for their appeal to children. But these features are not the stuff of permanent commitment. The Empire State Building is big, and Rambo is both fierce and ought to be extinct. The promotion of dinosaur mania to a career demands a matrix of ideas to absorb and channel that primal phenomenological oomph.

For me, a single book supplied that matrix—and also taught me the depth and value of "grown-up" documents. My parents belonged to a book club and forgot, one month, to mail back the "we don't want anything" card. One day, G. G. Simpson's *Meaning of Evolution* (first published in 1949, available today in a Yale University Press paperback edition, and still among the best general works on the subject) arrived in the mail. I knew nothing of evolution, but the cover sported tiny pictures of dinosaurs—just a little friezelike design in the spandrels around the title.

Earlier this year, the Dino-Store opened a few blocks from my home in Cambridge, Massachusetts—an entire emporium devoted to the reptilian paraphernalia of kiddie culture. But in 1951, I had only Alley Oop and Roy Chapman Andrews's *All About Dinosaurs* (already well digested)—so even a little cover design drew me within like a magnet. (In my grumpy Miniver Cheevy moods, I sometimes think you can have too much of a good thing—and leave no room for a child's imagination.)

So I read Simpson *faute de mieux*. And I discovered, via the best route of utterly unanticipated delight, that a fabric of ideas united me, and all of life, with dinosaurs. I grasped the concept of genealogical connection and transmutation—the two basic facts of evolution. More importantly, I peered (very darkly through a glass of childhood ignorance) at the fascination of what science, as a way of knowing, tried to do to pierce beyond basic facts to a notion of mechanism, and underlying generality (natural selection in this case) that could apply to me as well as to Tyrannosaurus, now almost a friend through the twin connections of actual genealogy and abstract theory.

Finally, I glimpsed at even greater distance the troublesome questions of "why" that science cannot answer, but that any person of passion and curiosity must contemplate—as Simpson did with all the generous humanism and uncompromising rationalism of a brilliant man humbled by the sweep of geological time.

I doubt that I understood even 10 percent of the book, but something drove me on. I had found the key that led from "big, fierce, and extinct" to what Darwin simply called "this view of life," and then characterized by a single noun—grandeur.

We must all, at some point, put aside childish things—and we remember the symbols of transition with poignancy (not with pain, if luck smiled upon us). I had two such moments. First, when I saw Olivia de Havilland playing Maid Marian and something I did not recognize stirred within me. (It is among my most treasured delights that I met this great lady, looking more beautiful than ever, a few months ago—and was able to tell her this little private secret face to face.) Second, when I lay on my bed late at night reading G. G. Simpson and found the real reason for lifelong fidelity to Tyrannosaurus.

A Tribute to My Mentors*

Joshua Lederberg

If I have any one message to convey, it is an account of my debts: to the individuals who gave so much of themselves as parents, teachers, colleagues, and friends, and to a system that has offered extraordinary nurture to whatever talent and ambition I could bring. That system, the social milieu of science, is under the microscope today, scrutinized for every aberration and pathology. Taken for granted, and thereby overlooked in the presentation of the scientific career to younger people, are its positive aspects of community and of the traditional (and reciprocal!) bonds of teachers and students, not to mention the unique thrills of discovery and the gratification of its application for human benefit.

Although I was born in Montclair, New Jersey, my early education was framed by the New York City public school system. A cadre of devoted and sympathetic teachers went far beyond their duty in encouraging a precocious youngster, despite his taunting them with questions they could not always answer. The culmination was Stuyvesant High School, which specializes in science. Stuyvesant also offered unusual opportunities for practical work in machine shops and analytical laboratories. Most important of all, it attracted a peer group (then unfortunately limited to boys) of the keenest young intellects: For the first time, I had a few intellectual sparring partners.

New York City in the 1930s had a network of institutions directed to enhancing the intellectual and social mobility of its melting pot youth. The numerous Nobel Prizes that have emerged from New York's science high

*Parts of this paper are reproduced, with permission, from the *Annual Review of Genetics, 27* © 1987 by Annual Reviews Inc.

schools and City Colleges are further witness of the encouragement given to the talent and ambition of its students, perhaps more than of the laboratory facilities or of the academic attainments of their faculty at the time.

The ambitions thus inspired were reinforced by a popular culture that idealized the medical scientist with novels and movies like *Arrowsmith, The Magic Bullet, The Life of Louis Pasteur,* and *The Symphony of Six Million.* In a mood born of the Great Depression, however, many of these works painted a bleak picture of the personal life of the scientist: Marriage and family were expected to be Baconian "hostages to fortune" (1625/1971).

Actual medical textbooks were not so readily available; nevertheless, I was able to read histology, microbiology, and immunology while in high school. Immunology, as then presented, was almost impenetrable to my efforts at orderly, scientific integration. (It took me two decades to realize that the fault was not mine.)

The library book that had the greatest influence on my further scientific development was Meyer Bodansky's *Introduction to Physiological Chemistry* (1934). The copy I received as a Bar Mitzvah present (1938) stands on my bookshelf today, the print almost worn off the pages.

Equally important as the schools to my own education was the local Washington Heights branch of the Carnegie-Astor New York Public Library system. These institutions symbolized and embodied the melting pot ideology. My father was an orthodox rabbi, born and educated in Israel, and thus had more prestige, higher intellectual aspirations for his children, and less income than most of his neighbors. Like many other first-generation Jewish youths in New York City at that time, I was recruited into an efficient and calculated system of Americanization, fostered by the rich opportunities and incentives of the educational system.

My earliest recollections aver an unswerving interest in science, as the means by which humankind could strive for understanding of its origin, setting, and purpose, and for power to forestall its natural fate of hunger, disease, and death. The Jewish reading in Genesis of the expulsion from Eden makes no presumptions of the benignity of Nature: "By the sweat of thy brow. . . ." This may have been the most acceptable deviation from the orthodox religious calling of my family tradition. These images were reinforced by the role of Albert Einstein and Chaim Weizmann as culture heroes whose secular achievements my parents and I could together understand and appreciate, regardless of the intergenerational conflicts evoked by my callow agnosticism. I could not then see how the monotheistic world view and the central teachings of the Old Testament, and their ethical imperatives for contemporary life, related to the tribal rituals shaped in the Diaspora. But the utopian-scientific ethic offered an acceptable resolution. My own career could advance our shared ideals in a modern, American idiom. Science would be a path to knowledge of the cosmic order. It would also be a means of alleviating human suffering. The Jewish tradition is remarkably tolerant of skepticism: I have in mind Maimonides' teachings of the unknowability of God. The agnostic set of mind thus permitted, together with my reaction to my father's orthodoxy, carried over into my reflex responses to other sources of authoritative knowledge.

The library was my university as I went through grade school and junior high school. Here was the universe of knowledge, huge but finite. The teenager, unencumbered by any informed guidance and tutelage, fantasized mastering all of it. There were few books (except perhaps musical scores) that were totally incomprehensible to me; most were merely difficult and would eventually yield to diligent study. At that age, of course, there is little sense of the finitude of human life. After 1938, I also had access to the Stuyvesant High School library, and more importantly Cooper Union, for its stacks gave access to scientific journals like the *Journal of the American Chemical Society* and *Science* magazine. The librarians did (and do) welcome me as their most enthusiastic patron; I loved nothing better than to scan the shelves, discipline by discipline, by no means confined to science, and try to find whatever work both challenged and was accessible to me. I was also a voracious reader of fiction from Dumas and Hugo to Thomas Mann, H. G. Wells, Ludwig Lewisohn, and Sinclair Lewis.

I did have some opportunity for "experiment" at home, what with the toxic and explosive chemicals that could then be purchased in chemistry sets, and over the counter at Eimer & Amend's, near Stuyvesant High School. Besides nearly destroying myself with looking for the threshold of explosion in throwing metallic sodium in water, most of these were elementary syntheses of azo dyes and the like, sometimes in concert with some high school chums. I did make a more original study of the reaction of ferric iron with thiosulfate, but got little satisfaction either from experiment or the published literature. Stuyvesant also sponsored a Biology Club, and one could beg the storekeeper to use the microtome to cut histological sections. I started a project to try to understand the effect on tissues of fixatives, chemicals intended to coagulate and solidify tissue proteins but hopefully leave the essential structures close to their living form. These chemicals, like formaldehyde, acetic acid, and picric acid, do alter the staining properties, and I thought this interaction should give clues to the chemical nature of tissue materials. Indeed it should; however, the use of specific enzymes, and then of antibodies, have provided much more specific probes for this kind of histochemistry.

The New York Museum of Science and Industry, and the New York World's Fair (starting 1939) were also wonderful stimuli, picturing science-technology utopias of the near future. They offered samples of polaroid optical sheets, and of the new Bakelite™ plastics that could be taken home for further experiments. Above all they left a vision of "Better Things for

Better Living Through Chemistry." (DuPont's slogan now leaves out the word chemistry—it has become a dirty word!)

One of the guides at the New York World's Fair was a young psychologist, Henry Platt. He had a vision of a means of encouraging young scientific minds, namely to offer them a laboratory where they could conduct authentic scientific research, with appropriate equipment and supervision. By lucky chance, he met Thomas J. Watson at the Fair, and persuaded him to support the project. This materialized as the American Institute of Science Laboratory, housed in an IBM showroom building on Fifth Avenue, in the shadow of the Empire State Building. I was lucky to be accepted into that program: Having graduated from high school in January, and being obliged to wait until September to start at Columbia University, it was a happy way indeed to occupy the interval.* And I was too young to work without running afoul of the child labor laws! The American Institute of Science Laboratory did indeed offer better facilities, and unbroken time, to continue the cytochemical work I started at Stuyvesant, and I began to focus on the chemistry of the nucleolus. It had many of the properties of a nucleic acid, but these were not consistent with pure DNA. (My own studies were hardly contributory to a solution: Brachet had already applied ribonuclease to the histochemical identification of RNA in nucleoli. His work, published in Nazi-occupied Brussels in November, 1940, was not communicated to the U.S. until some time later.) The work did confirm my fascination with the chemistry of the cell, and I determined to concentrate on such studies as soon as I could acquire the authentic and mature expertise that Columbia University could offer.

Before entering Columbia College, however, I had not yet met a working scientist. I can recall having attended popular lectures by Wendell Stanley on the chemistry of tobacco mosaic virus; earlier, as a 10 year old, how impressed I had been by the newspaper accounts of his having crystallized life (cf. Kay, 1986). These stories were among many accounts that pointed to the Rockefeller Institute as the sanctum sanctorum of biomedical science.

With these cardinal inspirations, my entry to Columbia that fall was motivated by a passion to learn how "to bring the power of chemical analysis to the secrets of life." I looked forward to a career in medical research where such advances could be applied to problems like cancer and the malfunctions of the brain.

I had applied to Cornell, on account of Lester W. Sharp's presence on the faculty—a name that I knew from his text, *Introduction to Cytology*. But

---

*My young colleagues at the American Institute Laboratory included many who have achieved great scientific distinction. In my own field, these are notably Barry Blumberg, Nobelist and Master of Balliol College, Oxford; and Charles Yanofsky, professor of biology at Stanford University.

Cornell was in practice open only to wealthy tuition-paying students, or to farm boys who could enroll in the New York State-funded College of Agriculture.

My application for a scholarship at Telluride House was rejected. I also had City College in mind, but thought of this as a last resort as it had limited graduate work and scarcely any research facilities. No one so much as hinted that I could seek work and scholarships at other state universities. Berkeley might have been a superb possibility, but California seemed like the other side of the moon. Financially, a commuter school like Columbia was almost the only feasible possibility, barring a scholarship. This perhaps did not exist for a Jewish boy from New York at that time.

Hitler had achieved power in Germany when I was eight years old, just old enough to have no doubt about the aims of his march across Europe. Eight years of fascinated horror at the unfolding of history followed—the persecution of the German Jews, the flight of intellectuals like Albert Einstein, the occupation of Austria, Munich, the Nazi-Soviet pact and partition of Poland, the fall of France, the victory of the RAF in the Battle of Britain, the Nazi invasion of Russia, the Japanese attack on Pearl Harbor. Then, in December, 1941, we knew that the War would dominate our lives until a painful victory was won.

Since 1945 the power to destroy has weighed in negative balance on the scientific conscience: We are no longer assured that net human benefit will be achieved as an automatic consequence of the enhancement of knowledge (cf. my related essays [1972 and 1973]). We are not abandoning the enterprise; the global competition, if nothing else, forfends a halt. Weighing the benefit of scientific research has become more complicated.

Today's popular portrayals of the scientific culture give short shrift to anything but fraud and competition. What contrast to the idealizations by Paul de Kruif and others that inspired my generation! This emphasis may stem in part from the reluctance of scientists to speak out in literary vein, with a few atypical exceptions: outstandingly, June Goodfield in her books and television series (1981, 1985), which are a renascence of the de Kruif tradition. The competitive stresses on young scientists' behavior today must be acknowledged. The role modeling and critical oversight of their scientific mentors have also warranted celebration (Kanigel, 1986; Zuckerman, 1977). These are now complicated by the disappearance of leisure in academic scientific life, the pressures for funding, and academic structures and a project grant system that give too little weight to the nurture and reassurance of the human resources of the scientific enterprise. The often contradictory demands on the scientific personality are ill understood: Antitheses such as imagination vs. critical rigor; iconoclasm vs. respect for established truth; humility and generosity to colleagues vs. arrogant audacity to nature; efficient specialization vs. broad interest; doing experiments vs. reflection; ambi-

357

tion vs. sharing of ideas and tools—all these and more must be reconciled within the professional persona, not to mention other dimensions of humanity (Eiduson and Beckman, 1973; Merton, 1976).

I have never encountered the extremities that Jim Watson painted in his self-caricature of ruthless competition in *The Double Helix* (1968), which is hardly to argue that they do not exist. Side by side with competition, science offers a frame of personal friendships and institutionalized cooperation that still qualify it as a higher calling. The shared interests of scientists in the pursuit of a universal truth remain among the rare bonds that can transcend bitter personal, national, ethnic, and sectarian rivalries.

References

Bacon, Francis. (1625). Of marriage and single life. In *The Essayes or counsels, civill and morall, of Francis Lo. Verulam* (pp. 36–39). London: Haviland. (Reprinted 1971, Menston, England: Scolar)

Bodansky, Meyer. (1934). *Introduction to physiological chemistry* (3rd ed.). New York: Wiley.

Eiduson, Bernice T., and Beckman, Linda. (Eds.). (1973). *Science as a career choice.* New York: Russell Sage Foundation.

Goodfield, June. (1981). *An imagined world: A story of scientific discovery.* New York: Harper and Row.

Goodfield, June. (1985). *Quest for killers.* Boston: Birkhaeuser.

Kanigel, Robert. (1986). *Apprentice to genius: The making of a scientific discovery.* New York: Macmillan.

Kay, Lily E. (1986). W. M. Stanley's crystallization of the tobacco mosaic virus. *ISIS, 77,* 450–472.

Lederberg, Joshua. (1972). The freedom and the control of science—Notes from the ivory tower. *Southern California Law Review, 45,* 596–614.

Lederberg, Joshua. (1973). Research: The Promethean dilemma. In R. J. Bulger (Ed.), *Hippocrates revisited—A search for meaning* (pp. 159–165). New York: Medcom.

Merton, Robert K. (1976). The ambivalence of scientists. In *Sociological ambivalence and other essays* (pp. 32–55). New York: Free Press.

Sharp, Lester W. (1934). *Introduction to cytology* (3rd ed.). New York: McGraw-Hill.

Watson, James D. (1968). *The double helix.* New York: Atheneum.

Zuckerman, Harriet A. (1977). *Scientific elite: Nobel laureates in the United States.* New York: Free Press.

Ways of Being Rational

Lorraine J. Daston

I must preface this personal account of my experiences as student and later teacher of the history of science with a *caveat lector*. In no field are the trajectories of entry so zigzag and divergent as those that lead to careers in the history of science. Because almost nothing human—or natural—is foreign to the history of science, and because colleges and universities rarely offer an undergraduate concentration in the subject, historians of science come from backgrounds in the sciences, history, philosophy, sociology, and even literature. Therefore, no one's experience counts as typical, least of all my own. Nor can the pedagogical generalizations based on that experience be safely generalized without canvassing a larger sample. With that caution, I offer my experiences and observations as that notoriously untrustworthy statistical device, the sample of one.

Like the vast majority of high school students (and for that matter, college students), I was taught only a few paragraphs' worth of history of science in the many hours I spent in chemistry, biology, physics, and mathematics classes. These were the sentences of praise and blame that were the stuff of textbook introductions. They were riveting, these condensed narratives of Galileo and Darwin, of Gauss and Mendel, for they were full of the heroism of the mind—a hopeful alternative for those of us who did better at algebra than at field hockey. But for just that reason, these brief tales of truth triumphant and error vanquished would hardly have inspired the best students to a career in the history of science, even if we had known that there was such a thing: We burned with the ambition to become the subjects of these legends, not their chroniclers; to become Holmes, not Watson.

My first intimation of a different, richer history came in an entirely different context: learning Euclidean geometry. The ideology of progress ran strong in the textbook paragraphs, and with it the unspoken but unmistakable implication that if newer was better, old was bad and older was worse. Yet here were arguments and conclusions—the most pellucid arguments and compelling conclusions I had ever seen—that were purportedly over two thousand years old. I looked up "Euclid" in the *World Book Encyclopedia* and read and reread the article until the pages were dog-eared and smudged. I spent the greater part of ninth grade trying to trisect an angle with ruler and compass. I drove my parents to distraction by insisting on demonstrations for statements like, "You must make your bed in the morning." Years later, I learned how many figures in the history of science and philosophy, both major and minor, had been bewitched by the arrow-straight, crystal-clear deductions of Euclid. For many of these thinkers, the clarity and certainty of geometry had stood in opposition to the vagaries and obscurity of history. But for me, the discovery of Euclidean geometry was the rehabilitation, if not the discovery, of history: It taught me that the present and future had no monopoly upon rationality.

What a genuine history of that kind of rationality (better, rationalities) we call scientific might look like was the discovery of my first few years in college. I had the great good fortune to attend one of the few institutions where history of science had been cultivated in its own right, with a small but lively department dedicated to its study, and where it had been integrated with rigor and imagination into many introductory science courses. As a freshman at Harvard, I had never heard of the discipline of the history of science, so it was simply blind luck that led me to Professor Owen Gingerich's introductory astronomy course. In Natural Sciences 9, as it was then called, we learned not only how to find latitude from the pole star and how to read stellar spectra; we also learned Aristotle's arguments for the quintessence, the admirable complexity of Ptolemy's devices, how Kepler had triangulated the orbit of Mars, the observations that had driven Planck to quantize energy, and the observations that had *not* driven Einstein to the theory of relativity. Professor Gingerich's lectures were studded with vivid anecdotes and with elegant laboratory demonstrations that always worked (not until I began teaching did I realize that this was a feat bordering on the miraculous: the laws of nature cannot be depended upon to hold in the classroom), but what most impressed me then and now were his explanations of past scientific theories, clear and convincing once certain plausible assumptions and unavoidable constraints were conceded. These explanations were fortified with exercises: We not only learned about equants and eccentrics, we applied them to the inequality of the seasons; we not only studied Kepler's triangulations, we repeated them; we not only listened to Professor Gingerich's proof that Newton's laws of force and Leibniz' conservation principles were technically equivalent, we solved problems using both. The explanations persuaded us of the reasonableness of past scientific theories, even if they hinged upon assumptions (for example, the unity and harmony of the cosmos) that we could no longer embrace. But it was the exercises that allowed us to see the world as the framers of these theories had seen it, for the exercises allowed us to categorize and manipulate that world as the framers had. The pedagogical lesson of Natural Sciences 9 was a behaviorist one: to think in a certain way, one must act in a certain way. Or to put it in different terms, world views begin with in-the-fingers knowledge.

Natural Sciences 9 won me over to the history and philosophy of science. In the course of my undergraduate and graduate studies, my fine teachers, including I. Bernard Cohen, Erwin Hiebert, Gerald Holton, and Dirk Struik, taught me the importance of situating scientific thought in its cultural and philosophical context and of a sort of anthropological sympathy with conceptions of nature other than our own. Without ever abandoning the precept that scientific theories must be reasoned accounts of nature, they showed their students how broad and deep the realm of reason could be and had been. In my own subsequent teaching, it is this catholic approach to the rational understanding of nature that has kindled the interest of the most thoughtful students, be they trained in the sciences or the humanities. Once the history of science ceases to be a history of error, it becomes a source of unsuspected insights into the role of methods and metaphysics for the scientists; once it ceases to be the history of inexorable, headlong progress, it becomes an example of the contingencies and complexities that fascinate the humanists.

Today's Jacks, Tomorrow's Giants

Francis J. Heyden, S. J.

Some of the scientists and teachers of tomorrow, like some of those of today, may be interested to follow the path of an 11 year old who made of a radio station a springboard toward his destiny as an astronomy teacher to hundreds of students from 6 to 30. This inquisitive Jack (who was I) was fortunate to find kindly giants, not ogres, to help him learn.

Later, as chair of the astronomy department at Georgetown University for almost 30 years, I watched over the progress of more than 90 graduate students. From my classes, they went on to prepare dissertations on topics from radio astronomy to the distribution of galaxies. Many of them left their student days for careers in science and teaching; most remain in touch as friends and colleagues.

Graduate students are independent, some of them knowing more than their mentors. At Georgetown, as well as at neighboring elementary and secondary schools, I discovered that guiding students as they began to do science meant work in many areas. My contributions built on those of parents, whose years of experience didn't always show in the advice they offered. I found myself helping young people who wanted to build a telescope or "make a star" for a science fair. And even though no one ever succeeded in making a star, some small would-be creators did eventually choose to take astronomy in college. Turning over the big stones of graduate topics often yielded caches of pebbles that became the raw materials of the "star."

How did I become an astronomer and a teacher? Remembering my own path makes me glad to work with students interested in unconventional approaches. In 1918, a friend and I decided to make a radio station. The technology was wireless then, so we made two transmitters. When we had our small setups working, we were able to help my father hear signals from armed forces ships defending us during the first world war. At this point, I thought I wanted to be an operator on a ship when I grew up, but by 1922, radio broadcasting was a growing concern.

The world was an exciting place.

So that we could talk to one another in code, we wound tuning coils on rolling pins and made condensers from the tinfoil in cigarette wrappers and some wax paper. Henry Ford generously provided the spark coils for the transmitter.

A Jack among benevolent giants (the regular hams with licensed government call letters), I was a ham operator within a year, adding my 25-mile range of transmittals to the cacophony produced by Buffalo's other operators.

The public library, with its three old books on wireless, was an important teacher. But, more significant for me, the giants were kind, letting me use their elaborate equipment and watch them work.

With their guidance, I improved my original cat whisker on galena to a radio tube; my primitive rolling-pin tuning coil multiplied and metamorphosed into plug-in honeycomb coils. My "Jack," which is still a common fixture in communications instruments, was then a simple way to change coils of wire in a radio set.

Now, I could draw circuits for receivers and transmitters faster than I could compute the algebra, which was supposed to be among my more pressing concerns. Although more conventional education eventually came my way, the generosity of the radio ham giants in 1918 to this small Jack influenced the course of my life. Remembering my beginnings, I never forget their example or brush off the demands of the Jacks (and Jills) of the future.

In 1931, a few months after I started teaching college physics in Manila at its fine observatory, my supervisor told me that, my interest in studying and teaching physics notwithstanding, I was needed not for physics but for astronomy. When I protested that I knew very little about astronomy, I received this answer: "But you know wireless. We need you in the Observatory for that."

And this former 11-year-old ham suddenly found himself in charge of radio time signals over the biggest transmitters the U.S. Navy had, including the one that kept the weather section in contact with the Philippines.

Astronomy was easier than all that. So, after the war, I took a Ph.D. in astronomy from Harvard University. Somehow, this qualified me to write the Manila Observatory's War Damage Claims in Washington, D.C., where I simultaneously worked to build Georgetown's offerings in astronomy and physics.

All of this happened because, by the age of 15, I had become a giant with my own little Jacks. My beanstalk never grew so high that I forgot that once I too had been a Jack, not a giant. I obviously had no trouble remembering that giants can be gentle. And my formal affiliation with graduate students in astronomy never kept me away from younger children with their specific, quirky interests in science and technology. I left the University at least 50 times a year to talk to elementary and high school students about their projects and questions.

On one such occasion, a principal asked if I could come on very short notice to address his school's science club, which had no speaker. I decided to talk about time scales, showing how scientists had discovered that the age of the earth and the universe was measured in billions, not thousands, of years.

Back came the inevitable question, "What does the Bible say about this?" A clerical collar on a stargazer must have seemed a contradiction. But my knowledge of Hebrew helped me explain that scientific and spiritual truths wear many shining faces.

Science clubs, science fairs, junior academies—and I still help to prepare and update texts on earth science—but the real coin of the realm is kindness. Many of my former students, now teachers of students at many levels and directors of research laboratories, assure me that they have neither forgotten my help nor, much more important, will they fail to extend their hands to others.

Sometimes a little handful of beans can make a big stalk grow.

Intimate Evolution of a Nature Lover

Lynn Margulis

As eldest daughter of a glamorous housewife and a liberated lawyer-turned-businessman, most of the free time in my earliest life was spent bossing around my younger siblings (given my father's second and third marriages, a total now of seven).

My major task was to modulate voice volume, except during dramatic performances—in which I generally served as director, producer, and star—in the basement of a three-story apartment building on the south side of Chicago, across the street from Lake Michigan. Zealous in my demands for promptness in rehearsals from my younger and smaller sisters, I inevitably grabbed the leading roles.

As I remember it, the rest of my uncommitted moments were dedicated to daydreaming, with or without props, such as books, live ant colonies, and dandelions gone to seed. A Mexican woman, who had lived for 30 years in the village of Tepoztlan, Morelos, before the installation there of electric lights, later described me as the moonlit girl with her head always in a book.

A Polish Jew brought up in Christian communities in Grand Rapids, Michigan, and Ridgewood, New Jersey, my father developed a great love for transformed Palestine: The banality of the busy Chicago downtown office where he worked was ameliorated by his zealous Zionism. Like my mother, I never shared his burning interest, retained to this day, in the machinations, tribalism, and fratricides of Middle Eastern politics.

With the partial exception of my immediate family, until I entered the University of Chicago as a student in the College, I knew no one whom I thought was leading an intellectually satisfying life: My fellow high school students and their "materialistic" moms and dads seemed concerned, respectively, with hoarding cashmere sweaters and tail-finned automobiles. Consumed with an ineffable passion for "something better," I held them all in great disdain. My other passion, as I grew older, was for young men, but I quickly learned that most of them were no different from the friends of my parents and the parents of my classmates.

No one in our family orbit knew anything more about science than the fact that the famous chemist Harold Urey's son John was in my sister's class at the Lab School of the University of Chicago. Once, my mother in a tizzy, the wonderful Ureys came to dinner. My father, literate and articulate though he is, had great difficulty explaining to me my ninth grade math assignments. I knew better than even to question my mother about such esoterica.

My conversion to science was bipartite: The major crucial component involved a college course called Natural Science 2. That the system conferring the degree of bachelor of arts at the University of Chicago's College of Arts and Sciences ever existed is hard for anyone to believe, especially a young person today: Classes had fewer than 24 students, labs were frequent, exams were only advisory (a 9-hour June final, alone, "counted" toward the final grade, and June always seemed to be a long time away from October). Furthermore, textbooks were considered anathema to learning: We read unabridged and uncompromised original works* by great scientists— Charles Darwin, Gregor Mendel, Hans Spemann, August Weismann, and the neo-Darwinists. The themes of Nat. Sci. 2 were "What is heredity?" "What links the generations?" "How do the materials in fused egg and sperm inspire whole animals?" These haunting questions about life drove those of us interested deeper into biology. They drive me still.

Science there—and it was a superb program—was quite simply the set of methods, honest, open, and energetic, that facilitated asking the really interesting questions of philosophy: "What are we?" "Where do we come from?" "How do we work?" "What is this universe?" I never for a moment doubt that I owe my choice of career in science to the wisdom of those Chicago educators. The materials, the great questions, the syllabi, the readings, and the policies of the College encouraged me to proceed on my own schedule; before graduating at age 19, I thrived in splendid courses such as Nat. Sci. 3 (Organization, Methods, and Principles of Knowledge or OMP, an acronym that describes an advanced course in philosophy, including philosophy of science).

I owe roughly 75 percent of my impetus toward the scientific life to the great minds I met in the books we read in Nat. Sci. 2 (those listed above plus Vance Tartar, Thomas Morgan, Hermann J. Muller, Theodosius Dobzhansky,

---

*In translation, when necessary.

A. H. Sturtevant, and the like) who spoke to me directly through the assignments in the syllabi. Genetics and evolution, geneticists and evolutionists fascinated me from the time I first read about them. This body of coherent scientific work—the American school of genetics at the beginning of this century—gave me a sense of matter, of place, of what chemistry should explain. However, I would be remiss if I failed to mention the influence of Carl Sagan, who, relative to my beginner status, was at that time advanced as a graduate student of physics and poised at the launching of his astronomical career. He shared with me his keen understanding of the vastness of time and space. More importantly, he was a living example that young people, with all their foibles and fumbling, could direct their energies toward the scientific enterprise. Even though our turbulent marriage ended six years later in divorce, it left me with two magnificent Sagan sons and a strong sense of my own potential as a scientist interested in large and serious questions.

With one notable exception, I have felt curiously uninfluenced by specific teachers in classrooms. As a graduate student at the University of Wisconsin, both in laboratory and in lectures, I was inspired by the erudition and honesty of cell biologist and chromosome expert Professor Hans Ris. Indeed, I have always learned most when I've had to teach others.

Finally, let me say I have always agreed with the sentiment of paleontologist George Gaylord Simpson when he remarked that evolution was not the most important science because he happened to work on it . . . rather he worked on it because it *is* the most important science. I heartily agree. Evolution as change through time—from cosmic and stellar evolution through biological (including human) evolution—brings to light the muddles of which we are the living legacy.

When filling out inevitable forms or writing required essays on my personal future, I always claimed that I wanted to be "an explorer and a writer," not then realizing that those are precisely the two activities of working scientists. If I were to begin my career over again, I would choose almost exactly the same route. Although I love the methods of microbiology and the power of molecular biological analysis, I suspect I'd major in the geological rather than the biological sciences because geologists go on spectacular field trips as part of their education. They never need to apologize for their love of nature.

Credit to Dorion Sagan, John Kearney, the National Aeronautics and Space Administration's Life Sciences Office, and the R. Lounsbery Foundation.

From Gifts to Talents
Through General Science

Robert A. Rice

Having devoted more than 50 years to science education, I would like to highlight some of the experiences and observations that may help science teachers grow and sustain their interest in teaching.

Of special pleasure to me is the continued contact I have with many former students. They include a police department administrator, a farmer, the custodian in my apartment building, the editor of a student science research journal *(BASE)*, a bank executive, the president of the University of California, and, of course, many teachers.

Starting in a small high school with teaching assignments in biology, chemistry, general science, physics, physical education, and, eventually, band forced me to develop many teaching skills and to make careful use of my time. I eventually became principal of the school. My experiences there taught me to see students—their interests and their capabilities—as individuals. Too often teachers prefer to teach only one or two subjects. Such teachers can become classroom bound, losing interest in teaching and students after a decade. This situation is negative for the students, the subject, and the school. Teachers working in smaller schools can gain respect from the community by taking advantage of the opportunities that abound for participation in recreational, health, musical, and other club activities outside the classroom.

Science teachers entering large urban schools tend to become tied to the classroom. On the other hand, when I joined the faculty at Berkeley High School (California), where the science faculty alone totaled 15, I overcame a classroom-bound tendency by assisting the coaches in basketball, football, and track. This was not only enjoyable but also established rapport with students, especially the difficult ones, in my science classes. I also helped with the marching band. Because I could find time for these additional activities, the principal was glad to give available time for science opportunities. So, I recommend that teachers work diligently with their principals, who may then be willing, in turn, to secure time for additional science activities.

During the 1930s and 1940s, there were plenty of education courses but few science courses to upgrade teachers' knowledge. Science-oriented students had even fewer encouraging activities available to them. But in the 1950s, industry, colleges, and universities became interested in schoolteachers and students. Shortly after becoming science department chairman at Berkeley High School, it was my good fortune to be selected as the first director of the San Francisco Bay Area Science Fair (SFBASF), sponsored by the Standard Oil Company of California. Soon after, other Bay Area corporations joined in. For six years, until 1959, I directed this new venture, which provided the first real opportunity for science teachers to meet, exchange ideas, see each other's work, and bring talented students together. Several science teacher and student associations sprang up from this venture. The Fair organization became the hub of communications for science teachers and students for many years; it is still an important and effective force in this region. In fact, organizations, such as those composed of antivivisectionists protesting the use of animals in experimentation, attacked the SFBASF as a means of reaching schools and science teachers. By working with groups such as the State Humane Association of California and the Society for the Prevention of Cruelty to Animals, the Fair was able to educate teachers on proper experimentation and care of animals in schools, thus effecting changes throughout the United States as national teachers' organizations took up the problem. My advice, then, to science teachers is to become involved with science fairs and other competitions.

In 1956, the University of California, Berkeley, decided to help improve science teaching through the National Science Foundation (NSF) Teacher Training Programs. Again, I was involved in planning and directing the first of these institutes at Berkeley and, over the next 15 years, supervised many others. I did this work while teaching my regular classes at Berkeley High School. For the first NSF institute, I selected several qualified science teachers who were also able to secure or make time to join me in this program. These teachers acted as section leaders. Today they are some of the leaders in science education. Participants came from nearly all over the United States as well as a few foreign countries and benefited from exchanging ideas, gaining new information, and generally upgrading their teaching techniques. Thus, more vistas were opened to develop science literacy and improve work with students. Important contacts with university personnel resulted, many of which unfortunately disappeared when NSF stopped the institute programs. Today, however, we have a partial return to similar programs funded by NSF and others. Teachers should take advantage of these opportunities as they develop.

Later, as president of the National Science Teachers Association, I had the opportunity to observe many of the problems science teachers face and to find out in what ways they need assistance. Again, therefore, teachers who want to improve and gain recognition must find time for science meetings, student competitions, institute programs, and the like. The teachers, students, and schools will benefit.

After developing the children's area of the U.S. Science Exhibit at the Century 21 World's Fair in Seattle, I returned to Berkeley to join a group of scientists and educators who were developing a new kind of center for science education, the Lawrence Hall of Science (LHS), located on the Berkeley campus of the University of California. Here teachers could focus on new ideas and experiments to better understand and explain the concepts of science. The LHS would also be a resource for teachers in search of assistance and for students looking for stimulation. The LHS is operational now, helping science and mathematics teachers. I continue to find great satisfaction in assisting students. A number of challenging programs encourage them toward careers in science. Some of these opportunities are the Westinghouse Science Talent Search, science fairs, symposia, olympiads, writing for journals such as *BASE,* and work experiences in science laboratories.

I have concentrated on two of the aforementioned activities, namely the Junior Science and Humanities Symposium (JSHS) and *BASE*. The JSHS provides a competition much like an adult scientific meeting in that it features communication of scientific information through public speaking— an important part of the student's presentation of a research project. This competition appeals to mature high school students and allows science and English teachers to combine their efforts. My second interest is in *BASE*, with one of my former students, a physician, acting as its editor. *BASE* presents the opportunity for young people to publish the results of their independent research.

While I have had a few more experiences than are available to many teachers, my background suggests that teachers must take advantage of and look for opportunities to get involved outside of the classrooms—they should get acquainted with nearby colleges, tech centers, and museums, which can offer considerable help and stimulation. Teachers are not only multipliers of information and learning techniques. They are stimulators in the development of young people. Most scientists attest to the fact that good teachers turned them toward a career in science. And, of course, we do not teach in class only. A teaching career goes beyond schooling. It turns its special knowledge, values, and skills to the entire field of education: that is, to the community at large, to the environment of which we are all a part.

Lessons I Have Learned

Gerald Skoog

During my 30 years in the teaching profession, I have learned some lessons that have been valuable to me. Some of them reflect common sense. Others are contradictory to common practice.

As I approached my first day of teaching, the advice dominating my inexperienced and apprehensive thoughts and shaping my preparation was to be tough and "not smile until Christmas." I took this advice seriously and acted accordingly. The students in the small Nebraska high school where I began teaching also acted accordingly. Many of the best students in my chemistry class dropped out. They did not want to put up with nonsense from an insecure teacher. Some of the more marginal students hung around and accepted me as a challenge. As I became more perceptive and secure as a teacher, I noted that students approached a new course with a certain amount of apprehension and a hope that this class might offer something different and useful. My "don't smile until Christmas" attitude fueled that apprehension and distracted from the value of the course. I eventually learned that the first day was crucial in setting the climate for the year, and I determined to open with involvement and excitement. Routines needed for effective instruction were started; threats and signs of hostility were laid aside. Clearly, students entering a class for the first time are looking for involvement, relevance, and a sense that the teacher is a reasonable person who is ready to help them learn an exciting subject.

I used to feel smug and satisfied when I noted that my students could score well on examinations in chemistry and biology without studying the text-book. I prided myself on the explanatory power of my lectures and the exploratory nature of the discussions. What I failed to see was that the students were relying on me as they "mastered" the subject. My teacher-centered practices were achieving immediate successes but were actually retarding the students in their development as lifelong learners. In a world of change, and one dominated by science and technology, students in science must be helped to be lifelong learners. I was slow to learn this important lesson.

When I see teachers' desks, bookshelves, and storage areas bare or containing only the "essentials," I wonder about the richness both of the classroom and of the teacher's life. I learned early as a science teacher to be a ravaging scrounger of things, including reading material, photos, and other people's ideas. As I read, I clip and file. At conventions, I take notes and comb the exhibitors' booths. I look for ideas or approaches that are generative: While it is useful to find a lab or discussion topic for one lesson, it is more profitable to find a method or approach that can help to teach several different topics. Also, certain instructional theories have rescued me from the pitfalls of trial and error and have allowed me to plan with some predictability. If I were to give a "best scrounger" and a "best teacher" award to a teacher in my life, it would go to Martin Grant, professor of biology at the University of Northern Iowa. His collection of books—including comic books!—periodicals, plants, and miscellany such as woodpecker tongues stimulated my curiosity and helped me raise and answer some interesting questions. Among the lessons I learned from Dr. Grant is a refusal (or an inability) to throw anything away or pass up something just waiting to be bagged; my "collection" has created storage problems but has given me a rich base for planning learning experiences.

Growing up in a small Nebraska town and attending a high school with 20 students and 2, sometimes 3, teachers, I had a great deal of spare time but only occasional glimpses of the world that extended more than 50 miles beyond my home. I learned, however, that books and periodicals could bring that world to me and fill the slow hours of the school day. By seventh grade, I was reading three or four books a week. This reading habit has served me well as a teacher. In particular, it helped as I faced four biology classes composed of those students in a high school of 3,600 who were not seen as capable of learning and thus were given to me, the new teacher, and assigned to a nonlaboratory setting. Having just completed my master's degree in a program sponsored by the National Science Foundation, I was full of new knowledge and ready to teach much "stuff." I learned early on that year that earthworms, grasshoppers, and dihybrid crosses would contribute neither to my survival in that situation nor to the students' present or future lives. But my reading led me to many topics and problems in science that today are called "science-related societal issues." These became the vehicles to gain the interest of the students and to teach important science concepts.

Today, if I were responsible for hiring a school district's science teachers, one of my first interview questions would be, "What do you read regularly?"

and, then, "What is the last book you read?" Also, I would encourage new teachers to make friends with the librarian, to start a paperback library for the students, and to read several periodicals regularly. Overall, my reading habits have opened up new interests and areas of thought that have continually enriched my teaching and my life.

If I have succeeded, my students have learned *something* from me. But I have also learned many lessons from them. During my first summer as a university professor in the early 1970s, I taught a small class of education students. It included a woman returning from the Peace Corps whose provocative and liberated viewpoints sparked controversy, a dedicated social worker, a teacher who raised hogs, a black female who was rightfully distraught over the racism she had experienced, and a conservative farmer, among other individuals. This lively mix of potential teachers led to passionate class dialogues; the textbook faded in importance; burning issues set the agenda. Probing questions about education and equality and the intensity of the students drove me to the library daily in search of ways to feed the searing discussions occurring and to help myself, individual students, and the class as a whole find meanings and answers to the questions that were emerging.

Throughout my career, I have been sustained and enriched by my colleagues. Despite certain public stereotypes, the teaching profession is filled with interesting people, and communities of scholars do exist. Lunchtime debates, peer observations, extended field trips, convention experiences, and other types of informal and formal interaction with my colleagues have opened new areas of interest, sharpened convictions, and, overall, helped me to continue to grow personally and professionally.

Growing up in a small town and beginning my teaching career in a small school forced me to become involved in activities and face challenges I might have escaped in more crowded settings. I learned early that commitments involve risks but provide opportunities for growth and satisfaction. For example, I was terrified when the curtain went up on a theatrical production I had directed. However, the production ended without disaster—even to applause!—the years passed, and much satisfaction has accrued. The involvement and risks experienced when I served as chair of a community board responsible for implementing and supervising some of the "war on poverty" programs in the early 1970s taught me some hard lessons about racism, poverty, politics, and self-interest that influenced my philosophy of teaching and my political direction. My continued involvement with the controversy surrounding the place of evolution and creationism in the public schools has given me extended opportunities to interact with scientists, theologians, activists representing a variety of different viewpoints, and occasional crackpots. As president of NSTA in 1985 and 1986, I welcomed my contact with teachers, science educators, corporate leaders, and scientists. Those were the richest years of my life—thus far. They also were demanding as I faced new tasks and responded to new challenges. From these and other experiences, I learned that it is important to allow oneself to be pushed and to stretch one's reach beyond previous levels of performance in ever-widening spheres of involvement in teaching, in today's complex society, in life. From this commitment comes much satisfaction and growth.

As a graduate of a small high school in a community of 150 and as someone who began his career as the only science teacher in the school, I have met each year in my professional and personal life challenges of which I never dreamed. While I suspect other work would also have offered challenges, I am satisfied in my choice of the teaching profession because "the lessons I have learned" have made my life interesting and fulfilling.

Letter to a Young Scientist

Glenn T. Seaborg

Dear Dianne:

I understand that you think you may be interested in a career in science. Perhaps it will be helpful if I share with you some of my thoughts on the value and rewards of such a career. I remember well the influences and considerations that led me to turn in this direction as a very young man.

My own history in this respect has both unusual and usual aspects. Up until the time I entered high school, I had no exposure to science and, therefore, little knowledge of its possibilities. I chose literature as my major subject, and I took no science until my junior year when, in order to meet the college entrance requirement, I took a chemistry course. Largely due to the enthusiasm and obvious love of the subject displayed by my teacher, Dwight Logan Reid of David Starr Jordan High School in Los Angeles, chemistry captured my imagination almost immediately. I had the feeling, "Why hasn't someone told me about this before?" From that point forward, my mind was made up. I felt I wanted to become a scientist and bent all my efforts in that direction. I have never been sorry, for I have found in science a life of adventure and great personal satisfaction.

In considering a career in science, you may ask yourself whether you really have the qualifications. You may feel—and many might try to tell you—that you need to be a genius. This is not true. While great advances have been made by our greatest minds—such as Einstein, Rutherford, Edison—the bulk of scientific discovery has been made by men and women who, while of better-than-average intelligence, were by no means in the genius category.

We have so many tasks which need doing in all phases of medicine, public health, agriculture, industry, and basic research, that we cannot hope to carry them out without help from people of many levels of ability. Further-

379

more, many discoveries are made by men and women whose scientific effectiveness came as a result of a combination of qualities. In a particular instance, manual dexterity, special experimental technique, a freshness of viewpoint, or an insight gained from past experience may be decisive. Science is an organized body of knowledge and a method of proceeding to an extension of this knowledge by hypothesis and experiment. By learning the fundamental principles, by mastering the elements of the scientific method, and by acquainting yourself with the experimental techniques available to the modern scientist, you can proceed with near certainty to significant scientific advances and to achievement which may exceed that of many mental giants of a generation ago.

My advice is this: Do not worry too much about your intelligence, about how you compare with your contemporaries, but concentrate on going as far as possible with the basic endowments nature has given you. Don't underestimate yourself. Some young people are probably somewhat more confident—or cocky—about their abilities than their years warrant, but if I may judge from my own experience in talking with young people, many lack self-confidence and are somewhat hesitant in visualizing themselves as potentially important scientists. You should have no hesitation at all about doing this. Set yourself a high goal of achievement and exert yourself to advance toward this goal. The development of your abilities will be most marked if you strike out steadfastly for a goal which may even be high enough that you never quite achieve it.

I would like to emphasize a particularly necessary element in the makeup of a good scientist: simple *hard work*. Many a person of only better-than-average ability has accomplished, just on the basis of work and perseverance, much greater things than some geniuses. Such a hardworking individual will succeed where a lazy genius may fail. Some scientific discoveries are made by armchair research, but most of them require considerable experimental work and represent a lot of perseverance and perspiration, as well as a properly conceived method of attack. Many people of quite superior promise never have that promise realized unless they are fortunate enough to be in an environment where they are continually prodded into activity. People differ enormously in this quality, as in other respects: Some are self-starters and have great physical endurance, some work best alone, and others are most effective in a team effort. You will have to evaluate your own characteristics and try to place yourself in the environment most likely, as a routine result, to draw hard work from you.

This matter of hard work runs counter to the trend of modern times, with its emphasis on leisure, shorter workweeks, and more leisure-time activities. I am in sympathy with these developments in society, generally, but I cannot feel that the 35-hour workweek has much relevance for a creative scientist. The greater effort expected of a scientist, however, is seldom extracted against her will. The great gift is the ability to secure employment that allows the opportunity to do work she genuinely loves; she does not work simply because it is necessary in order to live. We live in a money-oriented society, but I think that personal success in money matters is often overrated as the reigning monarch of our standard of values. I believe that every person has a deep psychological need to feel that what she is doing is of some importance, aside from the money paid for doing it.

The scientist has the satisfaction of this need built into her life, and this gives zest and motivation to her efforts over an indefinite period of time. The intellectual satisfactions, the thrill of discovery, and the sense of worthwhile effort are a rich reward and a strong stimulus to continued work. Scientists and engineers are definitely not clock-watchers. The majority of my personal acquaintances work in establishments where the doors of the laboratory are never locked and the lights frequently burn late into the night.

Scientists would feel a sense of purpose and inner satisfaction even if their efforts were not important to the world in which we live. In actuality, of course, there is no group of persons on whom society as a whole depends so heavily. Science has exciting challenges to meet. Great discoveries with great benefits to human beings everywhere are much closer than the far horizons, and the technology necessary to utilize these great discoveries for the better health and quality of life of mankind provides an immense field for your efforts. The scientific discoverer is the first to see or to know a really new thing: s/he is the locksmith of the centuries who has finally fashioned a key to open the door to one of nature's secrets. This age of discovery has changed to new frontiers in space, medicine, biology (i.e., the human genome), artificial intelligence, new sources of energy—the possibilities are almost limitless.

You can be part of it.

Part V
Bibliography

Bibliography

Evelyn Morholt
Linda Crow

This bibliography has been compiled under the constraint of three questions: What knowledges and values, what skills and practices, are useful in providing for the young who are inclined to a future as scientists? What might be the first instruments of teachers and supervisors who are devising programs for potential scientists? What compilation of works and references might be suitable for a first small but solid library on giftedness, one particularly useful for understanding the gifted young who have the talents to become the scientists of the future, one that at the same time would be fruitful in designing an environment that would fulfill the powers of young and teacher alike in the pursuit of excellence?

We are required to meet diverse needs: of a school principal planning to introduce a program for the gifted; of a teacher seeking manner and mode of identifying and providing environments of high appeal to the gifted; of a mentor guiding a gifted youth in proving recent work or in analyzing his or her conceptions against known researches, past and present.

Recognizing these needs, among others, the bibliography is structured into four parts as follows:

1. General Works: Early and Current Studies

2. Traits of Gifted Individuals

3. Programs: Strategies and Tactics, Needed Research

4. Journals and Publications of Selected Associations

Category 1. General Works:
Early and Current Studies

Early studies on giftedness probed the traits of eminent adults, attempting to analyze an apparent dichotomy in giftedness and talent, that is, between the arts and the sciences. Clusters of traits of eminently successful scientists were developed from questionnaires, biographies, interviews, and projective techniques. A relatively small number of works have been given over to researches devised on the basis of simulated or real-world environments created to augment opportunities for gifted children and adolescents to engage in independent investigations in science—experiences that were then utilized for appraisals of giftedness in science.

Arasteh, Josephine D. (1968, March). Creativity and related processes in the young child: A review of the literature. *Journal of Genetic Psychology, 112,* 77–108.

Offers a careful, comprehensive analysis of the literature of measurements of creativity and a variety of tests of intelligence and personality in preschoolers and children in elementary grades. The measures indicate that preschoolers possess individual differences in such factors as freedom of expression, nonconformity, curiosity, and playfulness. The bibliography includes 133 references, dating mainly from the 1930s through the 1960s, offering the reader an analysis of widely distributed works that are not readily available.

Arieti, Silvano. (1972). *The will to be human.* New York: Quadrangle Books.

On the basis of his own and other relevant researches into the psychodynamics of giftedness, Arieti analyzes and synthesizes conceptions of the factors that affect the development of an individual. Arieti develops his view that an individual's will, freedom, and dignity can be crushed in early development, education, and schooling, or by oppressive and deforming social conditions, especially by the misuse of authority. Further, he holds that individuals are determined by the choices they make in life in addition to gene-driven factors and socioeconomic influences.

Barbe, Walter Burke, and Renzulli, Joseph S. (Eds.). (1981). *Psychology and education of the gifted* (3rd ed.). New York: Irvington.

Provides an anthology of papers dating from the 1950s, 1960s, and 1970s, thereby offering a cross section of the areas of interest in giftedness over three decades. Selected papers probe the characteristics and modes of identification of the gifted young, as well as the development of their gifts.

Bibliography

Barron, Frank X. (1963). The disposition toward originality. In Calvin W. Taylor and Frank X. Barron (Eds.), *Scientific creativity: Its recognition and development* (pp. 139–152). New York: John Wiley and Sons.

Presents findings based on a series of tests, including the thematic apperception and Rorschach, anagrams, and unusual uses for objects, that were given to 100 captains in the U. S. Air Force. Barron's often-quoted listing of characteristics of "original" individuals includes their preference for complexity and some degree of apparent imbalance in phenomena; their tendency to be complex psychodynamically and to exercise wide personal scope; their exhibition of independence in judgments; their self-assertiveness and dominance; and their rejection of suppression as a mechanism for controlling impulses. Barron's findings have been reinforced by subsequent research studies.

Brandwein, Paul F. (1955). *The gifted student as future scientist: The high school student and his commitment to science.* New York: Harcourt Brace. (1981 reprint, with a new preface [Los Angeles: National/State Leadership Training Institute on the Gifted and Talented])

Describes provisions between 1945 and 1951 in developing a supportive environment within a public high school with a heterogeneous population. Based on his observations of scientists during his own tenure in laboratories as a research scientist, Brandwein developed a triad of factors observable in working scientists and applicable to the gifted with potential in science: a *genetic factor* (general intelligence, verbal and numerical ability); a *predisposing factor* (psychological, i.e., persistence and questing); and an *activating factor* (opportunities for advanced work and training in undertaking real-world investigations, i.e., performance, in science under the guidance of a mentor). Based on his and other researches, Brandwein proposes an interaction of these factors as necessary for the development of high ability (talent) in science. He describes a full program designed as an environment in which students "select themselves" (thus using self-selection as a mode of identification) for demonstration of talent in scientific research.

Briskman, Larry. (1981). Creative product and creative process in science and art. In Denis Dutton and Michael Krausz (Eds.), *The concept of creativity in science and art* (pp. 129–155). The Hague, the Netherlands: Martinus Nijhoff. (Distributed in the U.S. by Norwell, MA: Kluwer Academic Publishers)

Offers a critique of the researches of psychometricians who set out to identify (through factor analysis) a set of personality traits that might discriminate between creative and noncreative individuals. Briskman develops his idea of "transcendent product," that is, creative scientific or artistic products that transcend their early traditions.

Callahan, Carolyn. (1979). The gifted and talented woman. In A. Harry Passow (Ed.), *The gifted and talented: Their education and development, Part I* (pp. 401–423). (Seventy-eighth Yearbook of the National Society for the Study of Education.) Chicago: University of Chicago Press.
Offers a comprehensive analysis of the literature on the results of tests comparing men and women in certain abilities and achievements as well as in differences in personality. Analyzes the general cultural barriers and certain environmental factors that affect the success of all women, especially gifted ones. Stresses the need for special career planning and counseling of gifted young people that set aside cultural stereotypes.

Correll, Marsha. (1978). *Teaching the gifted and talented.* Bloomington, IN: Phi Delta Kappa Educational Foundation. (Fastback Series No. 119)
Probes current research concerning the traits of the gifted, instruments for identification (Renzulli and Hartman, and the Williams scale), special programs (enrichment, grouping, acceleration), the role of parents and need for community resources, the cost of effective programs, and the preparation of teachers.

Crockenberg, Susan B. (1972, Winter). Creativity tests: A boon or boondoggle for education? *Review of Educational Research, 42*(1), 27–46.
In an attempt to reconcile conflicting evidence regarding tests of "creativity," the author analyzes two tests of creativity representative of many: the Torrance Tests of Creative Thinking and the Wallach and Kogan Creativity Battery. A concern for the validity of these tests is in the interest of guiding those who devise programs supportive of the aspirations of young gifted individuals.

Dellas, Marie, and Gaier, Eugene L. (1970). Identification of creativity: The individual. *Psychological Bulletin, 73*(1), 55–73.
Presents a comprehensive review of the literature on giftedness, including testing instruments and validity studies. Dellas and Gaier offer a critical evaluation of work in the field up to 1970.

De Lone, Richard H. (1979). *Small futures: Children, inequality, and the limits of liberal reform.* New York: Harcourt Brace Jovanovich. (For the Carnegie Council on Children)

Reflects on inequality, egalitarian policy, and political and social reform over three eras; attends to child development and theories on the sources of inequality; and considers steps toward the fashioning of an egalitarian thrust in schooling, education, and social policy.

Dutton, Denis, and Krausz, Michael (Eds.). (1981). *The concept of creativity in science and art.* The Hague, the Netherlands: Martinus Nijhoff. (Distributed in the U.S. by Norwell, MA: Kluwer Academic Publishers)
Includes the contributions of philosophical as well as psychological inquiry into creativity (Harré, Koestler, Polanyi, and Sparshott, among others) and attempts to clarify the many-faceted areas to be found in the studies on giftedness and creativity.

Erikson, Erik H. (1982). *The life cycle completed: A review.* New York: W. W. Norton.
Recapitulates the stages in development of the maturing individual—a sequence of physical, cognitive, and social abilities. Erikson restates the significance of his widely known major steps in psychosocial development from infancy to old age. Here, in the briefest sum: in infancy—hope; in early childhood—will; in play stage—purpose; at school age—competence; in adolescence—fidelity; in young adulthood—love; in adulthood—care; in old age—wisdom. Each stage in the life cycle is carefully described to provide a summary of some of Erikson's earlier popular works.

Fox, Lynn, and Durden, William. (1982). *Educating verbally gifted youth.* Bloomington, IN: Phi Delta Kappa Educational Foundation. (Fastback Series No. 176)
Summarizes and analyzes current research in the field of the verbally gifted; a concise source of information on identification and provisions for sustaining gifts and resources.

Fox, Lynn, and Zimmerman, Wendy. (1985). Gifted women. In Joan Freeman (Ed.), *The psychology of gifted children: Perspectives on development and education* (pp. 219–243). New York: John Wiley and Sons.
Offers statistics to indicate that disparity between potential and achievement is much less for men than it is for women; for one example, most of the "16 percent of women rated as professionals are in traditionally female fields of education, nursing, health technology, and library science" (p. 219). In education, in 1978 only 14 percent of school principals were women. Fox and

Zimmerman analyze a vast literature, from which they extract the significant data concerning such areas as gender's relationship to orientation and differences in values, career interests, and educational planning.

Freeman, Joan. (1979). *Gifted children: Their identification and development in a social context.* Baltimore: University Park Press.

Reflects on the many meanings of giftedness and means for judging giftedness. Freeman describes the British Gulbenkian Project on Gifted Children and her role as director; the project is based on parents' recognizing giftedness and seeking help in creating supportive environments.

Gagé, N. L. (Ed.). (1976). *The psychology of teaching methods.* (Seventy-fifth Yearbook of the National Society for the Study of Education.) Chicago: University of Chicago Press.

Begins with a historical approach to the psychology of methods of teaching and offers contributors' papers on learning models of value to those planning curricular and instructional models. Among the numerous devices utilized in teaching included are learning theories, computer-assisted instruction, written instruction, tutoring, discussion dynamics, simulations and games, lecture methods, the use of television and films, and decision-making procedures.

Gallagher, James J. (1966). *Research summary on gifted child education.* Springfield, IL: Illinois State Board of Education, Gifted Program.

Stresses the need for summaries such as the one developed in his study to reduce the "redundant research" resulting possibly from a lack of a sense of history in the investigation of trends in education. Analyzes an extensive collection of studies relating to intellectual and personality factors, teachers' ratings, methods of identification (including ones for the underachiever), and intervention procedures for administrators (intervention involving new curricular and pedagogical approaches).

Gallagher, James J. (1985). *Teaching the gifted child* (3rd ed.). Boston: Allyn and Bacon.

Reflects on the broadening focus given to definitions of giftedness, administrative concerns about the nature of programs, and shifts in methods of teaching the gifted, such as independent inquiry. Gallagher offers a historical approach to the works on the gifted as well as practical approaches for the teacher and administrator.

Bibliography

Gardner, Howard. (1983). *Frames of mind: The theory of multiple intelligence.* New York: Basic Books.

Reinforces the idea of multiple intellectual competencies on the basis of Gardner's studies of prodigies and of gifted, normal, and brain-damaged individuals, as well as those from diverse cultures. Within the context of cognitive and developmental psychology, he describes multiple intelligences as including linguistic, musical, logical-mathematical, spatial, bodily-kinesthetic, and personal. In his view, "it should be possible to identify an individual's intellectual profile (or proclivities) at an early age" (p. 10) and thereby enhance a person's educational opportunities and options.

Goldberg, Miriam. (1960). Research on the gifted. In Bruce Shertzer (Ed.), *Working with superior students: Theories and practices* (pp. 41–66). Chicago: Science Research Associates. (North Central Association of Colleges and Secondary Schools Project)

Analyzes the research literature concerning the characteristics of the gifted, modes of identification, tests of creativity, as well as nonintellective factors. Goldberg offers suggestions for administrators of programs and possible ways to alter the method of teaching and the content of courses for the gifted.

Goldsmith, H. H. (1983, April). Genetic influences on personality from infancy to adulthood. *Child Development, 54*(2), 331–355.

Emphasizes the importance that behavioral geneticists studying personality and temperament give to the interrelationships between developmental processes and gene action. Goldsmith evaluates the extensive research—rather than theoretical assumptions—in genetic studies of adolescent and adult personality based on studies of twins and personality inventories of longitudinal studies of adult twins and families, as well as infant and childhood temperament and personality. He concludes, "With substantial confidence, it can be asserted that theories of personality development ignore the action of genetic factors at some risk" (p. 349). Among others, "task persistence" is suggested as a factor for psychobiological study.

Gould, Stephen Jay. (1981). *The mismeasure of man.* New York: W. W. Norton.

Develops an incisive analysis and formidable criticism of much of the early literature on measuring intelligence, that is, Broca and craniology; Binet and IQ; Goddard and the "feeble-minded"; Terman; and Yerkes. Gould especially focuses on the errors of Cyril Burt. Gould's book provides a rich reference source for all in the field of education of children.

Gowan, John Curtis, and Demos, George D. (1964). *The education and guidance of the ablest*. Springfield, IL: Charles C. Thomas.

Offers a rigorous analysis of research in the field into the 1960s, with comprehensive coverage given to scientific orientation. Gowan and Demos describe the need for a high general intelligence, predisposing factors (home background and personality), and environment (especially school experiences), as developed in the contributions of Ahrendt, W. Bloom, Brandwein, Guilford, Jablonski, Knapp, and Roe—pioneers in advancing the needs of the gifted, particularly in nurturing performance in science.

Gowan, John Curtis, Khatena, Joe, and Torrance, E. Paul (Eds.). (1981). *Creativity: Its educational implications* (2nd ed.). Dubuque, IA: Kendall/Hunt.

Brings together an anthology of the contributions of current researchers under headings: Theory and Practice; Developmental Characteristics; Can Creativity Be Increased by Practice?; Curriculum; Teachers and Parents; Identification and Measurement; and Guidance.

Guilford, J. Paul. (1968). *Intelligence, creativity, and their educational implications*. San Diego: Robert Knapp.

Presents selected papers concerning Guilford's "structure-of-intellect" theory and model. The model, representing Guilford's classification of intellectual abilities, is described along with implications for schooling.

Horowitz, Frances Degen, and O'Brien, Marion (Eds.). (1985). *The gifted and talented: Developmental perspectives*. Washington, DC: American Psychological Association.

Provides a comprehensive analysis of various definitions of giftedness by psychologists—theories specific to an understanding of the nature of domain-specific talents and multiple intelligences. While knowledge is considered domain-specific, the acquisition of and mechanisms for retrieval of knowledge may be general abilities that apply over many fields. Horowitz and O'Brien conclude that "psychologists know very little about the developmental course of giftedness or talent, about the nature of environmental opportunities that nurture their realization, and about the nature of the conditions that must change over time to ensure continued development of giftedness" (p. 45). They argue in favor of turning the focus away from intellectual and cognitive factors and investigating social and personality traits that may influence the expression of giftedness.

Bibliography

Hudson, Liam. (1966). *Contrary imaginations: A psychological study of the young student.* New York: Schocken Books.

Evaluates the data from wide-ranging studies, in a quest for understanding of "creativity." His subjects: two groups of students designated as "convergers" and "divergers" on the basis of their responses to Convergence-Divergence Inventories. His data: IQ tests, open-ended tests, attitudes concerning controversial statements, autobiographical scripts, and so on. Hudson evaluates his data in a quest for understanding of "creativity." He concurs with a thesis put forth by Cox, Roe, and MacKinnon and widely accepted by subsequent researchers that personal factors, not only intellectual ones, are crucial for optimum performance. Hudson stresses that

> the whole point of testing, in other words, lies in measuring those qualities which predispose a man to follow a particular bent. Some of these may be a matter of intellectual ability, but in all probability, the majority do lie—as Cox, Roe, and MacKinnon suggest—within the sphere of personality. (p. 109)

Hudson includes statistical data supporting his views and examples of the tests he gave to high school students.

Jackson, Douglas N., and Rushton, J. Phillippe (Eds.). (1987). *Scientific excellence: Origins and assessment.* Newbury Park, CA: Sage.

Focuses on many current, diverse studies of excellence in science. After a foreword by Harriet Zuckerman, this anthology of papers is divided into three parts. Part I on assessment is given over to Eugene Garfield's analysis, with papers on "Mapping the World of Science" and "Is Citation Analysis a Legitimate Evaluation Tool?" Part II's discussion of "personalogical" origins includes Philip E. Vernon's comprehensive, historical overview of research on scientific abilities. Sociocultural origins are covered in part III, which includes studies on avoiding bias in the publication review process, permitting creativity in science, examining the impact of graduate teaching and thesis supervision, investigating the psychology of women and scientific research.

Kail, Robert V., and Pelligrino, James. (1985). *Human intelligence: Perspectives and prospects.* New York: W. H. Freeman.

Evaluates theoretical perspectives of intelligence testing, cognitive psychology, and developmental studies (for example, those of Piaget). In addition, offers a rich yet concise description of new theories still developing in the 1980s—such as the triarchic theory of human intelligence and the theory of multiple intelligences.

Khatena, Joe. (1982). *Educational psychology of the gifted.* New York: John Wiley and Sons.

Refers to the past and current literature in the field of giftedness. Khatena develops in depth the areas of identification of giftedness, intellect, imagery, and creativity. He also addresses modes of guidance (educational models and support agents) as well as special problems of groups such as underachievers.

Kramer, Alan H. (Ed). (1981). *Gifted children: Challenging their potential.* Monroe, NY: Trillium Press. (World Council for Gifted and Talented Children)

Presents 33 papers selected from the proceedings of the Third World Conference on Gifted and Talented Children held in 1979 in Jerusalem. The editor notes that fewer than 10 out of the 23 countries represented at the conference had large-scale special programs for gifted children.

Kuhn, Thomas S. (1963). The essential tension: Traditions and innovation in scientific research. In Calvin W. Taylor and Frank X. Barron (Eds.), *Scientific creativity: Its recognition and development* (pp. 341–354). New York: John Wiley and Sons.

Stresses that creative scientists use both convergent and divergent thinking in posing problems and in planning and executing their investigations.

Link, Frances R. (Ed). (1985). *Essays on the intellect.* Alexandria, VA: Association for Supervision and Curriculum Development.

Compiles extensively the comprehensive contributions of current workers in fields of interest in schooling and education. Among a considerable number of areas are found researches in multiple intelligences, thinking and writing, critical thinking, and intellectual capacity.

Mansfield, Richard S., and Busse, Thomas V. (1981). *The psychology of creativity and discovery: Scientists and their work.* Chicago: Nelson-Hall.

Focuses on characteristics of creative scientists and offers theories to explain the creative process. Describes several strategies of research in the field of giftedness, especially a measure of real-life creativity (and relates it to other variables such as personality traits). Considerable attention is given in the appendix to research studies concerning criterion-related validity of several cognitive and perceptual tests and, among others, personality and vocational tests.

McClelland, David C., Baldwin, Alfred L., Bronfenbrenner, Urie, and Stodtbeck, Fred L. (1958). *Talent and society: New perspectives in the identification of talent.* Princeton, NJ: D. Van Nostrand.

Summarizes the issues in an early period of heightened interest in giftedness: identification, measurement of skill in social perception, achievement and social status, family interaction, values and achievement, and the role of an ability construct in a theory of behavior—mainly, that of adaptive behavior.

National Science Board Commission on Precollege Education in Mathematics, Science, and Technology. (1983). *Educating Americans for the 21st century.* Washington, DC: National Science Foundation.

Describes in this volume (and in a companion volume on source materials) a plan of action for improving mathematics, science, and technology education for all the nation's elementary and secondary school students in order that their level of achievement may be the highest in the world by 1995—a compilation of ideas, needs, and proposals for improving knowledge, values, and skills through schooling.

Newland, T. Ernest. (1976). *The gifted in socioeducational perspective.* Englewood Cliffs, NJ: Prentice-Hall.

Defines the gifted as described in the literature and delves into social, psychological, and philosophical considerations with reference to selected researches. Offers a comprehensive critique of both strengths and weaknesses of research on giftedness, providing a summary of work up to 1975.

Passow, A. Harry (Ed). (1979). *The gifted and the talented: Their education and development, Part I.* (Seventy-eighth Yearbook of the National Society for the Study of Education.) Chicago: University of Chicago Press.

Summarizes early studies of philosophy and judgments of some 30 contributors of papers such as Tannenbaum, Gallagher, Jackson, Callahan, Stanley, Gold, Renzulli, Feldman, Torrance, and Getzels, which range from the broad spectrum of policies and programs for the gifted, to practices for special populations of the gifted (artistic people, promising black and Mexican American students, gifted and talented women). Provides probes into counseling of the gifted young to facilitate the self-appraisal of unique qualities.

Powell, Gloria J. (Ed.), Yamamoto, Joe, Romero, Annelisa, and Morales, Armando (Associate Eds.). (1983). *The psychosocial development of minority group children.* New York: Brunner/Mazel.

Provides a background for understanding certain health (both mental and physical) and educational issues concerning minority group children, as well as some of their psychological, social, and familial patterns. Powell raises questions for research and social policy, suggesting that poverty is the severest handicapping condition of childhood.

Rothenberg, Albert, and Greenberg, Bette. (1976). *The index of scientific writings on creativity: 1566–1974*. Hamden, CT: Shoe String Press.

Presents a scholarly, significant search of the literature on creativity (1566–1974) subdivided into sections: Creativity—General; Creativity and Psychopathology; Developmental Studies (Life Cycles); Creativity in Fine Arts; Scientific Creativity; Creativity in Industry, Engineering, and Business; Creativity of Women; Means for Facilitating Creativity through Education.

Rubin, Louis J. (Ed.). (1969). *Life skills in school and society*. Washington, DC: Association for Supervision and Curriculum Development and National Education Association.

Analyzes the essential skills needed in a rapidly changing, technological society. Offers a thoughtful presentation of bold ideas for the times ahead.

Shore, Bruce M., Gagné, Françoys, Larivée, Serge, Tali, Ronald H., and Tremblay, Richard E. (Eds.). (1983). *Face to face with giftedness*. Monroe, NY: Trillium Press. (World Council for Gifted and Talented Children)

Develops, through the contributions of researchers over five continents, the curricular considerations as well as the attitudes of peers and parents regarding the concepts of giftedness and provisions made for the gifted. These selected proceedings from the Fourth World Conference on Gifted and Talented Children (held in Montreal in 1981) are concerned with social contexts, the meaning of giftedness, giftedness deflected, cultural perspectives, and curriculum considerations.

Sternberg, Robert J., and Davidson, Janet E. (Eds.). (1986). *Conceptions of giftedness*. New York: Cambridge University Press.

Organizes a diversity of conceptions of giftedness: educationally based concepts, cognitive-psychological views, developmental studies, domain-specific approaches (mainly in mathematics and music), totaling 17 different conceptions of giftedness in an attempt to provide a background for reassessing and unifying thinking in the field.

Bibliography

Tannenbaum, Abraham J. (1979). Pre-Sputnik to post-Watergate concern about the gifted. In A. Harry Passow (Ed.), *The gifted and the talented: Their education and development, Part I* (pp. 5–27). (Seventy-eighth Yearbook of the National Society for the Study of Education.) Chicago: University of Chicago Press.

Traces the changing values in the United States concerning the need for programs for gifted individuals following Sputnik in the 1950s, the unrest and campus revolts of the 1960s, more recent charges of elitism, the current focus on the needs of disadvantaged minorities, and the public's "devaluation" of science and cyclic surges of interest in science and technology. Many suggestions are offered for provisions used through the 1970s.

Tannenbaum, Abraham J. (1983). *Gifted children: Psychological and educational perspectives.* New York: Macmillan.

Provides a kind of vade mecum for the professional, requisite for understanding state-of-the-art science. Offers useful and comprehensive coverage of the history of the concern for the gifted and definitions of "giftedness" from a psychological perspective, with considerations for practical facilitation of programs. Provides a critical review of the literature on creativity.

Tannenbaum, Abraham J. (Guest Ed.). (1986, May). Reflections: Gifted child education—The last 25 years [Special Issue]. *Roeper Review, 8*(4).

Eleven contributors to the schooling and education of the gifted and talented reflect on their work of the past 25 years and select items and vignettes for emphasis. Included are Tannenbaum, Abraham, Passow, Goldberg, Gallagher, Brandwein, Rosenbloom, Torrance, Gold, Taylor, and Ward.

Taylor, Calvin W. (Ed.). (1972). *Climate for creativity.* New York: Pergamon Press. (Report of the Seventh National Research Conference on Creativity)

Summarizes a wide range of aspects of creativity that include predictors and criteria of creativity and uses of creative abilities in industry and scientific organizations, as developed in the Seventh National Research Conference on Creativity. A comprehensive review of the thinking of the 1970s.

Torrance, E. Paul. (1962). *Guiding creative talent.* Englewood Cliffs, NJ: Prentice-Hall.

Stresses the need for special counseling of gifted elementary school students so they come to recognize their unique gifts, yet develop their creativity without suffering social stresses from their peers.

Tuttle, Frederick B., Jr., and Becker, Lawrence A. (1980). *Characteristics and identification of gifted and talented students*. Washington, DC: National Education Association.

Reflects on the current work and offers new approaches to problems encountered by the gifted young. Describes a variety of identification procedures and activities for teachers of the gifted. Supplementary materials are offered by way of example.

Vasta, Ross. (1979). *Studying children: An introduction to research methods*. New York: W. H. Freeman.

Explains the goals of a vast number of research techniques useful in studies of children and adolescents. In part I, Vasta describes methods of scientific inquiry; part II compares and contrasts basic research designs (longitudinal/experimental/time series); part III focuses on the fundamental tactics of investigating and measuring children's behavior employed by researchers. Related issues, such as ethical considerations, complete the presentation in part IV.

Whitmore, Joanne. (1980). *Giftedness, conflict, and underachievement*. Boston: Allyn and Bacon.

Offers new approaches to identification of the causes of underachievement (including maladaptive behaviors and negative feelings toward school curriculums) among gifted elementary schoolchildren, based on Whitmore's work with the gifted who have a variety of learning disabilities. Whitmore offers extensive examples of practical early remedial methods and alternative educational programs for gifted underachievers. Many suggestions are applicable at secondary school as well as elementary school levels.

Witty, Paul (Ed.). (1951). *The gifted child*. Boston: D. C. Heath. (American Association of Gifted Children)

Provides a collection of papers of pioneers in studies of giftedness of many kinds, not solely superiority identified through intelligence tests. Among the many contributors are Terman, Oden, Strang, and Witty, who describe their probes into the identification of giftedness. Included also are comments on the Stanford Studies, the works of Leta Hollingworth, traits of teachers of the gifted, and the Science Talent Search.

Category 2. Traits of Gifted Individuals

A comprehensive literature exists describing the intellectual and affective as well as motivational factors to be found among gifted adults—geniuses of the past, eminently successful individuals—and the traits and behavior of gifted young. These studies, considered in the framework of the works described in the preceding papers, often reaffirm earlier studies and confirm earlier hypotheses in presenting a variety of models of intelligence, multifactorial hypotheses, and details of psychometric studies. Some few works refer to psychodynamic foundations of behavior (Arieti, Erikson, Hudson, Maddi, and others, as described earlier). More attention than in earlier years is given in current work to the role of the family and siblings as well as to the self-image of the gifted young.

Barron, Frank X., and Harrington, David M. (1981). Creativity, intelligence, and personality. *Annual Review of Psychology, 32,* 439–476.

Offers a comprehensive review of current research concerning the personality traits of intellectually gifted individuals.

Brandwein, Paul F., and Morholt, Evelyn. (1986). *Redefining the gifted: A new paradigm for teachers and mentors.* Los Angeles: National/State Leadership Training Institute on the Gifted and Talented.

Analyzes existing research and critical analyses leading the authors to posit that gifted individuals combine in themselves and bestow to society two gifts: a *gift of self* and a *gift of an originative work.* These are generated out of certain engines of their talent: a "standing quarrel" with the culture developed into a "lover's quarrel" with the culture and a "milieu," an interaction of the moral and intellectual dynamics of the individual in a fusion with the environment. The successful mentor in effect "sees" the gifted child in the persona of the gifted adult and thus is required to teach children to cope with the gift that is their mark. In any event, Brandwein and Morholt conclude, "Whatever tests validate a child's promise for doing originative work (present synonyms: giftedness, creativity, talent) in the school years, we will consider that promise unfulfilled unless it results in a gift to society" (p. 4).

Delisle, James. (1984). *Gifted children speak out.* New York: Walker.

Analyzes 6,000 completed questionnaires from gifted elementary school-children (ages 5–13) who reveal the burdens as well as rewards in assessments of self, parents, peers, and teachers. Delisle offers guides for discussions among teachers and discussion activities with children.

Freeman, Joan (Ed.). (1985). *The psychology of gifted children: Perspectives on development and education.* New York: John Wiley and Sons.
Presents contributions of works on an international level (eight countries on four continents). The first section defines the "gifted" and means for their identification; then follow the characteristics and behavior of the gifted with suggestions for their education. Freeman considers the term "gifted" as currently used to include about 24 percent of all children with varying abilities (social and creative skills); she details how different nations view the gifted by organizing programs that accommodate their needs—often in the light of national aims and customs.

Keller, Evelyn Fox. (1983). *A feeling for the organism: The life and work of Barbara McClintock.* New York: W. H. Freeman.
Describes both the personal and scientific "lives" of Barbara McClintock, the Nobel laureate, as she pursued her studies and discoveries mainly concerning the genetics of corn. Presents a view of a rigorous search for evidence and the obstacles and disappointments in the life of this persistent scientist. (McClintock didn't receive recognition until 1983 for her lifetime of contributions to genetics.)

Parloff, Morris P. (1972). Creativity research program: A review. In Calvin W. Taylor (Ed.), *Climate for creativity* (pp. 269–286). New York: Pergamon Press. (Report of the Seventh National Research Conference on Creativity)
Describes the kinds of personality patterns that are conducive to creative performance in science in the real world; recognizes that creative performance is a function of the interaction of personality structure, environmental influences, and cognitive capacities. Parloff describes a study of male candidates in the Westinghouse Science Talent Search. In sum, the more creative adolescents are differentiated from the less creative controls in that the former appear to be more efficient thinkers; are willing to experiment with new and unusual ideas; are more persistent, self-reliant, and independent in their thinking; are able to use their skills and resources in achieving their goals; and are less encumbered by distracting anxieties.

Roe, Anne. (1953). *The making of a scientist.* New York: Dodd, Mead.
Summarizes Roe's comprehensive, clinical study of eminent scientists (biologists, physicists, psychologists, and anthropologists). Roe's subjects were given long personal interviews, intelligence tests (specially constructed), and Rorschach and thematic apperception tests. Roe generalizes

her many observations in terms of common patterns of personal and professional behaviors. She considers major factors in the making of scientists to be the need and ability to develop independence and persistence to a high degree.

Subotnik, Rena. (1986). Scientific creativity: Westinghouse Science Talent Search winners' problem-finding behavior. In Arthur J. Cropley, Klaus K. Urban, Harald Wagner, and Wilhelm Wieczerkowski (Eds.), *Giftedness: A continuing worldwide challenge* (pp. 147–156). Monroe, NY: Trillium Press. (World Council for Gifted and Talented Children)
Analyzes a carefully planned 10-page questionnaire as well as interviews of 300 honors recipients in the Westinghouse Science Talent Search. Data collected compared high school students in their ability to find problems for investigation. Subotnik concludes that, given an enriched environment with opportunities to take a variety of science courses from knowledgeable teachers, high school students can demonstrate talent in science by designing and conducting their own experiments. Eighty-nine of the 1983 Westinghouse winners received adult assistance in finding a research problem; 62 percent of the 89 were assisted by a laboratory director or professor.

Tannenbaum, Abraham J. (1960). *Adolescents' attitudes toward academic brilliance.* New York: Teachers College, Columbia University. (Talented Youth Project)
Describes Tannenbaum's research studies of intellectually gifted youth. Using an attitude test, he recognized the negative feelings the gifted encounter from their peers.

Vail, Priscilla. (1979). *The world of the gifted child.* New York: Penguin Books.
Provides deep insight into the world of the gifted—their traits, progress in schooling, and their behaviors. Vail describes the upbringing of her own gifted daughter and her experiences as a teacher in this nontechnical, fluent, knowledgeable small book. Vail suggests solo and group activities for gifted youngsters and includes a section in which gifted young and adults (ranging in age from 6 to 75) describe their lives, their careers, their wishes.

Wallach, Michael A. (1985). Creativity testing and giftedness. In Frances Degen Horowitz and Marion O'Brien (Eds.), *The gifted and talented: Developmental perspectives* (pp. 99–123). Washington, DC: American Psychological Association.

Reviews the shifts in importance given to IQ tests and to tests of creativity as correlates of giftedness. Considers creativity as a "by-product of field-specific instruction rather than teachable skill in its own right" (p. 115). Wallach considers that "achievement during adolescence in a given field of endeavor tends to persist, making accomplishment in a particular field a good basis for predicting its continuance" (p. 116). Wallach holds that "excellence in a given field of accomplishment may offer a better basis for selecting individuals for educational benefits than does testing for such general dispositions as intelligence or creativity" (p. 113).

Category 3. Programs: Strategies and Tactics, Needed Research

There are few programs that deal specifically with modes of teaching the gifted or provide opportunities for independent research by those gifted who wish to pursue their interests in science. Many of the references listed in category 1 offer traits of the gifted, provisions needed for handicapped gifted, or help for those who need supportive guidance. In the following works, programs are often suggested. Current journals (see category 4) often describe specific programs in operation. Further research needed—often a part of these papers—calls for the support of hypotheses through actual research using individual young people, modes of conducting longitudinal studies, and redefinitions of "giftedness" and "creativity."

Anderson, Trudy. (1982, May). Real-world science. *The Science Teacher, 49*(5), 41–43.

Highlights the details and procedures involved in setting up a mentor program within a community—matching students with professionals in the area. Explains how some students continue after graduation to work with the same professionals and mentors in industry.

Brandwein, Paul F. (1952, February). The selection and training of future scientists III: Hypotheses on the nature of "science talent." *Science Education, 36*(1), 25–26. (Follows papers I and II [1947, 1951] with the same title)

Reflects that there is a tendency for educators to compare students' "gifts" without considering the differences in students' opportunities and incentives. The nature of high-level ability in science is a case in point; assumptions are made concerning the so-called trait "science talent," but the trait is not operationally defined. In all three papers, Brandwein offers certain hypotheses on the basis of his observations and work with adolescents of high ability. Is there a trait, "science talent," similar to talent in music? Or is "science talent" a component of general intelligence? Or are there other factors, such as specific opportunities within the school environment, which make a "science talent" apparent?

Brodshaugh, Jacqueline. (1986). The learning bank: Clearinghouse for education. In Arthur J. Cropley, Klaus K. Urban, Harald Wagner, and Wilhelm Wieczerkowski, (Eds.), *Giftedness: A continuing worldwide challenge* (pp. 397–401). Monroe, NY: Trillium Press. (World Council for Gifted and Talented Children)
Describes a vision of a community in which all educational resources are open to students as a public policy—a "coming together" of teachers, students, parents, art and museum directors, professional and business people to plan the use of community resources to offer good learning experiences for students. Brodshaugh describes how an educational clearinghouse—"a learning bank"—can be developed. Activities involve an arts program; a task force on agricultural education (plant genetics, robotics, etc.); career mentors, a mentor network for a region of some 7,000 students in grades 7–12, and career internship and exploration programs; the sharing of TV media; and early-entry college classes.

Clarizio, Harvey F. (1986). Psychometric limitations of Guilford's structure of intellect model for identification and education of the gifted. In Arthur J. Cropley, Klaus K. Urban, Harald Wagner, and Wilhelm Wieczerkowski (Eds.), *Giftedness: A continuing worldwide challenge* (pp. 70–79). Monroe, NY: Trillium Press. (World Council for Gifted and Talented Children)
Analyzes the literature on Guilford's structure of intellect (SOI) model in relation to psychometric characteristics: norms, reliability, validity, aids used in interpretation. He also evaluates the many implications of his analysis for the practitioner's role in identifying gifted students and creating programs for them.

Cropley, Arthur J., Urban, Klaus K., Wagner, Harald, and Wieczerkowski, Wilhelm (Eds.). (1986). *Giftedness: A continuing worldwide challenge.* Monroe, NY: Trillium Press. (World Council for Gifted and Talented Children)
Provides a wide spectrum of the background achievements and research needed to nurture the gifted in order to develop their special talents in

leadership, music, dance, mathematics, and the sciences. The selected proceedings of the Sixth World Conference on Gifted and Talented Children, held in Hamburg in 1985, stressed the need for disseminating worldwide current thinking of leaders in the field of the study of the gifted.

Keller, J. David. (1980, November). Akron's exploratory program: A program for gifted and talented in mathematics and science. *School Science and Mathematics, 80*(7), 577–582.

Describes an approach used to identify gifted and talented students who took mathematics and science courses designed specifically for the gifted. Describes a modularized inquiry approach in which mentors in the community guided students in pursuit of their research interests; results indicated a significant growth in students' achievement in science and mathematics.

LaSalle, Donald. (1979, March). On Talcott Mountain. *Science and Children, 16*(6), 27–29.

Describes a program for the gifted and talented that included investigations in astronomy, geology, and climatology; notes that the predominant trait of the participants was their persistence.

Maker, C. June. (1982). *Teaching models in education of the gifted.* Rockville, MD: Aspen Publishers.

Offers a comprehensive review of many teaching-learning models useful in implementing a curriculum for the gifted. Theoretical models include those of Bloom, Bruner, Guilford, Kohlberg, Parnes, Renzulli, Taba, C. Taylor, Treffinger, and F. Williams.

Martin, Kathleen. (1979, December). Science and the gifted adolescent. *Roeper Review, 2*(2), 25–27.

Describes gifted adolescents as idealistic and critical thinkers who often respond immediately and directly to their world; emphasizes that science should be taught not only as a collection of facts, but also as a reverence for life.

Moorman, Carolyn Kay. (1979, April). Expeditions—Journeys with a purpose. *American Biology Teacher, 41*(4), 217–218.

Provides curriculum and identification guidelines, time lines, and methods for implementation and evaluation used in an independent biological study program designed for seventh and eighth graders.

Passow, A. Harry. (1983). The four curricula of the gifted and talented: Toward a total learning environment. In Bruce M. Shore, Françoys Gagné, Serge Larivée, Ronald H. Tali, and Richard E. Tremblay (Eds.), *Face to face with giftedness* (pp. 379–394). Monroe, NY: Trillium Press. (World Council for Gifted and Talented Children)

Stresses the total learning environment comprising the curriculums *all* students experience: general education, specialized, subliminal/covert (school climate and environment), and nonschool educative settings.

Renzulli, Joseph S., Reis, Sally M., and Smith, Linda H. (1981). *The revolving door identification model.* Mansfield Center, CT: Creative Learning Press.

Describes the important role of the teacher as a "talent spotter" recognizing those individuals who might benefit from advanced work beyond the level of the regular course of study. In essence, this guidebook describes the revolving door identification model. The "model" is based on Renzulli's *The Enrichment Triad Model* (1977), which comprises three factors: above-average ability, task commitment, and creativity. Chapters describe procedures for implementing and evaluating this revolving door model; many examples are given using diagrams and charts to assist both teachers and administrators (as well as parents). Relevant information is available in many appendixes: descriptions of available creativity tests, achievement tests, intelligence tests, and tests of special talents, checklists, parent nomination forms, and other child-screening questionnaires for kindergarten and early childhood education. Samples of "exploration" scholarship programs for older students are included.

Rice, Joseph P. (1985). *The gifted: Developing total talent* (2nd ed.). Springfield, IL: Charles C. Thomas.

Indicates that this revised edition is intended to summarize elements for setting up educational programs for the gifted. Criticizes the exclusive use of general tests of intelligence in selection of students for school programs and provides a comprehensive classification of talent. Lists the characteristics Barron and MacKinnon describe relating to personality traits of the creative person; offers a chart of Frank William's "search" for the creative teacher.

Romey, William D., with Hibert, Mary L. (1988). *Teaching the gifted and talented in the science classroom* (2nd ed.). Washington, DC: National Education Association.

Presents an overview of characteristics of the gifted and offers a rationale for each of several kinds of activities in which children in elementary school may participate. For example, among science activities described for the gifted, the authors suggest specific exercises in perception and awareness, fantasy, classifying, drawing pictures, playacting, singing songs, divergent thinking, finding science in the everyday environment (including TV—in "soap operas" and situation comedies, for example).

Schnur, James, and Stefanich, Greg. (1979, December). Science for the handicapped gifted child. *Roeper Review, 2*(2), 26–28.

Reflects that too often the handicapped receive little or no exposure to science, which may restrict the number of handicapped individuals who enter science fields. Describes the lack of proper physical environments and programs, as well as the attitudes of counselors, which may deter handicapped gifted students from pursuing careers in the sciences.

Sherman, Marie. (1984, October). A biochemistry course for high-ability secondary students. *Journal of Chemical Education, 61*(10), 902–903.

Describes a course in biochemistry designed for students who have completed three years of science study; the goals of the course as well as descriptions of many science fair projects are offered.

Siegler, Robert S., and Kotovsky, Kenneth. (1986). Two levels of giftedness: Shall ever the twain meet? In Robert J. Sternberg and Janet E. Davidson (Eds.), *Conceptions of giftedness* (pp. 417–435). New York: Cambridge University Press.

Describes a variety of definitions of giftedness and traits of the gifted. Proposes several possible, fruitful approaches to research, for example: a focus on people "in the process of becoming productive, creative contributors to a field—high school students who win Westinghouse Science Competition prizes, who publish articles in nationally circulating magazines, or who have their drawings shown in major exhibits" (p. 434). States further that such students "already have made creative contributions—they have not just learned to perform well on tests—but they are still in the process of becoming eminent" (p. 434).

Stanley, Julian C. (1979). The study and facilitation of talent for mathematics. In A. Harry Passow (Ed.), *The gifted and talented: Their education and development, Part I* (pp. 169–185). (Seventy-eighth Yearbook of the National Society for the Study of Education.) Chicago: University of Chicago Press.

Summarizes the philosophy and practice of Study of Mathematically Precocious Youth (SMPY), which began in 1971 at Johns Hopkins University, and the annual SMPY Mathematics Talent Search. Implications of the study on programs of mathematics in schools are considered.

Swami, Piyush, Schaff, John F., and DeBruin, Jerome E. (1979, April). Research for gifted students: Cultivating a national resource. *The Science Teacher, 46*(4), 28–29.
Provides details of a summer research program in science for high school students. Students worked with research scientists and had opportunities to write research papers and have them published; these opportunities for increasing students' scientific knowledge and skills often motivated them toward careers in research science.

Thomas, S. C., and Kydd, R. A. (1983, January). A university science enrichment program for gifted high school students. *Journal of Chemical Education, 60*(1), 27–28.
Offers details of the selection of students for a program in college-level research projects.

Whitmore, Joanne, and Maker, C. June. (1985). *Intellectual giftedness in disabled persons*. Rockville, MD: Aspen Publishers.
Suggests practical guidelines to aid disabled gifted youth by way of planned programs and a variety of career opportunities.

Young, Richard. (1979, December). The science gifted. *Roeper Review, 2*(2), 23–24.
Reflects that gifted students in their pursuit of science possess an ability to think in terms of mental images with a high level of ability in problem solving; indicates these students should be given opportunities to work in activities that involve understanding, analyzing, and organizing the world around them.

Category 4. Journals and Publications of Selected Associations

Current works on giftedness, curriculums, and programs aimed at improving the environment for all learners, as well as the gifted, appear in many journals and in the published proceedings of conferences. The following listing is only a sampling of the many publications available.

Journals

Child Development
Educational Leadership
Exceptional Children
Gifted Child Monthly
Gifted Child Quarterly
*Gifted Child Today**
Gifted International
Journal of Chemical Education
Journal of College Science Teaching
Journal of Creative Behavior
Journal of Research in Science Teaching
Phi Delta Kappan
Psychological Bulletin
Review of Educational Research
Roeper Review: A Journal on Gifted Education
School Science and Mathematics
Science and Children
Science Education
Science Scope
The Journal of Educational Research
The Science Teacher

Government Sources

U.S. Department of Education

Office of Educational Research and Improvement (OERI), 555 New Jersey Ave., NW, Washington, DC 20208 ([800] 424-1615), supports and conducts research on education, collects and analyzes education statistics, and disseminates information, among other functions. Its work is carried out by the Office of Research, Center for Statistics, Programs for the Improvement of Practice, Library Programs, and Information Services. Of particular interest to science educators and those interested in working with the gifted in science or in other areas are the following projects and services:

---

*Originally titled *G/C/T [Gifted/Creative/Talented]* from 1978 to 1986.

1. Available through Information Services are free statistical publications and referrals to appropriate sources, including those below (described in *Institutional Projects* [April, 1988]).*

• The Educational Resources Information Center (ERIC), Washington, DC 20208, "is a national education information system responsible for developing, maintaining, and providing access to the world's largest education research database" (p. 40). The ERIC Clearinghouse on Handicapped and Gifted Children (Council for Exceptional Children, 1920 Association Drive, Reston, VA 22091) attends to the education and development of these two groups. The ERIC Clearinghouse on Science, Mathematics, and Environmental Education (Ohio State University, 1200 Chambers Road, Room 310, Columbus, OH 43212) offers information on education in these fields as well as engineering.

• The National Research and Development Centers "are university-based projects that focus research on topics of national significance to educational policy and practice" (p. 16).

• The Research Synthesis Center for the Teaching, Learning, and Assessment of Science, affiliated with the Biological Sciences Curriculum Study, directs its attention to science education through The Network, Inc., 290 South Main Street, Andover, MA 01810.

2. Through the Office of Programs for the Improvement of Practice,

• The National Diffusion Network makes exemplary programs available for adoption by educational institutions at all levels. For a description of the Network and of *Science Education Programs That Work* (1988), write to the Superintendent of Documents, Washington, DC 20402.

• Nine regional educational laboratories provide research, development, and technical help to the states under their jurisdiction.

3. Among the Library Programs is the Educational Research Library (Washington, DC 20208), which lends books and other materials.

Superintendent of Documents
The U.S. Government Printing Office, Washington, DC 20402, makes available publications on educational statistics.

Other Sources

Science Service. This organization judges and selects winners of scholarships and awards provided by the Westinghouse Educational Foundation (the

---

*For this publication, write the Information Office at the address above (Room 300) OERI.

Westinghouse Science Talent Search). Also offers science reports and publications of interest to teachers and the public generally.

World Council for Gifted and Talented Children. Proceedings of the biennial world conferences on gifted and talented children are available from Trillium Press, P. O. Box 209, Monroe, NY 10950.

Related Journals. Several publications not necessarily devoted to schooling or education of gifted children per se (for example, the NSTA journals and books, *Parents' Magazine,* and *The Instructor*) often treat various aspects of giftedness as well as listing and explicating helpful programs.

Part VI

Biographical Notes on Contributors

Biographical Notes on Contributors

Sigmund Abeles. Science consultant for the Connecticut State Department of Education, Abeles has had an interest in programs for the gifted and talented for many years. Formerly supervisor of education for the gifted for the New York State Education Department, he presently serves on the Connecticut task force for the gifted and talented. Since 1961, he has been involved in curriculum development in science with particular emphasis on the physical sciences and technology. He has written many articles on topics and issues in science education. Besides his work as administrator of the federal math-science bill in Connecticut, he is currently involved in revising Connecticut's "Guide to Curriculum Development in Science," a document designed to assist school districts in the development and implementation of their science programs.

Bill G. Aldridge. Executive Director of NSTA for the past nine years, Aldridge previously served three years as a program manager in the Division of Science Education Development and Research, Science Education Directorate, of the National Science Foundation. He has an undergraduate degree in physics, as well as advanced degrees in physics and educational evaluation, from the University of Kansas. He holds a degree in science education from Harvard University. He has taught high school physics and mathematics and college physics. Aldridge has written two textbooks, nine monographs, and numerous magazine and journal articles. He has received awards and recognition from the National Science Foundation and from the American Association of Physics Teachers. As director of the largest science education organization in the world, Aldridge has the opportunity to pursue a myriad of activities and interests associated with science and science education.

413

Isaac Asimov. Born in Russia in 1920, Asimov arrived in the United States in 1923 and has been an American citizen since 1928. Educated in New York's public schools, he went on to Columbia University, where he earned all his degrees, including a Ph.D. in chemistry in 1948. He joined the faculty of Boston University's School of Medicine and is a professor of biochemistry there, although he has done no work (and received no pay) since mid-1958.

Asimov became interested in science fiction in 1929, began writing in 1931, submitted his first story on June 21, 1938, made his first sale on October 21, 1938, and has since written 351 short stories (and published them, of course) and has 6 more in press. He has also published 391 books and has some 22 in press. He has written at least 2,000 nonfiction essays, mostly about science. In fact, for 50 years virtually all he has done is write, on almost every conceivable topic and for almost any conceivable audience. He expects to continue this until he is quite, quite dead.

Isaac Asimov is married to Janet O. Jeppson, psychiatrist and writer, and has two children by a previous marriage (a boy and a girl).

Paul F. Brandwein. As a research biologist, Brandwein probed host-parasite relationships in cereal fungi; he taught biology and ecology in schools and in university. Certain of his researches in teaching and learning probed the traits of gifted and disadvantaged young; as editor-in-chief at Harcourt Brace Jovanovich he directed preparation of instructional programs for national and international schooling and education. Brandwein has written books in science and in education and is now at work on *Walden III: The Coming Birth of an Educational System.* He is listed in *American Men and Women of Science* and in *Leaders in Education.*

Linda Crow. An earth scientist and science educator in high school and in colleges, Crow has been teaching for the past 10 years at an urban community college. She has written numerous journal articles and has contributed to three monographs. Her major research has been in the area of factors affecting student science achievement. In 1985 she won an NSTA Ohaus Award for innovation in college science teaching. Her college has also acknowledged her as an outstanding teacher and recognized her as a master teacher. Recently, she has received a federal grant to develop college materials that enhance critical thinking in the sciences.

Lorraine J. Daston. Dibner Associate Professor of history and history of science at Brandeis University, Daston has also taught at Harvard, Columbia, and Princeton Universities. She is the author of *Classical Probability in the Enlightenment* (Princeton: Princeton University Press, 1988) and co-editor of *The Probabilistic Revolution* (Cambridge, MA: MIT Press, 1987), and she has written many articles on the history of mathematics, psychology, and natural history.

Robert A. Day. Currently professor of English at the University of Delaware, Day has been involved with scientific writing, editing, and publishing for many years. He has served as managing editor of the *Journal of Bacteriology* and eight other journals published by the American Society for Microbiology. He later served as director of ISI Press and vice president of its parent company, the Institute for Scientific Information. Active in many organizations, he has been chairman of the Council of Biology Editors and president of the Society for Scholarly Publishing. His book, *How to Write and Publish a Scientific Paper,* now in its third edition, is used as a text in hundreds of colleges and universities.

Charles R. Eilber. Director of the North Carolina School of Science and Mathematics from its organization in 1979 until his resignation 10 years later, Eilber helped found the nation's first residential, tuition-free school for 11th and 12th grade students academically talented in science and mathematics. Before his work there, he served in Michigan's public schools for 10 years as a teacher of science and mathematics and as a secondary school administrator. He has been a Fulbright Exchange Teacher in Great Britain and a National Science Foundation Fellow at Harvard, where he received his master's degree in education. He is a member of many local, state, and national organizations and frequently serves as speaker or consultant to state and national groups interested in public and private schooling and education as well as other fields.

Robert L. Ellison. An industrial/organizational psychologist, Ellison is research director of the Institute for Behavioral Research in Creativity and adjunct professor of the University of Utah's department of psychology. He earned his Ph.D. in psychology from the University of Utah. He has written many papers and technical reports, principally on his research in the areas of talent identification and development, assessment of individual and organizational performance, and organizational development.

Deborah C. Fort. After receiving her Ph.D. in comparative literature in 1974, she taught literature and writing in colleges and universities in the Washington, D.C., area for a number of years. Fort has singly and cooperatively designed writing programs both in and out of the academy and presented papers about teaching writing. In recent years, she has turned to free-lance writing and editing, specializing in science education and publishing occasionally in the field. She is mother to two daughters.

415

Vincent G. Galasso. Galasso, who received a bachelor of science degree from the City College of New York in 1961 and a master of science degree from New York's St. John's University in 1967, has done further graduate work at New York University and Long Island University. He has been teaching in the New York City school system since 1962 and was appointed chairman of the biology department at The Bronx High School of Science in 1981.

Galasso has made numerous presentations relating to science and education for the gifted and talented locally, nationally, and internationally. He also has worked in development programs for teachers and supervisors and as an adjunct professor at The City University of New York and Columbia University.

Stephen Jay Gould. A paleontologist, evolutionary biologist, teacher, and writer, Gould is based at the Museum of Comparative Zoology at Harvard University, where he is Alexander Agassiz professor of zoology. Gould's scientific field studies have been largely of the Bahamian land snail *Cerion.* He has written numerous books, essays, and reviews and received many grants and fellowships, literary and academic medals and awards, and honorary degrees. Gould serves on the editorial and/or advisory boards of several journals and scientific organizations. A frequent and popular speaker on science and other matters, since 1973 he has also written a monthly column, "This View of Life," for *Natural History.* He is father to two sons.

Francis J. Heyden, S.J. Heyden, director of the solar/optical division of the Manila Observatory (Philippines), majored in philosophy, theology, and astronomy, earning his Ph.D. in the latter at Harvard University. For 26 years, he served as chairman of Georgetown University's astronomy division; during this period, he also prepared the Manila Observatory's war damage claims. As a teacher, Heyden helped some 95 graduate students prepare dissertations and offered some 50 lectures annually to precollege students ranging from fifth graders to high school seniors.

David W. Johnson. A University of Minnesota professor of educational psychology (emphasis in social psychology), Johnson has master's and doctoral degrees from Columbia University. He has written 17 books, including: *The Social Psychology of Education; Joining Together: Group Theory and Group Skills; Learning Together and Alone: Cooperation, Competition, and Individualization; Educational Psychology;* and *Circles of Learning: Cooperation in the Classroom.* He has published over 250 research articles in leading psychological journals.

In 1972, he received a national award for outstanding research from the American Personnel and Guidance Association; in 1981, he received another award for outstanding research on intergroup relationships from Division 9 of the American Psychological Association. He is currently listed in *Who's Who in the World*.

For the past 20 years Johnson has served as an organizational consultant to schools and businesses in such areas as management training, team building, ethnic relations, conflict resolution, interpersonal and group skills, prevention of drug abuse, and evaluating the affective outcomes of school systems. He is an authority on experiential learning and is a recent past editor of the *American Educational Research Journal*.

Roger T. Johnson. A professor in the University of Minnesota's department of curriculum and instruction (emphasis in science education), Johnson holds an M.A. degree from Indiana's Ball State University and an Ed.D. from the University of California, Berkeley. His public school teaching experience includes teaching in kindergarten through eighth grade in self-contained classrooms, open schools, nongraded situations, cottage schools, and departmentalized (science) schools.

Johnson is an authority on inquiry teaching and has worked at the national level in science education. He is a member of the NSTA Search for Excellence Team and was a member of the Project Synthesis group, which explored discrepancies between what science education ought to be and what is actually happening in science classrooms.

Johnson has served on many major task forces of the Minnesota Environmental Education Board. He has written numerous articles, several book chapters, and has cowritten (with David W. Johnson) *Learning Together and Alone* (2nd ed., Prentice-Hall, 1987) and *Circles of Learning* (Association for Supervision and Curriculum Development, 1984). He has received several national awards, including the Helen Plants Award from the American Society for Engineering Education in 1984.

Milton Kopelman. Kopelman received a bachelor of science degree from The City University of New York in 1946; a master of science degree from Columbia University in 1949; and has done further graduate work at the University of Colorado, Pennsylvania State University, and Columbia University. He has been teaching in the New York City school system since 1948 and served as the chairman of the biology department at The Bronx High School of Science from 1961 until 1977. He was then selected as principal of The Bronx High School of Science and has held that post since 1977. He has also been an adjunct professor at The City University of New York and Columbia University.

Joshua Lederberg. In 1946, Lederberg, then a 21-year-old medical student, discovered that a form of sexual reproduction—gene recombination—occurs in bacteria, opening the door to many applications of bacterial genetics in medicine and in biotechnology. He received the Nobel Prize for this work in 1958, along with Edward L. Tatum and George Beadle.

Lederberg was educated in New York City public schools, including Stuyvesant High School, and then at Columbia University and its medical school, Columbia College of Physicians and Surgeons, where he worked with his prime mentor, Francis J. Ryan. He then taught at the University of Wisconsin and Stanford University, where he collaborated with E. A. Feigenbaum in artificial intelligence, pioneering the development of "expert systems," before becoming the president of New York City's The Rockefeller University in 1978, the post he holds today.

He worked on the Mariner and Viking missions to Mars, and he has served as a consultant to the Arms Control Disarmament Agency, to many other government agencies, and to the World Health Organization. From 1966–1971, he wrote a weekly syndicated column for the *Washington Post* on the social impact of scientific progress.

Lederberg has had a lifelong insatiable curiosity about every aspect of science, and in his present role is particularly concerned about understanding and enhancing the creative potential of young investigators.

Lynn Margulis. She is Distinguished University Professor of botany at the University of Massachusetts at Amherst. Previously she taught at Boston University, where she has worked since 1966. Since 1958, Margulis has written or been coauthor of 7 books and over 200 shorter publications—among them research and nontechnical chapters and articles. She has directed the Ph.D. or master's degree theses of over 20 graduate students and retains an interest in the development of hands-on science education for young people. Margulis, who works largely in cell biology and microbial evolution, is currently studying the symbiotic origin of cell organelles and living microbial mats as analogs of ancient microbial communities. Parent to four children, she now publishes frequently with her son Dorion Sagan.

Evelyn Morholt. A former teacher, Morholt has been editor of *The Teaching Scientist* (Federation of Science Teachers, New York City), chair of a New York City high school science department, and acting examiner for the New York Board of Education. She has written nine books used in the schools, the most recent with Paul F. Brandwein, *A Sourcebook for the Biological Sciences* (3rd ed.), 1986.

Joseph D. Novak. A Fulbright scholar and professor of education and biological sciences at Cornell University, Novak began his research into the process of learning when he tried to discover why so few students were effective at problem solving. Among his contributions are 17 books, including *A Theory of Education* (Cornell University Press, 1977), as well as television programs on learning techniques produced by Australian and U.S. public television. He has served as president of the National Association for Research in Science Teaching.

A. Harry Passow. Passow is the Jacob H. Schiff professor of education at Teachers College, Columbia University, where he has been a faculty member since 1952. He initiated the Talented Youth Project at Teachers College in 1954 and served as its director for 12 years. Editor of *The Gifted and the Talented: Their Education and Development, Part I,* the 78th Yearbook of the National Society for the Study of Education, Passow is the current president of the World Council for Gifted and Talented Children, Inc.

Steven J. Rakow. Assistant professor of science education at the University of Houston—Clear Lake, Rakow has written over 50 publications. His interests include computers in education, the status of minorities in science education, reading and science education, and the role of science in middle and junior high school. Rakow is the field editor of *Science Scope,* the NSTA publication for middle/junior high schoolteachers. He currently spends his time teaching science methods courses for elementary and secondary teachers, offering teacher inservice presentations, and—his greatest love—teaching a class of gifted and talented fourth graders.

Robert A. Rice. Currently a science consultant at the Lawrence Hall of Science, Rice is a science educator with more than 50 years of experience in teaching, supervising, and administering science activities for students and teachers. He was one of the developers of the children's area, United States Science Exhibit, Century 21 Exposition, Seattle, Washington; he was responsible for the development of numerous teacher and student training programs and institutes at the University of California, Berkeley; and he was one of the developers of the Lawrence Hall of Science. His major interest is in science fairs and symposia for students and a student research journal of science and technology, *BASE.* This journal provides opportunities for students worldwide to publish their research work.

Annemarie Roeper. Cofounder of the nationally esteemed Roeper City and Country School and coeditor of the Roeper Review, Roeper taught graduate courses at Oakland University, Michigan, and Wheelock College, Boston. She has published four books for children, numerous articles, and a book titled *Education for Life.* She received an honorary doctorate of education from Eastern Michigan University. Roeper is listed in *Who's Who in the World* and *Who's Who of American Women.* With her husband, George Roeper, she has received citations from the Michigan Board of Education and the Michigan Senate and the key to the city of Detroit.

Gifted Young in Science: Potential Through Performance

Irving S. Sato. Director of the National/State Leadership Training Institute on the Gifted and the Talented since 1972, Sato is a speaker and consultant to numerous organizations concerned with the schooling and education of the gifted and talented. He has conducted workshops, conferences, and meetings, made presentations, or given speeches in all 50 states as well as addressed forums in many other countries worldwide. Sato is author or editor of numerous books and papers and has conducted six research projects on various aspects of giftedness. He belongs to many professional organizations and advisory boards and has received many honors and awards.

Madeline Schmuckler. Having completed her undergraduate work at City College of New York, she did graduate work in education at Yale University and in educational administration at St. John's University (New York). In 1976, she began teaching at The Bronx High School of Science, from which she is presently on leave. She is mother to two children.

Glenn T. Seaborg. Seaborg is University Professor of Chemistry at the University of California, Berkeley, associate director of the Lawrence Berkeley Laboratory, and chairman of the Lawrence Hall of Science. He joined the Berkeley campus faculty in 1939, after receiving his Ph.D. there in 1937, and served as chancellor from 1958 to 1961. From 1961 to 1971 he chaired the U.S. Atomic Energy Commission. Winner of the 1951 Nobel Prize in chemistry (with E. M. McMillan) for work on the chemistry of the transuranium elements, Seaborg is one of the discoverers of plutonium (element 94). During World War II he headed the group that devised the chemical extraction processes used in plutonium production for the Manhattan Project. He and his coworkers have since discovered 9 more transuranium elements: elements 95–102 and 106. Among his codiscoveries are many isotopes with practical applications in research, medicine, and industry (such as iodine 131, technetium 99m, cobalt 60, cesium 137, americium 241, plutonium 238), as well as the fissile isotopes plutonium 239 and uranium 233.

Sid Sitkoff. An instructional specialist in science, Sitkoff has written and directed numerous science programs including "Stones and Bones," disseminated through the National Diffusion Network. His major interest is in science education. He serves as science specialist for the Los Angeles Unified School District and is a lecturer at Loyola Marymount University, Los Angeles.

Gerald Skoog. Skoog has taught science in schools ranging in number of students from 150 to 3,600 and has been a science educator at Texas Tech University since 1969. His research interests focus on the coverage of evolution in high school biology textbooks. Skoog was NSTA president in 1985–1986.

Gunther S. Stent. Born in Berlin in 1924, Stent graduated from Chicago's Hyde Park High School in 1942 and studied physical chemistry at the University of Illinois (B.S. 1945; Ph.D. 1948). He then attended the California Institute of Technology as a postdoctoral fellow to join Max Delbruck's "Phage Group," the fountainhead of the discipline that, a few years later, came to be called "molecular biology." Stent has been on the University of California, Berkeley, faculty since 1952, as professor of molecular biology since 1959. His current research concerns the embryological development of the nervous system. Besides contributing to the scientific literature, Stent has also published on the history and philosophy of science and, in 1986, edited for posthumous publication Delbruck's philosophical testament *Mind from Matter?* He is a member of the National Academy of Sciences, the American Academy of Arts and Sciences, and the American Philosophical Society.

Pinchas Tamir. Professor at the School of Education, Hebrew University, Jerusalem, Tamir works at the Israel Science Teaching Center there. He received his Ph.D. at Cornell University in 1968. Since 1969 he has been the director of the Israel High School Biology Project. He has published numerous chapters in books and more than 130 research papers in journals specializing in science education, curriculum, and educational psychology, mainly in the U.S., Canada, and the United Kingdom. In addition, he has published many papers in Hebrew and edited several books.

His research deals with topics such as curriculum development, evaluation, education of teachers, instruction, cognitive preferences, and various aspects of science education. He received the 1977 Palmer O. Johnson Award of the American Educational Research Association for the best research paper. He has presented papers in numerous professional and international conferences and served as visiting professor in Australia, Mexico, Canada, Norway, and a number of universities in the U.S.

Abraham J. Tannenbaum. Tannenbaum's Ph.D. is in social psychology. He is professor emeritus of education and psychology at Teachers College of Columbia University, where he directed programs in the education of the gifted and of the behaviorally disordered. Upon assuming his faculty post in the Department of Special Education in 1965, he accepted the responsibility for designing and implementing master's and doctoral programs in the areas of giftedness and of behavioral disorders, and he maintained those responsibilities continually until his 1987 retirement.

From 1975 to 1980, Tannenbaum directed the federally sponsored Graduate Leadership Education Project, which involved seven universities in concerted efforts to produce future leaders in the field of educating the gifted. He has served as president of the Metropolitan Association for the Study of the Gifted (1964–1965) and president of the Council for Children with Behavioral Disorders (1975–1976). In 1981, he received the Hollingsworth Award for research on the gifted; in 1985, the National Association for Gifted Children's Distinguished Scholar Award. His most recent major work in the field is *Gifted Children: Psychological and Educational Perspectives* (Macmillan, 1983).

Calvin W. Taylor. A native of Utah, Taylor obtained B.A. and M.A. degrees in psychology at the University of Utah and a Ph.D. in psychology at the University of Chicago. Taylor is a member of Phi Beta Kappa, Sigma Xi, and Phi Kappa Phi. He wrote his doctoral dissertation under the supervision of L. L. Thurstone, author of *The Vectors of Mind* and *Multiple Factor Analysis.*

Taylor did basic and applied research underlying the U. S. Employment Service, helped establish personnel research as a permanent army-wide function in the military, and has produced more than 15 books and research project volumes and more than 100 articles in a variety of professional and educational journals. Taylor received the Richardson Creativity Award from the American Psychological Association and is an elected member of the Education Hall of Fame in the U. S.

Stephen J. Warshaw. Head of the Science Department of the North Carolina School of Science and Mathematics, Warshaw received his Ph.D. in biology from Yale University. He has taught in public school and worked in the private sector and in the U.S. Navy as a research biologist. He has written many publications and received the NSTA Science Teaching Achievement Award in 1986.